Advances in Intelligent Systems and Computing

Volume 1371

The series "Advances in Intelligent Systems and Computing" contains publications on theory, applications, and design methods of Intelligent Systems and Intelligent Computing. Virtually all disciplines such as engineering, natural sciences, computer and information science, ICT, economics, business, e-commerce, environment, healthcare, life science are covered. The list of topics spans all the areas of modern intelligent systems and computing such as: computational intelligence, soft computing including neural networks, fuzzy systems, evolutionary computing and the fusion of these paradigms, social intelligence, ambient intelligence, computational neuroscience, artificial life, virtual worlds and society, cognitive science and systems, Perception and Vision, DNA and immune based systems, self-organizing and adaptive systems, e-Learning and teaching, human-centered and human-centric computing, recommender systems, intelligent control, robotics and mechatronics including human-machine teaming, knowledge-based paradigms, learning paradigms, machine ethics, intelligent data analysis, knowledge management, intelligent agents, intelligent decision making and support, intelligent network security, trust management, interactive entertainment, Web intelligence and multimedia.

The publications within "Advances in Intelligent Systems and Computing" are primarily proceedings of important conferences, symposia and congresses. They cover significant recent developments in the field, both of a foundational and applicable character. An important characteristic feature of the series is the short publication time and world-wide distribution. This permits a rapid and broad dissemination of research results.

Indexed by DBLP, INSPEC, WTI Frankfurt eG, zbMATH, Japanese Science and Technology Agency (JST).

All books published in the series are submitted for consideration in Web of Science.

More information about this series at http://www.springer.com/series/11156

Arti Noor · Abhijit Sen · Gaurav Trivedi
Editors

Proceedings of Emerging Trends and Technologies on Intelligent Systems

ETTIS 2021

Springer

Editors
Arti Noor
Education and Training Division
Centre for Development of Advanced Computing (CDAC)
Noida, Uttar Pradesh, India

Abhijit Sen
Computer Science and Information Technology
Kwantlen Polytechnic University
Surrey, BC, Canada

Gaurav Trivedi
Department of Electronics and Electrical Engineering
Indian Institute of Technology Guwahati
Guwahati, Assam, India

ISSN 2194-5357 ISSN 2194-5365 (electronic)
Advances in Intelligent Systems and Computing
ISBN 978-981-16-3096-5 ISBN 978-981-16-3097-2 (eBook)
https://doi.org/10.1007/978-981-16-3097-2

This Springer imprint is published by the registered company Springer Nature Singapore Pte Ltd.
The registered company address is: 152 Beach Road, #21-01/04 Gateway East, Singapore 189721, Singapore

Preface for ETTIS-2021

We are pleased to present the proceedings of the International Conference on "Emerging Trends and Technologies on Intelligent Systems" (ETTIS-2021) being organized during March 4–5, 2021, at the Centre for Development of Advanced Computing (CDAC), Noida, in association with the Automatic Control, Computers and Electronics Department, Faculty of Mechanical and Electrical Engineering, Petroleum-Gas University of Ploiesti, Romania, as academic partners. ETTIS-2021 is the first edition of the conference in the series targeting the research in the area of intelligent systems. With the proliferation of artificial intelligence, intelligent systems have permeated our lives invariably in every aspect. Intelligent systems solutions are now being deployed almost in all spheres to solve a variety of problems and enhance the productivity of the systems. The sectors such as agriculture, education, smart cities and health care, to name a few, are utilizing the benefits of intelligent systems.

As the theme of the conference is recent, it attracted academicians, scientists, researchers and experts from the different domains to showcase their research ideas and share information about cutting-edge developments in the field on a common single platform. The accepted papers were divided into six technical tracks based on the main theme, namely (i) network security and emerging technologies, (ii) deep learning, fuzzy and neural systems, (iii) AI and ML, (iv) NLP and speech understanding, (v) information security and (vi) hybrid intelligent systems.

Although the conference was conducted in virtual mode due to the COVID-19 pandemic, the response to the conference was overwhelming. A total of 31 papers were accepted for presentation under six different tracks after double-blind review process by experts. These papers represented the recent developments in the subfields of intelligent systems. Selected papers after the further revision are now being published in the book series *Advances in Intelligent Systems and Computing* (AISC) by our publishing partner Springer.

The conference featured invited talk from eminent experts of the different countries besides the regular paper presentations by authors. We are highly thankful to our keynote speakers: Prof. Keshab Parhi, University of Minnesota; Prof. Bernhard Pfahringer, Waikato University, New Zealand; Prof. Gaurav Trivedi, IIT Guwahati; Prof. Abhijit Sen, Kwantlen Polytechnic University, Canada; Dr. S. S. Aggarwal,

Director General, KIIT Group of Colleges, Emeritus Scientist CEERI/CSIR; Prof. Emil Pricop, Petroleum-Gas University of Ploiesti, Romania, for sharing the new developments in the field and adding value to the conference.

At this point, we take this opportunity to thank Sh. Vivek Khaneja, Executive Director and Patron of the ETTIS-2021, for supporting the conference in all aspects to make it a mega event. We would also like to thank Sh. V. K. Sharma, Senior Director and Group Coordinator (Education and Training) and also Co-Patron, of the conference for guiding us at every step so that the conference could maintain a high standard of quality even though being the first edition of the series.

We are also grateful to Dr. Hemant Darbari, Director General, CDAC, and Dr. B. K. Murthy, Scientist G and Group Coordinator, CDAC, for embracing the inaugural session and enlightening the audience with the innovative research activities being taken up by CDAC in the area of artificial intelligence for societal benefit and initiatives taken by MeitY to promote original research in AI and allied fields.

Last but not least, we would like to express our sincere appreciation to all authors and participants of ETTIS-2021. We also express our sincere gratitude to Springer for being our publication partner.

We hope that the accepted papers of the conference will attract good citations in future and stimulate the minds of the young researchers to develop innovative solutions for applications in terms of effective techniques and algorithms. We feel honored and privileged to be a part of this journey and hope that ETTIS conference series will continue in future also serving its stipulated purpose.

Surrey, Canada — Prof. Abhijit Sen
Noida, India — Dr. Arti Noor
Guwahati, India — Dr. Gaurav Trivedi

Contents

About the Editors

Dr. Arti Noor is presently working as Senior Director at C-DAC, Noida. She has done her Ph.D. from IIT, BHU, in 1990. She has 20 years of teaching VLSI design-related courses to M.E. students of BITS, Pilani, and C-DAC, Noida. She has guided six Ph.D. and guided 200 student's projects of B.Tech./M.Tech./M.E. and examined 100 M.Tech. theses. She has published 81 research papers in journals and conferences including monographs.

Dr. Abhijit Sen is currently Professor of the Computing Sciences and Information Technology at Kwantlen Polytechnic University, Canada. He holds a Ph.D. from McMaster University, Hamilton, Ontario, Master of Science degree from University of California, Berkeley, USA, and B.Tech. in Electrical Engineering from IIT Kharagpur. He has over 20 years of academic and administrative experience at Kwantlen Polytechnic University. At Kwantlen Polytechnic University, he has taught a wide range of courses in network and communications, distributed systems, software development, programming, Internet application development, and emerging technologies. He serves as a member of accreditation team of Canadian Information Society (CIPS). He brings in considerable industrial and managerial experiences having worked in organizations such as Canadian Aviation Electronics, Montreal, Canada, and Microtel Pacific Research, Burnaby, Canada. He also worked as a consultant to Canada Post, Montreal, and InfoElectonics, Montréal, Canada. He is also actively involved in professional activities and presented seminars and papers in number of conferences. He is a member of Institute of Electrical and Electronics Engineers Association (IEEE), New York Academy of Sciences, and Canadian Information Processing Society.

Dr. Gaurav Trivedi is currently the Associate Professor at IIT Guwahati. He holds a Ph.D. and M.Tech. from IIT Bombay, and Bachelor of Engineering from Shri Govindram Seksaria Institute of Technology and Science (SGSITS), Devi Ahilya University, Indore, Madhya Pradesh. His research interests include circuit simulation, VLSI CAD, electronics system design, computer architecture, semiconductor devices, hardware security, embedded systems, IoT, high performance computing, large-scale optimization, and machine learning. He has published several papers in national/international conferences and journals.

Application Layer DDOS Attack Detection and Defense Methods

Sadhu Sreenivasarao

Abstract Nowadays in the cyberworld, the Internet has dramatically revolutionized many different fields and different public services globally. Due to any reason, unavailability of these services leads to enormous cost implications and it even affects society. A distributed denial of service (DDOS) attack is a major cybersecurity threat designed to deny services to legitimate users. These days, application-layer distributed denial of service attacks are the main threat on web servers. This paper addressed application layer DDOS attacks typical architecture, common detection mechanisms and defense methods. Numerous Application layer DDOS attack detection techniques have been developed, these can generally be described as detection methods based on the signature, anomalies and hybrids. In anomaly detection, we discussed statistical and machine learning-based methods by researchers. We classified defense mechanisms as defense methods and a combination of detection and defense methods.

Keywords DDos attacks · Application layer · Statistical · Machine leaning · Detection · Defense · Hybrid mechanism

1 Introduction

Denial of Service (DoS) is an active type of cyber-attack targeting availability service which is one of the network security principles. DOS attack disrupts the network or system unavailable to legitimate clients. The attack is performed by flooding the target with traffic or sending information that triggers a crash. Distributed denial of service is a DOS type attack where multiple attackers attempt to make it and to accomplish the same goal. It is a distributed, coordinated, and massive attack. The number of attackers involved sets the DDOS attack apart from the DOS attack. A DOS attack usually involves a small number of attackers, at times even a single

S. Sreenivasarao (✉)
C-DAC, Hyderabad, India
e-mail: sadhur@cdac.in

A. Noor et al. (eds.), *Proceedings of Emerging Trends and Technologies on Intelligent Systems*, Advances in Intelligent Systems and Computing 1371,
https://doi.org/10.1007/978-981-16-3097-2_1

attacker. DDOS attacks are larger and may involve hundreds or even thousands of attackers.

Based on web server resource consumption, DDOS attacks are categorized as volume-based DDOS attacks, protocol-based DDOS attacks, and application-layer DDOS attacks.

Volume-based DDOS attacks attempt to saturate the bandwidth of targeted systems. During this attack, the attackers flood the victim with a lot of packets. The intensity of the attack can be measured in bits per second (bps). UDP floods, ICMP floods (Ping) are examples of volume-based DDOS attacks. In the UDP attacks, the main target of the attacker is the starvation of system resources, the attacker sends a huge quantity of UDP packets to the random ports of the victim's system to launch the attack. This causes the system to repeatedly verify that the application is listening on this port and (when no application is found) replies with an 'Unreachable Destination' ICMP package. It leads to system resource starvation. In an ICMP flood attack, the attacker overwrites the targeted resource with Internet Control Message Protocol (ICMP) request echo (ping) packets, it consumes both outgoing and incoming bandwidth, so it leads significantly to saturate and slow down the victim's network infrastructure.

The primary goal of protocol-based DDOS attacks is to exhaust server resources or "intermediary communication devices" such as firewalls and load balancers instead of bandwidth. As part of this attack, the attackers overwhelm systems and server resources by sending more protocol requests for consuming available resources. The attack strength can be determined by packets per second (PPS). The Ping of Death and SYN flood are examples of Protocol-based DDOS attacks. The initial attack on the Ping of Death is more unusual today. A related attack referred to as the ICMP flood attack is more widespread. SYN flood attacks work by tapping into a TCP connection's handshake process. An SYN flood is designed to disable a server for legitimate traffic by utilizing all available server resources. When sending multiple Initial Connect Request (SYN) packages, the attacker has the ability to overwrite all available ports on a targeted server, make sure the target device responds to legitimate traffic at a slow or no pace.

The main goal of Application layer attacks is to take out or crash an application, an online service, or a website. The purpose of these attacks is to attack the application itself, focus on specific vulnerabilities or problems of the application, which makes the application not available for legitimate users. The attack strength can be determined in requests per second (RPS). HTTP flood attacks and Slowloris are examples of application-layer attacks. In an HTTP attack, the attacker attacks web servers using legitimate HTTP GET or POST requests. HTTP attacks are designed to allow the affected server or application to assign as many resources as possible in direct response to each request. This makes it possible for the attacker to overload the system or application, flooding as many requests as possible. Slowloris uses partial HTTP requests to open connections from a system to an intended web server and maintain those connections open for as long as possible, crushing and slowing the target. It requires minimal bandwidth for operation and only affects target web applications, leaving all remaining services and ports unaffected.

The first DDOS attack occurred in 1996 using the SYN flood attack on one of the oldest internet service providers Panix that was shut down for several days. Application layer DDOS attack was discovered first in 2009, using slow rate HTTP post-attack. 4.83 million DDoS attacks occurred in the first half of 2020, a rise of 151% over the same period in 2019. The number of application-layer attacks seeing a sharp rise in comparison to volume-based attacks in 2020. The Cisco Visual Networking Index (VNI) estimates that there will be 14.5 million DDOS attacks by 2022. Amazon Web Services, the 800-pound gorilla of everything cloud computing, was affected by a massive DDOS attack in February 2020.

2 Application Layer DDOS Attacks

Network and application layer DDOS attacks are important threats. Network Layer DDOS attacks are intended to target network bandwidth. Application Layer DDOS attacks are designed for web server-based applications.

Application layer DDoS attacks have become very popular in today's network world. Attackers create complete TCP connections to the victim server. Attackers create complete TCP connections to the victim server. Application layer DDOS attack architecture [1] is composed of the attacker, and the web server or victim machine. Each botnet contains handlers and zombies. Handlers pass on attack control instructions to the zombies. Zombies directly attack a victim system or web server. Handlers contain a group of zombies. Handlers also transmit information from the zombies to the attacker regarding the victim machine. Handler devices and zombies are compromised machines in an attacker-controlled communication network. Users of these machines do not know that their machines are used in connection with a certain botnet. Zombies send a large number of attack requests to the server. The attacker has a tremendous impact on the victim server by increasing its size (Fig. 1).

3 Application-Layer DDOS Attack Detection Techniques

Detecting application layer DDOS attacks gets harder because more resources and techniques are available to attackers these days. Application Layer DDOS attack detection techniques are usually classified under signature, anomaly and hybrid detection techniques.

Signature detection techniques use patterns and knowledge to find known attacks. Advantages of signature-based detection techniques are simple design and detect the known attacks with minimum false alarms. Drawbacks are unable to detect zero-day attacks and frequent updates are required for new type attack signatures. Anomaly techniques can detect new types of attacks but they can face problems with high positive false alarms and unclassified alerts. It needs initial training. Hybrid detection techniques are a mixture of signature and anomaly methods.

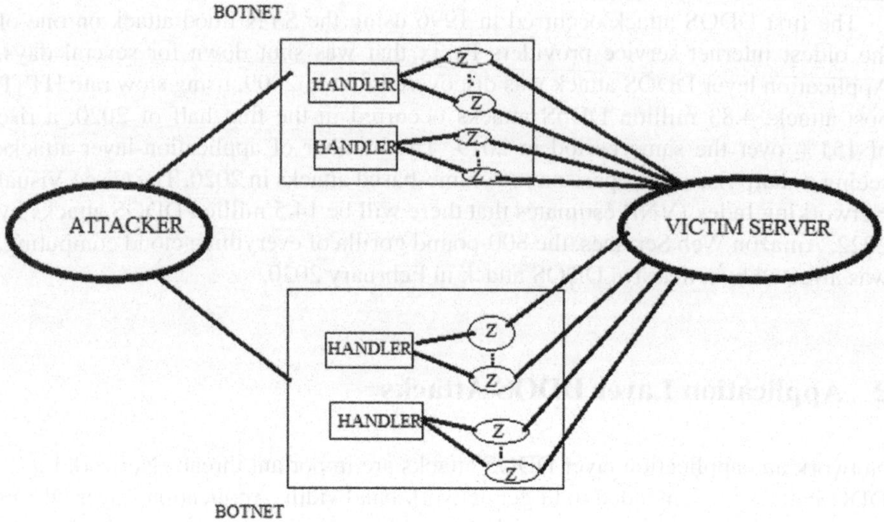

Fig. 1 Typical application layer DDOS attack architecture

3.1 Signature Detection Techniques

Signature-based detection techniques detect the attacks based on patterns, Rules or knowledge, and state. Techniques detect attacks by identifying characters, forms, and patterns within the data. These techniques are easily implemented and hash functions may be used for identification. Rule-based techniques construct a set of rules that specify known attacks. It detects the attacks by using attack signatures in network traffic. It works at a low false-positive rate and has a very high detection rate. It takes a lot of rules to identify all the possible attacks. Rules need pattern matching, so the computational cost for a rule-based system is very high. In state-based techniques attack patterns are represented with state transition diagrams. It examines a stream of events for potential attacks. It operates at a low rate of false positives and uses probabilistic and self-learning techniques.

Yu et al. [2] proposed a Trust Management Helmet (TMH) detection method based on a signature against application-layer DDOS attacks. This is a lightweight mitigating mechanism that differentiates legitimate users from attackers. The user's trust is assessed on the basis of the user's access history. Based on users' trust, schedule the users to their request to access the server.

Xu et al. [3] proposed a novel methodology to detect user behavior for asymmetric application layer DDOS attacks, which is referred to as a random walk model. User browsing behavior is described by page request sequence, based on this build a user browsing behavior model. According to the sequence of page requests, construct the random walk graph. This provides a probability of page transition. Based on the likelihood of page transition predict the sequence of the next page request of the user.

Based on the comparison of the predicted page query sequence with the observed page request sequence, we can measure whether the user is legitimate or attacked.

3.2 Anomaly Detection Techniques:

Anomaly detection techniques calculate anomaly behavior in-network or user traffic in two ways such as using statistical methods and using machine learning methods periodically. A significant discrepancy between the observed pattern and the typical pattern is considered an anomaly.

A statistical method constructs a probability distribution for the behavioral pattern of a user. Low-probability events are viewed as potential attacks. Statistical measurements used to compute discrepancies. Here monitor each packet rather than inspecting data traffic, which means a footprint of the flow. Statistical methods are used for the determination of any significant differences between observed behavior and typical behavior. Statistical methods use univariate, multivariate, and time series models.

Lin et al. [4] proposed RM (rhythm matrix) a statistical model for DDOS attacks of the application layer concentrated on Web sites. It characterizes user access behavior. Anomalous change rate outliers in the rhythm matrix are used to detect DDOS attacks from the application layer and identify malicious hosts. The experiments are tested on three simulated DDOS Application Layer attack modes generated by the LOIC and HOIC tools.

Yang et al. [5] proposed an anomaly-based detection method for low-rate DDOS attacks. This method uses probability distribution functions to differentiate the network traffics. Legitimate and attack traffic is differentiated by the use of widespread entropy and information distance metrics. Experiments show that it efficiently detects the first DDOS attacks at low throughput. It also decreases the false-positive rate.

Khundrakpam et al. [6] proposed an effective fuzzy-GA approach to detect application layer DDOS attacks and differentiate flash events from normal customers according to their access behavior. This method analyzes the statistical parameters of incoming packets. These parameters are included in the fuzzy classification. The genetic algorithm (GA) provides a range of values optimized for the input parameters. These optimized values are applied to fuzzy logic to recognize flash events on the web accessing client behavior. Experiments performed on the CAIDA dataset and results showed that this model provides 98.4 and 97.3% accuracies in the detection of DDOS attacks and flash events accordingly.

The process of retrieving information out of the huge amount of data and performing operations by itself is called machine learning. Machine learning contains several algorithms for finding data patterns and predicting the behaviors of data. Anomaly-based detection techniques using machine learning algorithms to find anomalies in network traffic.

Machine Learning [7] methods are two kinds: supervised and unsupervised. Supervised learning types are divided into classification and regression types. A few examples of supervised learning algorithms are linear regression, logistics regression, decision trees, K-Nearest Neighbors (KNN), and Naive Bayes. PCA, apriori, and K-means are unsupervised algorithms.

Training and testing steps are involved in supervised learning. The class labels and associated features are identified during the training phase. The supervised learning algorithms learn from the data samples during the training phase. In the detection of network traffic anomalies, general class labels are anomalies or not. Supervised algorithms classify the data into associated classes according to training data. Feature selection is also an important step in machine learning to remove unnecessary features. There are many supervised learning techniques, each having its own pros and cons. The classification algorithms are support vector machines, decision trees, neural networks, nearest-neighbor, and naïve Bayes.

Unsupervised learning allows intriguing information to be obtained from entry data sets without a class label. Input data points are generally considered to be a collection of irregular factors. In unsupervised learning, data are grouped into different classes through the learning process. Under the anomaly detection method, unsupervised learning grouped users into normal users and attackers. The main unsupervised learning methods are principal component and cluster analysis.

Ye et al. [8] proposed clustering-based model describes the web user's browsing behavior against DDOS attacks of the application layer. Cluster the client's sessions by extracting four features from the user's behavior. Bot's browsing behaviors are considered abnormal behavior. Normal user browsing behavior is achieved by using a sequence of huge requests of legitimate users. Experiments showed that detection rates are 93.16% and 90.59% on ADS1 and ADS2 data sets accordingly.

Chwaliński et al. [9]. proposed a detection method using clustering and likelihood analysis against HTTP-GET attacks. Likelihood analysis distinguishes attacker requests between legitimate request sequences. Introduced method formatting newer traffic models from web servers. The anomaly of the web sessions is measured according to the probability of behavior of the web sessions. Experiments have been performed on CLARKNET, NASA, and ESHOP datasets. Results showed that it classifies the legitimate and illegitimate users 90% accurately.

Liao et al. [10] proposed the SVD-RM algorithm-based detection method against the four different application layer DDOS attacks. This method uses the L-Kmeans cluster algorithm and SVD-RM classifier. The sparse vector contains the user's request frequency sequence. It is classified by the classification algorithm SVD-RM.

Satyajit et al. [11] proposed a detection method using feature construction and logistic regression. This method builds a system that differentiates between a legitimate user and attacker behavior. This method shows that it effectively classifies attack traffics from legitimate user traffic.

Indraneel et al. [12] proposed a Bio-Inspired anomaly-based HTTP-Flood Attack Detection (BIFAD) technique against HTTP-flood attack. In this approach, the Bat algorithm is used for quick and early identification of HTTP-flood attacks [13]. This algorithm contains two key steps, first defining feature selection to identify the traffic behavior is legitimate or not, and second, for training and testing customize the Bat algorithm. Experiments were carried out on the CAIDA dataset.

3.3 Hybrid Detection Techniques

Traditional detection techniques can't identify new malicious attacks. Even anomaly-based detection techniques show low accuracy and high false alarms. Hybrid detection techniques use both signature and anomaly-based techniques to improve accuracy and minimize false-positive rates.

Jiang et al. [14] proposed a hybrid method based on user traffic behavior against DDOS attacks of the application layer. It addresses two key issues such as accuracy and the time and space complexity of detection algorithms. This detection is a hybrid two-layer detection method, it works on both traffic features and user behavior features. It extracts these two features from the weblog. Neural networks are adopted for anomaly detection for better performance. CICIDS-2017 data set used for experiments. Experiments showed that this method improved accuracy by up to 99.23% while reducing 90% of time cost and recall increased by about 3.70%.

Mikhail et al. [15] proposed an approach to detect anomalies against DDOS attacks of the application layer. In this approach, by utilizing encrypted protocols extract statistical values from network packets. Here, network traffic is analyzed without its decryption. Analyze conversions between webserver and clients and group them into clusters. These clusters are examined with resulting clusters using stacked autoencoders. It is a deep learning class algorithm to classify normal and anomaly behaviors.

Rizwan et al. [16] proposed a clustering-based detection approach against application-layer attacks such as HTTP-POST, HTTP-GET and Slow Rate attacks. Clustering algorithms used to group users who have similar behavior patterns. Correlation is applied to cluster groups to detect attacks. This method maintains two log files for flow monitoring module and user behavior monitoring module during the attack detection process. Flow monitoring module analyzes data flow information and user behavior monitor analyzes user behavior. The performance is measured by using detection rate and accuracy. This method results compared with K-medoid and K-means clustering algorithms. The experiments showed a 95.9% detection rate and 96.5% accuracy.

Table 1 summarizes the application layer DDOS attack detection techniques:

4 Application Layer DDOS Attacks Defense Techniques

Application layer defense mechanisms classified into two types: defense techniques and a combination of detection and defense techniques. Defense techniques only defend the victim server from attacks. Detection and defense techniques do both

Table 1 Application layer DDOS attack detection techniques

Detection technique type	Subcategories of detection type	Name of the detection method/algorithm	Proposed by	Detection based on
Signature detection techniques		Trust management helmet (TMH)	Yu et al.	User's access history
		Random walk model	Xu et al.	User browsing behavior
Anomaly detection techniques	Statistical methods	RM (rhythm matrix)	Lin et al	Anomalous change rate outliers in the rhythm matrix
		Low-rate DDOS attacks detection method	Yang et al	Probability distribution functions to differentiate the network traffics
		Fuzzy-GA approach	Khundrakpam et al	Analysis of the statistical parameters of incoming packets
	Unsupervised machine learning methods	Clustering-based model	Ye et al	Cluster the client's sessions
		Clustering and likelihood analysis against HTTP-GET attacks	Chwaliński et al.	Likelihood analysis distinguishes attacker requests
	Supervised Machine learning Methods	SVD-RM algorithm	Liao et al.	User's request frequency sequence classification
		Feature construction and logistic regression	Satyajit et al.	Classification of legitimate and attacker behavior
		Bio-Inspired anomaly-based HTTP-Flood(BIFAD)	Indraneel et al.	Defining feature selection to identify the traffic behavior
Hybrid detection techniques		Hybrid detection method	Jiang et al.	Works on both traffic features and user behavior features
		Detect anomalies against DDOS attacks	Mikhail et al.	Utilizing encrypted protocols extract statistical values from network packets

(continued)

Table 1 (continued)

Detection technique type	Subcategories of detection type	Name of the detection method/algorithm	Proposed by	Detection based on
		Clustering-based detection approach	Rizwan et al.	Correlation is applied to cluster groups to detect attacks

attack detection and defend the victim servers from the attack. Most of the defense techniques are deployed at the victim-end, some are at the distributed-end.

4.1 Defense Techniques

Ranjan et al. [17] proposed a defense technique for web servers against DDOS attacks of the application layer which are called DDOS-Shield. Reverse proxy integrated into this method. Based on the rate-limiting strategy, detect HTTP session characteristics by using statistical methods. It contains a DDoS-resilient scheduler and a suspicion assignment mechanism. The suspicion assignment mechanism measures each client's suspicious value. The suspicious value assigned based on the users' session history. Based on suspicious values DDOS-resilient schedulers accept session requests.

Ye et al. [18] proposed a defense method for DDOS attacks in the application layer using clustering. In this method, extract statistical features to cluster user behavior. This method uses a hierarchical clustering. When requests come from the users, calculate the deviation between user request and normal cluster which is used to determine the session is normal or suspicious.

Liu et al. [19] proposed a Defense system Against Tilt attacks (DAT). It contains defenders Ingress filtering and Behavior analyzer. The responsibility of Ingress filtering is to filter the packets from identified attackers and handle the abnormal connections. Behavior analysis verifies Tilt attacks. Behavior analyzer analyzes users' behavior and reduces the effect of tilt attacks. DAT defense asymmetric attack.

4.2 Detection and Defense Techniques

Yu et al. [20] proposed a Defense and Offense Wall (DOW) technique to detect asymmetric application-layer attacks such as session flooding, request flooding. This technique does the process of both detection and currency technologies. This detection technology is anomaly-based and utilizes k-means clustering. This method detects and filters asymmetrical attacks and flood requests. Detection technology used to stop suspicious sessions, and currency technology used to encourage legitimate sessions,

Table 2 Application layer DDOS attack defense techniques

Defense technique type	Name of the defense method/algorithm	Proposed by	Defense based on
Defense techniques	DDOS-Shield	Ranjan et al.	Based on the rate-limiting strategy and detect HTTP session characteristics by using statistical methods
	Clustering based AL-DDOS defense	Ye et al.	Uses a hierarchical clustering
	Defense system against tilt attacks (DAT)	Liu et al.	Ingress filtering is to filter the packets and behavior analysis verifies Tilt attacks
Detection and defense techniques	Defense and offense wall (DOW)	Yu et al.	Detects and filters asymmetrical attacks and flood requests
	Signature-based detection and blocking method	Sivabalan et al.	Web page request rate analysis

which are used to defend against session flooding. In this method, the client session rate is treated as a currency in the encouragement model. So, DOW guarantees the legitimate user's service requests can be processed in time.

Sivabalan et al. [21] proposed an approach to detect and defend DDOS attacks of the application layer using signature-based detection and blocking methods of the AYAH or CAPTCHAs web page. In this method, based on web page request rate analysis, the webserver signature generator generates each client signatures and upgrades the signature database. At the time of the user's request, the server first verifies the signature with a signature database and determines if the client is legitimate or not. It sends CAPTHAs to the user when the load on the server goes above the threshold, based on user response, it determines dynamically that the client belongs to legitimate or suspicious.

Table 2 summarizes the application layer DDOS attack defense techniques:

5 Conclusion

This paper attempted to provide a survey of proposed Application layer DDOS attack detection techniques and defense techniques. This survey concludes the hybrid-based detection techniques are so effective and efficient. It also concludes application layer DDOS attack defense with detection is efficient. This survey is useful for researchers looking into application layer DDOS attacks. We plan to develop a hybrid detection and defense method using deep learning algorithms to counter the DDOS attacks of the application layer.

References

1. Dasari, K.B., Nagaraju, D.: Distributed denial of service attacks, tools and defense mechanisms. IJPAM **120**(6), 3423–3437 (2018). ISSN: 1314-3395
2. Yu, J., Fang, C., Lu, L., Li, Z.: Mitigating application layer distributed denial of service attacks via effective trust management. IET Commun. **4**(16). https://doi.org/10.1049/iet-com.2009.0809
3. Xu, C., Zhao, G., Xie, G., Yu, S.: Detection on application layer DDoS using random walk model. In: IEEE ICC 2014, Communication and Information Systems Security Symposium, pp. 707–712, Sydney, NSW, 10–14 June 2014
4. Lin, H., Cao, S., Wu, J., Cao, Z., Wang, F.: Identifying application-layer DDoS attacks based on request rhythm matrices. IEEE Access **7** (2019). https://doi.org/10.1109/ACCESS.2019.2950820
5. Xiang, Y., Li, K., Zhou, W.: Low-rate DDoS attacks detection and traceback by using new information metrics. IEEE **6**(2) (2011)
6. Johnson Singh, K., Thongam, K., De, T.: Detection and differentiation of application-layer DDoS attack from flash events using fuzzy-GA computation. In: IET Information Security, E-First on 21st June 2018. ISSN 1751-8709. https://doi.org/10.1049/iet-ifs.2017.0500
7. Indraneel, S., Vuppala, V.P.K.: HTTP flood attack detection in application layer using machine learning metrics and bio-inspired bat algorithm. Appl. Comput. Inform. (2017)
8. Ye, C., Zheng, K., She, C.: Application layer DDoS detection using clustering analysis. In: IEEE, 2nd International Conference on Computer Science and Network Technology, 10 June 2013. https://doi.org/10.1109/ICCSNT.2012.6526103
9. Chwaliński, P., Belavkin, R., Cheng, X.: Detection of application layer DDoS attack with clustering and likelihood analysis. In: Globecom 2013 Big Data Workshop
10. Liao, Q., Li, H., Kang, S., Liu, C.: Application layer DDoS attack detection using a cluster with a label based on sparse vector decomposition and rhythm matching. Secur. Commun. Netw. **8**, 3111–3120 (2015). https://doi.org/10.1002/sec.1236
11. Selvakumar, S., Yadav, S.: Detection of application-layer DDoS attack by modeling user behavior using logistic regression. In: IEEE 4th International Conference on Reliability, India, 15 Sept 2015
12. Sreeram, I., Praveen Kumar, V.: HTTP flood attack detection in application layer using machine learning metrics and bio-inspired bat algorithm. In: IET Inform. Secur. **15**(1) (2019)
13. Choi, J., Choi, C., Ko, B., Kim, P.: A method of DDoS attack detection using HTTP packet pattern and rule engine in the cloud computing environment. Soft Comput. **18**(9), 1697–1703 (2014)
14. Jiang, J., Yu, Q., Yu, M., Li, G., Chen, J., Liu, K., Liu, C., Huang, W.: ALDD: a hybrid traffic-user behavior detection method for application layer DDoS. In: Published: 17th IEEE International Conference on Trust, Security and Privacy in Computing and Communications/12th IEEE International Conference on Big Data Science and Engineering
15. Zolotukhin, M., Ham Al Linen, T., Kokkonen, T., Siltanen, J.: Increasing web service availability by detecting application-layer DDoS attacks in encrypted traffic. In: Published: 23rd International Conference on Telecommunications (ICT), Date Added to IEEE Xplore, 30 June 2016. ISBN: 978-1-5090-1990-8. https://doi.org/10.1109/ICT.2016.7500408
16. Rahman, R., Tomar, D.S., Jijin, A.V.: Application-layer DDOS attack detection using hybrid machine learning approach. IJSAA **11**
17. Ranjan, S., Robinson, J., Chen, F.: DDoS-Shield: DDoS-resilient scheduling to counter application layer attacks. IEEE/ACM Trans. Networking **17**(1), 26–39 (2009). https://doi.org/10.1109/TNET.2008.926503
18. Ye, C., Zheng, K., She, C.: Application layer DDOS detection using clustering analysis. In: 2nd International Conference on Computer Science and Network Technology, Date Added to IEEE Xplore: 10 June 2013. https://doi.org/10.1109/ICCSNT.2012.6526103
19. Liu, H.I., Chang, K.C.: Defending systems against tilt DDoS attacks. Telecommun. Syst. 22–27 (2011)

20. Yu, J., Li, Z., Chen, H., Chen, X.: a detection and offense mechanism to defend against application layer DDoS attacks. In: 3rd International Conference on Networking and Services (ICNS 07), p 54, 19–25 June 2007
21. Sivabalan, S., Radcliffe, P.J.: A novel framework to detect and block DDoS attacks at the application layer. IEEE TENCON Spring Conference, Sydney (2013)

Identification and Analysis of Threat Vector for Security Evaluation of LAN

Simranjeet Kaur Rehan and Rekha Saraswat

Abstract *Introduction*: Security of LAN in today's world is a challenge we face on daily basis, despite applying authentication algorithms or strong firewalls on networks to counteract the internal as well as external threats. The analyses of the LAN network from within for any threats can be prevented from any exploitable link. To understand this, the security of LAN network must be evaluated using security evaluation algorithms including attack graph approach, intelligent algorithms, etc. Security evaluations entertain the factors that can act as threat vectors. This study presents an approach to identify such threat vectors and to evaluate security from within a wired LAN. *Methods*: The threat vectors of LAN were identified and categorized as per their weightage, i.e., priority. Threat vectors were based on the level of priorities associated with their risk of exploitation toward the network and their respective control which was evaluated with a criterion. The most exploitable or risky path was determined using attack graph approach once all the nodes with network were provided respective weightage/risk factor. *Results and Conclusion*: The internal working of network and its vulnerabilities are identified, and associated risk factors are observed to monitor the most and least impactful vector so that the overall security of internal network can be enhanced.

Keywords Security evaluation · Threat vectors · Attack graph · Traffic light method · 9-scale comparison technique

1 Introduction

Today's world creates zettabytes of data every day it is imperative to safeguard it being infringed and used for nefarious purposes. This evolved the field of "security." Despite the availability of several antiviruses, algorithms, digital techniques, authentication techniques, integrity constraints, there is always a threat for breach in the

S. K. Rehan (✉) · R. Saraswat
C-DAC, Noida, Uttar Pradesh, India

R. Saraswat
e-mail: rekhasaraswat@cdac.in

A. Noor et al. (eds.), *Proceedings of Emerging Trends and Technologies on Intelligent Systems*, Advances in Intelligent Systems and Computing 1371,
https://doi.org/10.1007/978-981-16-3097-2_2

security. There is an unmet need to attain an absolute sense of security. As we are developing ways to secure our networks, researchers/hackers are devising strategies to breach them, at a similar pace. Therefore, it is imperative to periodically evaluate, vigil and upgrade network security such that even the weakest link is protected from being exploited.

Basically, security gives two types of protections [1, 2]. The first secures our data from getting accessed or manipulated by unauthorized source. Another provides computer security which helps to secure data using various encryptions techniques and stop attackers/hackers to invade our privacy [3]. The evaluation of nodes in the network through which unauthorized access is feasible, needs to be identified so that the security and the efficiency level of network are raised. This necessitates the knowledge of the uprising field of "security evaluation."

Security evaluation is a crucial step that every organization must incorporate in their security guidelines. It provides so many algorithms or techniques to test the various types of exploit present within the network. These exploits are differentiated on the basis of the nature of network, i.e., wireless or wired results in internal or external exploits [4], respectively. These exploits further bifurcate according to the nodal structure of the organization. Once exploit is identified, in response its solution is developed and incorporated into the respective security guideline. This enhances the efficiency and safety of the network and makes the network easily detectable for any exploits.

1.1 Methods

Despite the application of present security evaluation techniques to secure communication and data, there exists loophole through which attackers can breach. In the present paper, a top-down approach to observe threat vectors present in wired LAN bifurcated till nodal levels was followed to examine threat vector, and their relationship toward other vectors was observed in terms of risk that they may represent. Once these vectors were analyzed, they were weighed [5] and prioritized and then using cumulative risk function their risk was calculated. Attained variables were accommodated as inputs in a security evaluation technique. The outputs were observed in terms of most vulnerable path and likely to be exploited by the attacker.

2 Literature Review

Several researchers have proposed and devised algorithms to mitigate exploitation of the network. Below follows a brief of such work that is taken as baseline to implement the research presented in the paper.

i. Yousefi et al. [6] proposed an amalgamated framework depending on two different techniques, attack graph and Q-learning techniques. To minimize high complexities of attack graph, Q-learning technique was preferred. It assembles the current states of the network to model a reward-based environment. The framework used attack graph to show the interdependencies of the vulnerabilities in system and simplified them into a transition graph. The vulnerabilities are prioritized and tackled accordingly.
ii. Yi et al. [7] presented a LAPA framework to detect and analyze potential security risks and attacks. The attack tracks and graphs are built to model how zero-day vulnerability would affect the network. Reasoning rules were formulated to form logical attack graphs for various attacks, to deduce the performance and evaluation efficacy of the framework.
iii. Ming [8] proposed a comparative study on evaluating a network for the vulnerabilities it possessed. The study used two algorithms based on quantum-based particle swarm algorithm. To analyze the optimal points of training data, the fitness functions are used, and results are analyzed for both algorithms.
iv. Kumar et al. [9] evaluated a security network based on MulVAL framework and CVSS to yield the vulnerability values for 3 host machines. Using the scanned data attained from host machines, logical attack graph was formed to deduce exploitability and impact values. A metric was also established to score the vulnerabilities and to access the scope and range of impact caused on network risk.
v. Lan et al. [10] established an indexed evaluation system for cyberspace defense systems. The concept of thermodynamic entropy was used along with fuzzy theory to quantify the effectiveness of 2-layered index. The index was weighed and normalized to attain entropy coefficients and subordinate degree values, to indicate the effectiveness levels of systems.
vi. Lu et al. [11] established a hierarchical security evaluation indicator system. The indicators are quantified using expert scoring method, to attain quantification and permission values for further processing. A method was modeled to establish a comparison standard and dimensionless gray process, to conduct evaluation on network using indicator values. The results were based on the types of permission values accessed by network users.
vii. Francis et al. [12] used an IAA framework and four applications to set up the modeling environment with purpose of implementing a security model. Model consisted of a client–server component having peer-to-peer fashion communication and three FLOSS applications that worked as text editors. A baseline graph was created to quantify the attack surface of the purpose-driven application at each step successively, using multipliers for each exploit. Each path of the baselined graph was traversed for each exploit until the multipliers are turned to zero.
viii. Liu et al. [13] used Delphi technique to formulate two parameters for security indicator system. Once the data were collected using genetic algorithm,

it was converted into chromosomes and based on threshold and fitness function, population probability was selected whose crossover and mutation were determined.

ix. Zhang et al. [14] used fuzzy AHP model to evaluate the network in which factor and comment sets are used to determine the weight of each factor from nine-scale comparison table which produced a fuzzy evaluation matrix. Then, results were evaluated to determine the security state of the factor.

x. Ge et al. [15] proposed a security evaluating model which depends on three parameters of security, i.e., value of asset loss, threat the attack may cause and lastly the importance of asset. Based on this, the network's attack graph was formulated, and risk associated was analyzed.

3 Implementation

This paper is based on a wired LAN with a defined network consisting of routers, firewalls, switches and systems like pc or laptops, based on which an indexed structure of nodes and edges representing the components of network and flow of data between them was created. The following five steps were applied to evaluate the network.

3.1 *Analyzing Threat Vectors*

Using Indian risk surveys published by ficci [16–18], six primary threat vectors are identified, which are further bifurcated into 44 tertiary threat vectors based on checklists and risk reports of SANS [19], CISCO [20], ISO 27001 [21] and many government organizations. Table 1 represents evaluation index with further classification of 6 primary threat vectors into secondary vectors.

3.2 *Weightages of Threat Vectors*

Weightages of threat vectors are quantified based on 9-scale comparison table [14] as depicted in Table 2, to construct the comparison judgment matrix (M) [14] like Fig. 1. Figure 1 shows values $r_{11}, r_{12}, r_{13}, \ldots, r_{nn}$ depicting the relative significance between m_i and m_j.

After constructing the matrix, eigen values and normalized eigen vectors are calculated, which are attributed as respective weightages.

$$W = (w_1, w_2, w_3, \ldots, w_n)$$

Table 1 Evaluation index

Identifier	Primary index	Identifier	Secondary index
Υ_1	Security misconfiguration [16, 17]	Υ_{11}	Firewall misconfiguration [19]
		Υ_{12}	Router misconfiguration [20]
		Υ_{13}	Switch misconfiguration [22]
Υ_2	Logical security [16, 17]	Υ_{21}	Backing up of data [23]
		Υ_{22}	Data recovery criteria [24]
		Υ_{23}	System logs [25]
Υ_3	Nodal misconfiguration [16, 17]	Υ_{31}	Patch management [26, 27]
		Υ_{32}	Open ports [28]
		Υ_{33}	Security settings [23, 29]
Υ_4	Security awareness [16, 17]	Υ_{41}	Agreements [21, 30]
		Υ_{42}	Security safeguards [31, 32]
Υ_5	Node value	Υ_{51}	Information contained [33–35]
		Υ_{52}	Connection/linkage [33–35]
		Υ_{53}	Hierarchy of node [23, 34, 35]
Υ_6	Internal attacks [16, 17]	Υ_{61}	Authorization restriction [23, 29]
		Υ_{62}	Policies of organization [23, 28]

Table 2 9-Scale comparison table

9-Scales	Description
1	Significance of m_i equal to m_j
3	Significance of m_i slightly more than m_j
5	Significance of m_i clearly more than m_j
7	Significance of m_i strongly more than m_j
9	Significance of m_i vastly more than m_j
2, 4, 6, 8	Depends upon the neighboring values

Fig. 1 Judgment matrix (*M*)

M	m_1	m_2	m_3	…	m_n
m_1	r_{11}	r_{12}	r_{13}	…	r_{1n}
m_2	r_{21}	r_{22}	r_{23}	…	r_{2n}
m_3	r_{31}	r_{32}	r_{33}	…	r_{3n}
⋮	⋮	⋮	⋮		⋮
m_n	r_{n1}	r_{n2}	r_{n2}	…	r_{nn}

Table 3 Switch misconfiguration threat vector's index

Secondary index	Identifier	Tertiary index	Identifier
Switch misconfiguration [22]	m	Vlan table contains unassigned vlan	m_1
		Unused port administratively up	m_2
		Access port with portfast disabled	m_3

Table 4 Matrix for switch misconfiguration vector

m	m_1	m_2	m_3
m_1	1	3	9
m_2	1/3	1	3
m_3	1/9	1/3	1

Table 5 Risk evaluation set

Risk element	R_1	R_2	R_3	R_4	R_5
Risk level	Extremely low	Low	Moderate	High	Extremely high
Value	1	2	3	4	5

To have a better interpretation, let us take an example of weightages calculations for switch misconfiguration threat vector's index depicted in Table 3.

Firstly, a comparison matrix depicted in Table 4 was constructed based on the switch configuration summit archive report published by SANS in 2019 for "ae solutions" [22], and values are provided using the 9-scale comparison table as conveyed in Table 2. Then, eigen values are calculated ($\lambda = 0, 0, 3$) and normalized eigen vector is observed and conveyed as respective weightages of $[m_1, m_2, m_3]$. For this matrix, the value of normalized eigen vector was: [0.94345, 0.31448, 0.10482].

3.3 Risk Calculation Criterion

In a network, different nodes could possess same vector with various level of risks depending on how well the vector controls[1] are implemented. Therefore, a set of controls were generated and were used as control checklist to analyze the respective risk of exploitation of vectors. For this a risk evaluation set [14] was established to define the different level of risk that could be possessed as depicted in Table 5.

The set values depend on how vector controls are marked using traffic light technique where green, amber and red color blocks are used to evaluate whether they are properly implemented or not.

[1] Each vector has various aspects through which they could be monitored whether if they are properly checked and implemented or not.

Using weightages and established risk evaluation set, risk associated with each vector was calculated using the following risk function,

$$f = [W] \cdot [E]^T \tag{1}$$

And cumulative risk function,

$$F = \sum f \tag{2}$$

3.4 *Risk Calculations*

Using risk evaluation set, the risk values for vectors were observed, along with weightages, derived risk values for all threat vectors, using (1) and (2).

3.5 *Applying the Algorithm*

After the risk of each vector was calculated, their accumulative node values were determined which were used as weights of nodes to the structured network created in the beginning of the approach to which security evaluation technique, i.e., attack graph algorithm [15], was applied. In this algorithm, a weighed network was used to find which path in the network holds maximum risk of exploitation.

4 Case Study

For better insight of the approach, let us take an example of a wired network as shown in Fig. 2. The network consists of 9 nodes, comprising personal computers, switches, routers and firewalls, depicting communication between PC1 and PC2. Using the above evaluation index and weightage calculations, risk of each node was calculated. After the nodes are weighed, the attack graph algorithm [15] was applied to the structured and weighed network shown in Fig. 3. The result observed was in terms of the most vulnerable path that attacker may exploit with node-to-node risk value.

Most vulnerable path: {I (4.888940) – G (0.958630) – E (0.733780) – C (0.405230) – A (0.202660)}.

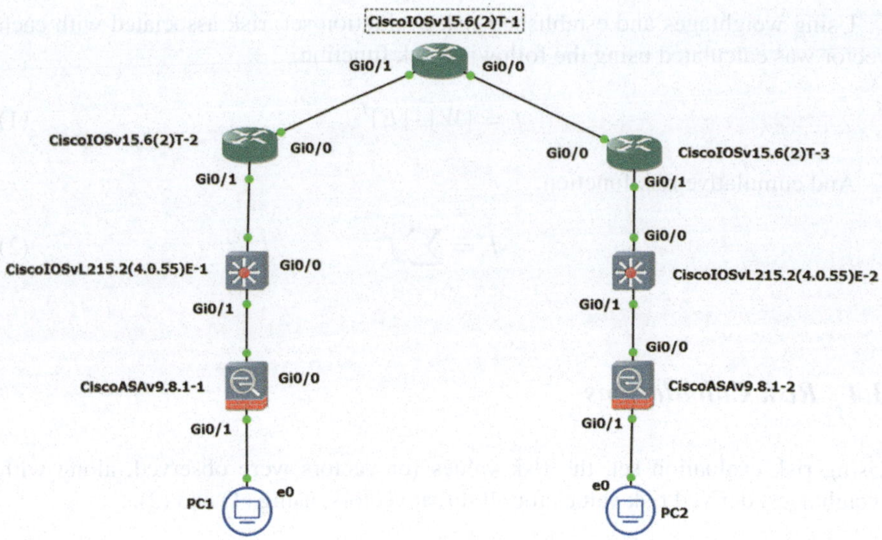

Fig. 2 Wired LAN network

Fig. 3 Structured and weighed network

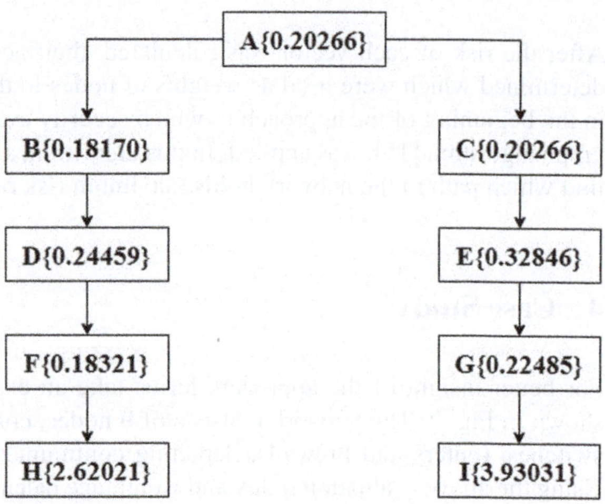

5 Observations and Results

- Using the defined evaluation index, the internal working of threat vectors could be easily studied. It could be easily understood how vectors work internally, i.e., how the network risk will be affected if one of the vectors respective controls are not implemented properly. It was observed that the relationship between the risk and threat vector implementation is linear and node-to-node risk evaluation could be easily retrieved.

- Different scenarios were created with permutation and combination of implemented controls and evaluation values, to observe the vulnerabilities associated with these threat vectors correlated with each other. It was observed that threat vectors like security misconfiguration and logical security can hold more risk toward the security of networks. Similarly, there are other vectors which when exploited can harm information contained within the network.

Using the risk values obtained for each threat vector, dependencies between them can also be analyzed. For example, what will happen when threat vector "node value" of node A and "logical security" of node B of a network are exploited simultaneously? The risk of the network will increase depending on how these exploits are made, what is their risk level and how their controls implemented. Depending on all these factors, the total path risk from node A to node B will be determined, so that necessary measure can be mitigated.

With thorough understanding, the most vulnerable threat vectors which can make network vulnerable to many exploits could be identified.

6 Conclusions

The paper depicts an approach to calculate overall security risk of a network and risk associated with nodes in a network using various controls and vectors.

In this paper, the threat vectors are properly analyzed and are classified till node levels, into 44 tertiary vectors. Using these vectors, a control implementation checklist was prepared to know how well these vectors are controlled so that exploitation can be minimized. From this the risk corresponding to each node is observed so that node-to-node risk evaluation can be performed.

Using this approach, the most impactful threat vector and most vulnerable node of our network could be identified. These advancements provide a systematic approach to keep the vulnerabilities in check and to enquire how much attention a vector requires, to increase the overall network security.

References

1. Prabhakar, S.: Network security in digitalization: attacks and defence. Int. J. Res. Comput. Appl. Robot. **5**(5), 46–52 (2017)
2. Witzke, E.L.: Computer network security: then and now. In: 2016 International Carnahan Conference on Security Technology (ICCST), pp. 1–7. IEEE, Orlando, FL, USA (2016). https://doi.org/10.1109/CCST.2016.7815710
3. Awad, F., Al-Refai, M., Al-Qerem, A.: Rogue access point localization using particle swarm optimization. In: 2017 8th International Conference on Information and Communication Systems (ICICS), pp. 282–286. IEEE, Irbid, Jordan (2017). https://doi.org/10.1109/IACS.2017.7921985

4. Shuping, Y., Yingyan, G.: Network security situation quantitative evaluation based on the classification of attacks in attack-defense confrontation environment. In: 2009 Chinese Control and Decision Conference (CCDC), pp. 6014–6019. IEEE, Guilin, China (2009). https://doi.org/10.1109/CCDC.2009.5195279
5. Qu, Z., Wang, X.: Application of comprehensive fuzzy evaluation in Lan security. In: 2009 International Conference on Networks Security, Wireless Communications and Trusted Computing, pp. 209–213. IEEE, Wuhan, China (2009). https://doi.org/10.1109/NSWCTC.2009.44
6. Yousefi, M., Mtetwa, N., Zhang, Y., Tianfield, H.: A reinforcement learning approach for attack graph analysis. In: 2018 17th International Conference on Trust, Security and Privacy in Computing and Communications/12th International Conference on Big Data Science and Engineering (TrustCom/BigDataSE), pp. 212–217. IEEE, New York, NY, USA (2018). https://doi.org/10.1109/TrustCom/BigDataSE.2018.00041
7. Yi, F., Cai, H.Y., Xin, F.Z.: A logic-based attack graph for analyzing network security risk against potential attack. In: 2018 IEEE International Conference on Networking, Architecture and Storage (NAS), pp. 1–4. IEEE, Chongqing, China (2018). https://doi.org/10.1109/NAS.2018.8515733
8. Ming, S.: Computer network security evaluation based on intelligent algorithm. In: 2017 Sixth International Conference on Future Generation Communication Technologies (FGCT), pp. 88–91. IEEE, Dublin, Ireland (2017). https://doi.org/10.1109/FGCT.2017.8103737
9. Kumar, S., Negi, A., Prasad, K., Mahanti, A.: Evaluation of network risk using attack graph based security metrics. In: 2016 14th International Conference on Dependable, Autonomic and Secure Computing, 14th International Conference on Pervasive Intelligence and Computing, 2nd International Conference on Big Data Intelligence and Computing and Cyber Science and Technology Congress (DASC/PiCom/DataCom/CyberSciTech), pp. 91–93. IEEE, Auckland, New Zealand (2016). https://doi.org/10.1109/DASC-PICom-DataCom-CyberSciTec.2016.30
10. Lan, Y., Liu, S., Lin, L., Ma, Y.: Effectiveness evaluation on cyberspace security defense system. In: 2015 International Conference on Network and Information Systems for Computers, pp. 576–579. IEEE, Wuhan, China (2015). https://doi.org/10.1109/ICNISC.2015.120
11. Lu, Z., Zhou, Y.: The evaluation model for network security. In: 2014 Fourth International Conference on Communication Systems and Network Technologies, pp. 690–694. IEEE, Bhopal, India (2014). https://doi.org/10.1109/CSNT.2014.145
12. Manning, F.J., Mitropoulos, F.J.: Utilizing attack graphs to measure the efficacy of security frameworks across multiple applications. In: 2014 47th Hawaii International Conference on System Sciences, pp. 4915–4920. IEEE, Waikoloa, HI, USA (2014). https://doi.org/10.1109/HICSS.2014.602
13. Liu, S., Fang, Y.: Application research in computer network security evaluation based on genetic algorithm. In: 2012 International Symposium on Instrumentation and Measurement, Sensor Network and Automation (IMSNA), pp. 468–470. IEEE, Sanya, China (2012). https://doi.org/10.1109/MSNA.2012.6324623
14. Zhang, R., Huang, L., Xiao, M.: Security evaluation for wireless network based on fuzzy-ahp with variable weight. In: 2010 Second International Conference on Networks Security, Wireless Communications and Trusted Computing, pp. 490–493. IEEE, Wuhan, China (2010). https://doi.org/10.1109/NSWCTC.2010.122
15. Ge, H., Gu, L., Yang, Y., Liu, K.: An attack graph-based network security evaluation model for hierarchical network. In: 2010 International Conference on Information Theory and Information Security, pp. 208–211. IEEE, Beijing, China (2010). https://doi.org/10.1109/ICITIS.2010.5688764
16. India risk survey 2017. http://www.ficci.in/pressrelease/2806/india-risk-survey-ficci-press-release.pdf, last accessed 2020/3/1
17. India risk survey 2018. http://ficci.in/Sedocument/20450/India%20Risk%20Survey%20-%202018.pdf, last accessed 2020/3/1
18. India risk survey 2019. http://ficci.in/Sedocument/20487/India-Risk-Survey-2019-ficci.pdf, last accessed 2020/3/1

19. SANS Firewall checklist: https://www.sans.org/media/score/checklists/FirewallChecklist.pdf, last accessed 2020/1/11
20. ISO 27001: Router security audit checklist. http://tajdini.net/ebook/Itsec-router-auditchecklist.pdf, last accessed 2020/1/11
21. ISO 27001-2013: Auditor checklist. https://www.rapidfiretools.com/wpcontent/uploads/2018/04/ISO-27001-Auditor-Checklist.pdf, last accessed 2020/1/11
22. Switch misconfiguration. https://www.sans.org/cyber-security-summit/archives/file/summit-archive-1553007190.pdf, last accessed 2020/2/1
23. Security controls. https://nvd.nist.gov/download/800-53/800-53-controls.xml, last accessed 2020/2/7
24. Patch management controls. https://www.Us-cert.Gov/sites/default/files/recommended_practices/rp_patch_management_s508c.Pdf, last accessed 2020/2/18
25. Syslog. https://nvlpubs.Nist.Gov/nistpubs/legacy/SP/nistspecialpublication800-92.Pdf, last accessed 2020/2/20
26. Data backup. https://www.library.cmu.edu/datapub/dms/data/secure, last accessed 2020/2/14
27. Patch management. https://www.infosec.gov.hk/english/technical/files/patch.pdf, last accessed 2020/2/14
28. Open port configuration. https://pentest-tools.com/public/sample-reports/openvas-scan-sample-report.pdf, last accessed 2020/2/1
29. Security policies. https://www.exabeam.com/information-security/information-security-policy, last accessed 2020/2/6
30. Security agreements. https://ecfsapi.fcc.gov/file/6516186048.pdf, last accessed 2020/1/13
31. Safeguards. https://www.hhs.gov/sites/default/files/ocr/privacy/hipaa/administrative/securityrule/techsafeguards.pdf, last accessed 2020/2/7
32. Security safeguards. https://www.hhs.gov/sites/default/files/ocr/privacy/hipaa/administrative/securityrule/techsafeguards.pdf, last accessed 2020/3/1
33. Data recovery. https://security.Uci.Edu/security-plan/plan-control.Html, last accessed 2020/2/24
34. Security structure for nodal information. https://nvlpubs.nist.gov/nistpubs/SpecialPublications/NIST.SP.800-12r1.pdf, last accessed 2020/2/24
35. Security structure for nodal information. https://nvlpubs.nist.gov/nistpubs/Legacy/SP/nistspecialpublication800-115.pdf, last accessed 2020/2/24

Review of Stack-Based Binary Exploitation Techniques

Vanita Jain, Bhanupratap Singh, and Swapnil

Abstract Binary exploitation focuses on making use of flaws and oversights in the program design to accomplish a task that the program wasn't intended toward. Some flaws are fairly common programming errors, while some are not so obvious and give rise to complex exploitation techniques. This paper addresses some of the techniques extensively used to test the vulnerabilities and security holes in the system, and their mitigation techniques. The goal of these techniques is to hijack the target binary's execution flow.

Keywords Binary exploitation · Stack overflow · Buffer overflow · Format string · Exploitation · Return-oriented programming

1 Introduction

A program is a set of instructions, and thus, the execution of a compiled binary performs a specific task. However, the programmer is susceptible to making logical and semantic errors, and thus, the result of the code doesn't always correspond to what the programmer intended it for. Some programming errors are prevalent and can be found almost anywhere if the code review is not done correctly or the programmer/reviewer is not aware of the unintended functionality of a library function. A pattern is established, in regards to which, some generalized exploit techniques [1] were devised that take advantage of such mistakes. These techniques have evolved over the years and are employed in various situations.

Most exploits make use of memory corruption [2] for the execution of arbitrary code injected into the memory. This arbitrary code can be targeted to achieve some

V. Jain (✉) · B. Singh · Swapnil
Bharati Vidyapeeth's College of Engineering, New Delhi, India
e-mail: vanita.jain@bharatividyapeeth.edu

B. Singh
e-mail: bhanupratapsingh.it2@bvp.edu.in

Swapnil
e-mail: swapnilkumar.it2@bvp.edu.in

A. Noor et al. (eds.), *Proceedings of Emerging Trends and Technologies on Intelligent Systems*, Advances in Intelligent Systems and Computing 1371,
https://doi.org/10.1007/978-981-16-3097-2_3

specific tasks, such as information leak. Such instances mostly exist due to improper handling of some cases which work well until exploited. The program may then crash, or the attacker may gain control of the execution flow in a carefully controlled environment. Another instance that may lead to exploitable programmer errors is the quick modification of a program to extend its functionality. Even though this extension allows support for a larger functionality, it also increases the program's complexity, which makes an oversight in the program more likely to occur.

The purpose of the paper is to review various exploitation techniques in use by the attackers, which allows them to take control of a program and use it to their advantage. The paper first discusses what stack-based exploitation actually means and the importance of stack in Virtual Memory and Computer Architecture. We then move to the two techniques which allow data to be overwritten on the stack and its effects. We then discuss the various mitigations to prevent exploitation of bugs and the bypass for those mitigations. We then finally discuss the case study of CVE-2019–18634 in which we develop our exploit for a buffer overflow.

2 Stack-Based Exploitation

In the context of computer science, a stack [3] is a data structure that can be visualized as a stack of plates, holding an array of elements. It supports a push method to add an element at the top and a pop method to remove the element at the top.

The stack data structure as shown in Fig. 1 is used as a temporary storage area in virtual memory that allows the processor to store and retrieve data in memory. When a process starts on the system, the process loader sets up the stack in the virtual memory [4, 5]. This stack is used by the function to store local variables, the parameters, and the return address to the caller function.

When function X calls function Y, X pushes the address of next instruction on the stack and jumps to function Y. When function Y finishes execution of the program, it pops the return address and jumps to this address to resume execution back in function X [7]. During the exploitation of a binary, the attacker attempts to change the control flow [8] of the program by modifying either the return address of the function or use logic errors to alter variable values which allow us to guide the program to execute arbitrary code [9] on the system.

This paper discusses both the vulnerabilities and the exploitation technique for each case. However, it should be noted that a vulnerability need not be exploitable, and it requires creativity to convert a vulnerability into an exploit. The basic principle of exploitation is to gain/create a write-what-where primitive, also known as w-w-w primitive [10]. A write-what-where primitive allows the attacker to write arbitrary data wherever they want in the memory. This primitive can be used to overwrite the return address or corrupt values in memory to guide the program in the direction the attacker wants.

Fig. 1 Stack layout [6]

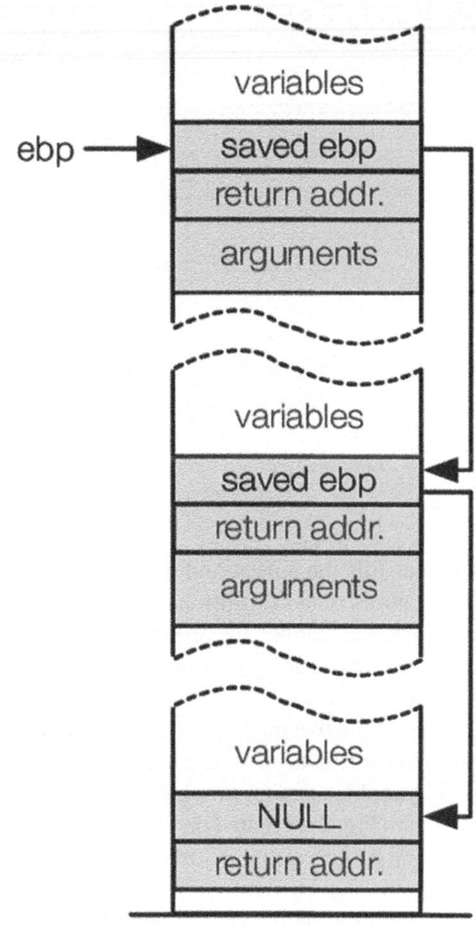

3 Buffer Overflow

Buffer overflow [11] is one of the most common vulnerabilities that has been around as long as the computers have and are still found to be part of major vulnerabilities discovered today. C [12] being a high-level programming language makes it the programmer's responsibility to verify data integrity since compiler integrated data integrity checks would significantly slow down the execution because of the integrity checks on each variable. Therefore, if the programmer isn't careful, the program might be vulnerable to memory leaks and buffer overflow. Thus, there are no in-built safety measures to ensure the contents of a variable are in accordance with the allocated memory.

Buffer overflow in action is the insertion of more bytes into the memory than the number of bytes space has been allocated for. In such a scenario, the required number

Fig. 2 Stack frame [13]

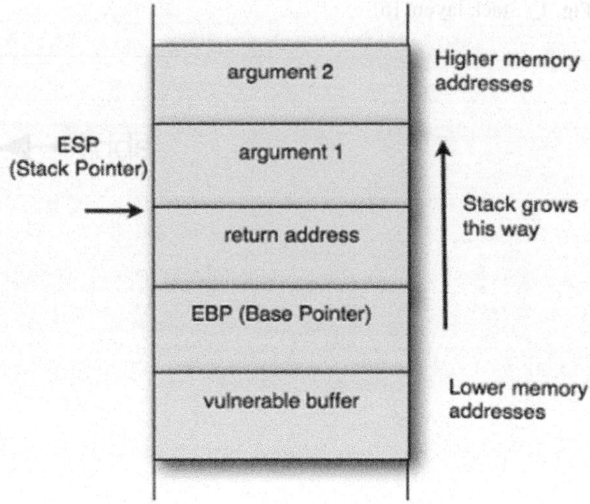

of bytes fills the buffer, and the extra bytes overwrite the bytes following the buffer. This causes data corruption in memory. If any critical bytes get overwritten in an uncontrolled manner, the program is bound to crash. Within a controlled environment, a crafted input can be used to exploit this overflow and take control of the execution flow in the binary.

One of the five memory segments used by programs is the stack. It is a LIFO (Last-In-First-Out) data structure used to support execution flow and provide context for local variables and arguments when functions are called. A new stack frame [14] as shown in Fig. 2 is created on the stack when a function is called. This process is called function prolog as the instruction pointer jumps to the first instruction of the called function. A stack frame consists of local variables declared in that function along with the return address. This address is popped off of the stack into the instruction pointer when the function is done.

Buffer overflow can be exploited to overwrite the said return address. The execution flow will try to use this value to return the function. If it is filled with garbage value, the program crashes. However, if the overflow is controlled and the return address is replaced by a valid address, the execution flow will jump to that specific location. This location can be targeted from the binary itself. A shellcode [15] is a piece of code which is the payload for our exploit. This shellcode can act as a backdoor for us after exploiting the system. If the attacker controls the data on the stack, they can put a shellcode on the stack, and the return address of the function can then be changed to the address of the shellcode to execute our payload. The other common sections are *.data* and *.bss*, which are used to store initialized and uninitialized global and static objects. In case of an overflow in the buffer stored in *.bss*/*.data* section, it can allow modification of internal flags that guide the program execution, as demonstrated in the *sudo* vulnerability (CVE-2019–18634) which led to local privilege escalation [16]. As these sections do not contain return addresses, it is difficult to create a w-w-w primitive. For exploitation of such vulnerability, it is important to understand how the data stored in the *bss/data* section affect the program flow and require creativity to convert the vulnerability to an exploit.

4 Format Strings

The basic format strings, as shown in Table 1, are used extensively by formatting functions like *printf()/sprintf()*, are simply evaluated and perform value substitution anytime a format parameter is encountered. Every format parameter is expecting a variable, so if there are two format parameters in given format string, then two more arguments specifying the variables need to be provided to the function after the string to resolve those parameters. The function expects these values on the stack following the format string to be the arguments passed to the function and thus reads those values. A security hole [17–19] occurs when either the programmer does not provide the right number of arguments, or if a user input string is directly passed to a function that will evaluate this string. In such a case, the values preceding the format string on the stack do not correspond to the arguments, but to the values stored on the stack by other instructions. This can be used to leak the stack and obtain useful information regarding the current process.

4.1 *Reading from an Arbitrary Memory Address*

The '*%x*' and the '*%s*' format parameters, for hexadecimal and character formats, respectively, are extensively used to leak stack data when a format string vulnerability is encountered. This gives a fair idea of the location of the addresses on the stack. The user input may include a memory address followed by the aforementioned format parameters so as to read from this memory address.

Table 1 Format parameters [20]

Format string	Output type
%d	Integer
%u	Unsigned integer
%x	Hexadecimal
%n	Bytes written so far

4.2 Writing to an Arbitrary Memory Address

The '*%n*' format parameter works a bit differently; it writes the number of bytes to a variable provided in the specified argument, in contrast to other format parameters that read and display data. This functionality can be extended to write data to a memory address. The field-width/padding of the format parameters can be manipulated to reach a certain value, which is equivalent to the number of bytes written, that can then be written to the specified memory address.

4.3 Short Writes

The '*%n*' format specifier writes a whole byte, which isn't always necessary. A length modifier can be used instead to write two-byte shorts by adding an '*h*' between the format parameter.

4.4 Direct Parameter Access

To avoid stepping through each parameter argument sequentially, we can use Direct Parameter Access to access the parameters directly. In the format parameter itself, specify the number of the parameter followed by a '*$*' sign. A backslash might be required in order to tell the command shell to avoid interpreting it as a special character.

4.5 .dtors

It's a special table section made for destructors, and these functions execute just before the *main()* function exits. This section is an array of function addresses terminated by a NULL address. In this section are the addresses of all the functions that have destructor attribute set. Since the write permission is set on this section, the addresses in this section can be overwritten with a memory address directing the program's execution flow to that address when the *main()* returns.

4.6 Overwriting Global Offset Table [21]

The procedure linkage table [22] (PLT) is a section which consists of multiple jmp instructions, each corresponding to an address of a function defined in the loaded library. It works like a trampoline, each time a function from a library needs to be called; its control will pass through the PLT. These jmp instructions jump to pointers to addresses. These addresses exist in the section called the global offset table (GOT), which generally has write permission. If an address of a function, say *exit()*, is overwritten with the address of a shellcode, then the program will actually redirect execution to the shellcode when the function is called.

5 Mitigation Techniques

The two vulnerabilities (buffer overflow and format strings) are the stack-based vulnerabilities that can provide w-w-w primitive. Modern operating systems have mitigation techniques enabled, which make it harder to exploit these bug classes. This paper discusses two operating system level mitigation techniques supported by all modern OS, DEP (Data Execution Prevention) also known as NX (No Execute) and ASLR (Address Space Layout Randomization).

5.1 NX/DEP

NX/DEP [23, 24] was introduced in Linux kernel 2.6.8 on August 14, 2004 and is one of the pillars of modern exploit mitigation technologies. Before its introduction, the permission of heap and stack was read-write-execute. That means if user-controlled data on stack contain valid instructions, the CPU will execute these instructions. As discussed above, this allowed user write-able sections to execute arbitrary code leading to compromise of the system. DEP prevents this from happening by setting the default permissions of sections where data are to be stored to be read–write and not executable. A simple buffer overflow can no longer be used to jump to the payload which contains instructions as the OS will never execute those instructions.

5.2 Bypassing NX/DEP Using ROP

Return-oriented programming [25, 26] is a technique to use existing instructions in the binary to achieve the same goals as a shellcode. For performing an ROP attack, special instructions called gadgets are needed.

Gadgets are the set of instructions which have a ret instruction at the end, and they lie in the code section of the binary. A ret instruction effectively is a *pop eip* instruction, which pops the data from the stack to the program counter. In a buffer overflow, a chain of these return addresses is put on the stack.

When the vulnerable function returns, it pops the first address and jumps to it. The address contains instructions which work just like the shellcode that could have been

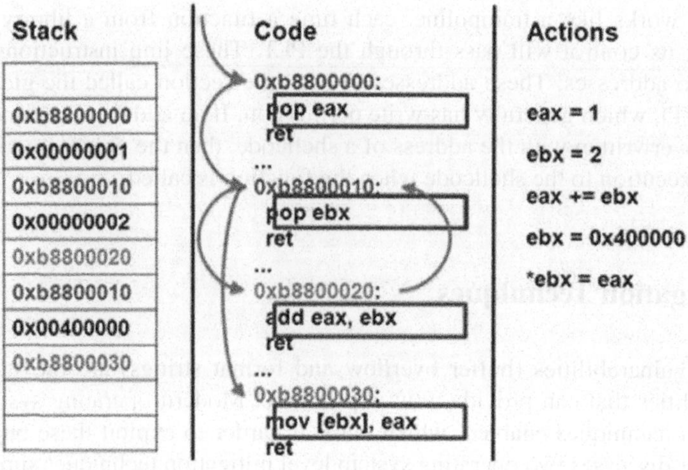

Fig. 3 Working of ROP [27]

placed on the stack. As shown in Fig. 3, once the first gadget finishes, the return at the end pops the address of the second gadget from the stack and the program jumps on the second gadget. This keeps going until the attacker has achieved its goal.

5.3 ASLR

ASLR [28, 29] or address space layout randomization was introduced in Linux kernel 2.6.12 on June 17, 2005, and its main goal is to prevent the attacker from knowing the memory address of the program. As shown in Fig. 4, if the attacker cannot reliably know the address of the code, stack and other segments, he cannot write an exploit to corrupt a specific memory location. As the name suggests, ASLR randomizes the location of these segments and prevents the attacker from reliable guessing the memory location. The address space in 64-bit is so large that a brute force attempt is

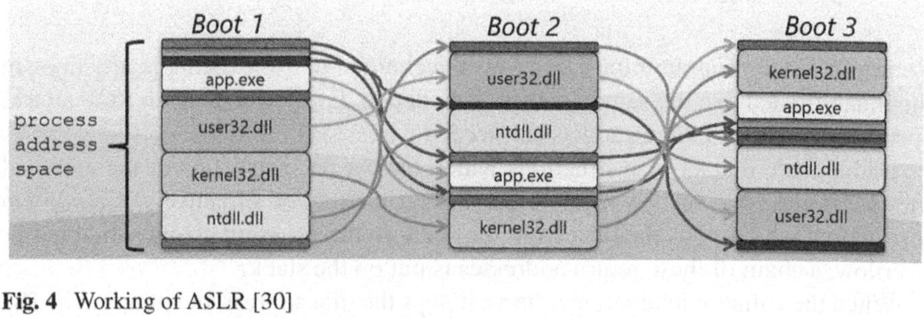

Fig. 4 Working of ASLR [30]

Table 2 Brute force probability percentage in multiple runs [32]

Unknown	0.024	6.25	100
0xb7	56	b	132
0xb7	58	e	132
0xb7	5e	5	132
0xb7	54	d	132
0xb7	5c	f	132

impossible, whereas in 32-bit, brute force can be attempted and has a low probability of succeeding. There are two primary methods to bypass ASLR, brute force and information leak.

5.4 *Bypassing ASLR Using Brute Force [31]*

In 32-bit systems, the address space is a maximum of 232 bits or 4 bytes. Since the pages in virtual memory are aligned at 4 KB or 4096 or 212bytes, the last 3 nibbles always remain constant. As shown in Table 2, the 4th nibble has an exploit reliability of 6.25% while brute forcing, whereas the 3rd byte has an exploit reliability of 0.024%. In 64-bit systems, brute force is not a viable option as the probability is very low and a higher level of randomization and entropy could be provided using the extra 4 bytes.

5.5 *Bypassing ASLR Using Information Leak*

Since the randomization occurs at the page level, if the pointer can be leaked [33] in any section, then all the other addresses can be referred to as relative to this address. This allows us to have 100% exploit reliability as that one address can now be used to find the addresses of other instructions and functions. Information leak also referred to as death by the pointer. Information leak is considered the best method to bypass ASLR, and modern exploits are multi-staged exploits. This first stage used the vulnerability to leak some address, and then the second stage uses that address to reliably calculate the addresses of the instruction it needs getting 100% reliability during the last stage.

6 A Case Study of CVE-2019–18634

Sudo [34] is a tool available for Linux systems to allow authorized users to execute a command as a superuser or as another user on the system, as specified in security policy. CVE-2019–18634 [35] was disclosed on the website of the developers of sudo utility on January 30th 2020. It described the vulnerability in the piece of code that handled password feedback. By default, sudo does not provide any feedback when a password is typed in the system, but if password feedback has been enabled then for each character typed, sudo outputs an asterisk character. The advisory [36] described the bug as a stack-based buffer overflow which can crash the sudo utility. After further research, the bug turned out to be a buffer overflow in the *.bss* section and not on the stack. As the overflow was not on the stack but in the bss section, it was not possible to get a write-what-where primitive and to exploit this bug a creative approach was needed. During analysis, it was found that the bss section contains flags which tell the program how to take input for the password. sudo can take input from either the standard input or take it from a program which provides the password. This program is run as the user who invoked the sudo program, and the details of the user are also present in the bss section. With this information, the overflow can be used to overwrite the user details so that it runs our program as root user, and when our program runs as root, it can open a shell as root giving us local privilege escalation.

7 Conclusion

Many mitigation techniques have been employed, and automated testing helps find bugs to some extent, but it's practically next to impossible to rid a program of all bugs and vulnerabilities, thus giving a window to programmers with malicious intent to cause harm to an organization by either gaining access to their system or by leaking their data. Bug bounty hunting is now a full-fledged profession aimed at finding bugs in software and servers in exchange of monetary awards. Organizations enroll their products on popular platforms to encourage people towards finding security holes. The payouts depend on the severity, and the impact of the bug found, but it can be safely assumed that any sort of monetary reward isn't as big of a loss in contrast to an attacker gaining control of the system, in which case the company loses its credibility and reputation, along with any lawsuits and fines that might follow. Basic techniques described in the paper can be used by programmers to check for security holes in their applications during development and testing to at least make their applications secure against common exploitation techniques that an unusually large number of applications fall prey to. A look into the exploitation techniques and their causes encourages the programmers to adopt best programming practices and pay close attention to the functions and definitions while referring to the documentation.

References

1. Jonathan, P., Baker, B.: Beyond stack smashing: recent advances in exploiting buffer overruns. IEEE Secur Privacy **2**(4), 20–27 (2004)
2. Saito, T., et al.: A survey of prevention/mitigation against memory corruption attacks. In: 2016 19th International Conference on Network-Based Information Systems (NBiS). IEEE (2016)
3. Drimak, E.G.: Stack mechanism for a data processor. U.S. Patent No. 3,889,243 (1975)
4. Denning, P.J.: Virtual memory. ACM Comput Surv (CSUR) **2**(3), 153–189 (1970)
5. Gorman, M.: Understanding the Linux Virtual Memory Manager. Prentice Hall, Upper Saddle River (2004)
6. Stack layout. https://www.researchgate.net/figure/Stack-layout-with-the-frame-pointersfig3 305423832
7. Harman, P.: Computer system with virtual memory and paging mechanism. U.S. Patent No. 6,970,991 (2005)
8. Abadi, M., et al.: Control-flow integrity principles, implementations, and applications. ACM Trans. Inf. Syst. Secur. (TISSEC) **13**(1), 1–40 (2009)
9. Sommestad, T., Holm, H., Ekstedt, M.: Estimates of success rates of remote arbitrary code execution attacks. In: Information Management Computer Security (2012)
10. Write-what-where condition. https://cwe.mitre.org/data/definitions/123.html
11. Elias, L.: Smashing the stack for fun and profit. Phrack.org: Issue 49 (1996)
12. Kernighan, B.W., Ritchie, D.M.: The C Programming Language (2006)
13. Stack frame. https://www.researchgate.net/figure/Stack-frame-before-return-to-libc-attackfig 1221505572
14. Rodes, B.: Stack layout transformation: Towards diversity for securing binary programs. In: 2012 34th International Conference on Software Engineering (ICSE) (2012)
15. Mason, J., et al.: English shellcode. In: Proceedings of the 16th ACM Conference on Computer and Communications Security (2009)
16. Provos, N., Friedl, M., Honeyman, P.: Preventing privilege escalation. In: USENIX Security Symposium (2003)
17. Lhee, K.-S., Chapin, S.J.: Buffer overflow and format string overflow vulnerabilities. Softw. Pract. Exp. **33**(5), 423–460 (2003)
18. Shankar, U., et al.: Detecting format string vulnerabilities with type qualifiers. In: USENIX Security Symposium (2001)
19. Newsham, T.: Format String Attacks (2000)
20. Scanf documentation. http://www.cplusplus.com/reference/cstdio/scanf/
21. Cabrero, J.E., Holland, I.M.: System and method for providing shared global offset table for common shared library in a computer system. U.S. Patent No. 6,260,075 (2001)
22. Procedure Linkage Table. https://docs.oracle.com/cd/E26505_01/html/E26506/chapter6-1235.html
23. Gao, Y.C., Zhou, A.M., Liu, L.: Data-execution prevention technology in windows system. In: Information Security Communications Privacy (2013)
24. Van de Ven, A.: New Security Enhancements in Red Hat Enterprise linux v. 3, update 3. Red Hat, Raleigh, North Carolina, USA (2004)
25. Roemer, R., et al.: Return-oriented programming: systems, languages, and applications. ACM Trans. Inf. Syst. Secur. (TISSEC) **15**(1), 1–34 (2012)
26. Prandini, M., Ramilli, M.: Return-oriented programming. IEEE Secur. Privacy **10**(6), 84–87 (2012)
27. Honap, A.M., Lee, W.: Hiding Kernel level rootkits using buffer overflow and return oriented programming. In: International Conference on Information Systems Security. Springer, Cham (2017)
28. Lu, K., et al.: ASLR-Guard: stopping address space leakage for code reuse attacks. In: Proceedings of the 22nd ACM SIGSAC Conference on Computer and Communications Security (2015)

29. Snow, K.Z., et al.: Just-in-time code reuse: on the effectiveness of fine-grained address space layout randomization. In: 2013 IEEE Symposium on Security and Privacy. IEEE (2013)
30. Address Space Layout Randomisation. Available: https://miro.medium.com/max/1400/1*1fNJ0GYaDY5smouTFNszHQ.png
31. Müller, T.: ASLR smack laugh reference. In: Seminar on Advanced Exploitation Techniques (2008)
32. Modern Binary Exploitation. Homepage: http://security.cs.rpi.edu/courses/binexp-spring2015/
33. Serna, F.J.: CVE-2012–0769, The Case of the Perfect info leaks Blackhat Conference (2012)
34. sudo documentation. https://www.sudo.ws/sudo.html
35. CVE-2019-18634. https://nvd.nist.gov/vuln/detail/CVE-2019-18634
36. Effects of pwfeedback set in sudo. https://www.sudo.ws/alerts/pwfeedback.html

OntoBlogDis: A Knowledge-Centric Ontology Driven Socially Aware Framework for Influential Blogger Discovery

V. Adithya, Gerard Deepak, and A. Santhanavijayan

Abstract Blogging in present-day times has become a common form of voicing opinions and explaining them. Therefore, commercial interventions to advertise products and political interventions, to share a party's agendas, etc., have been started to come upon blogs. The bloggers also make money on this, and a lot of upcoming bloggers are now finding high-grossing blogs and further write about the same topic or paraphrase it to increase their views in a very short time. This does not help in the growth of this mode of media in a long run. Thus, by finding the influential blogger these problems can be solved. This paper proposes a statistical knowledge engineering-based approach to provide an effective solution to this problem. The proposed OntoBlogDis uses TF-IDF, Synonymization, NPMI for Term Co-occurrence, Static Ontology Matching using ANOVA, Jaccard similarity, and knowledge harvesting from Wikidata. OntoBlogDis when compared to the baselined approach is inferred to be superior and recorded an overall F-measure and FNR are 96.84% and 0.02.

Keywords Influential blogger · Jaccard similarity · Knowledge harvesting · Static ontology matching · Synonymization

1 Introduction

In this world of today People on Online Social platforms, interact with each other, exchange intelligence, information, ideas, and knowledge that propagates quickly affecting those in the same platform for taking different decisions. Blog sites are one of those online social platforms that enables its users to build, write articles, and publish ideas, thoughts on different topics of concern or linking to them in the form of

V. Adithya
Department of Computer Science and Engineering, SRM Institute of Science and Technology, Ramapuram, Chennai, India

G. Deepak (✉) · A. Santhanavijayan
Department of Computer Science and Engineering, National Institute of Technology, Tiruchirappalli, India

A. Noor et al. (eds.), *Proceedings of Emerging Trends and Technologies on Intelligent Systems*, Advances in Intelligent Systems and Computing 1371,
https://doi.org/10.1007/978-981-16-3097-2_4

blog posts, attracting interest, and encouraging readers to carry out a range of tasks. Centered on blog documents material shared by users, they are becoming famous. Bloggers influence people and other bloggers, finding that the most influential blogger will help us to get an awareness for lots of analytic purposes on finding the mindset of the people and also to find the first blogger who has influenced other bloggers on the same topic which in turn will help a lot in finding and classifying credibility of specific news, help people think on a new topic which increases much more knowledge, etc.

Nowadays, blogs have started to influence people and manipulate them in the wrong ways also. At the same time, these upcoming bloggers tend to copy the content of some of the top bloggers in current topics and then publish them to become famous and also to earn money by monetizing their websites. To monetize blog posts, a specific number of views and likes are needed. To achieve this in a short period, budding bloggers just copy the content of the top bloggers whose articles are very famous and paraphrase them and thereby use the social platforms in a selfish way rather than to share authentic and new information. At the same time, some bloggers also influence other bloggers on the topic which makes the influenced bloggers also write about it. In case the influential blogger shares wrong news, which influences a lot of bloggers, which a lot of readers also tend to read; these unauthentic contents may cause lots of everlasting repercussions. In such cases, finding the credibility and authenticity of every blog content on that topic is tough. If we can find the influential blogger then we can use that as a key to solve these problems and also to authenticate the upcoming blogs on this content has some novelty from this or not and based on it can be accepted for publishing.

Finding the influential blogger, that can act as primary key information that can be used to solve all the above-mentioned problems. This paper proposes a knowledge-centric approach to find influential bloggers. The statistical knowledge engineering approach can be used to find the solution to this problem. The top blogger profiles are loaded from the Social Web and it is pre-processed and an ontology model is created for this, and then, statistical computations are performed on the data to identify features, and then, its similarity between the otology and the data is found out, and based on the similarity score influential score is calculated and they are categorized based on this score. The proposed OntoBlogDis architecture performance is compared to the baseline approaches and it is inferred that this model's performance is superior when compared to the baseline approaches. The proposed model records an accuracy of 96.12%.

The forthcoming part of the paper consists of the following sections. Section 2 consists of the research works done related to the topic. Section 3 briefly explains the proposed architecture. Section 4 contains the implementation and performance evaluation of the architecture, and the paper is concluded in Sect. 5.

2 Related Works

Vasanthakumar et al. [1] have briefly explained the need to find the influential blogger in the business point of view and have put forth an efficient approach to solve this problem with the help of a simple text processing approach by using TF-IDF, synonym substitution, and cosine similarity; the limitation of this approach is that it is not a knowledge centric. Agrawal et al. [2] have addressed the challenge of finding the influential blogger by conducting experiments with real-world blog sites data; they have evaluated multiple faces of identifying the problem of finding the influential bloggers. They have proposed some desirable attributes related to blogpost using statistics, and then, they have defined a model using these statistics to find the influential blogger. Since this is a statistical approach, these attributes may vary between datasets so it may not give a generic solution. Zanette et al. [3] have together performed exploratory research to find the business or commercial intervention in their blogs which they publish. They have concluded their research which provides many ways which are both advantageous to bloggers and companies to provide a win–win situation; this is exploratory research and does not give a state of the art solution. Khan et al. [4] have put forth together a novel weighted metric to find the influential blogger; their approach consists of three metrics, namely productivity of the blog, popularity of the blogs, and Blog Rank which is calculated and these are used as features with some weights which proves to be efficient when compared against their baseline approaches but these three features alone cannot be solely dependent on finding the influential blogger and all the features cannot be determined so easily and finding those will be very time-consuming and may not generalize to real-world data. Gilwa et al. [5] had together proposed a novel method for finding the influential bloggers which use the idea of an association of relation that exists by the people's comments in the threads and the blogger or blog. Several statistical features have been used to accomplish the solution to the problem, but these are not automated solutions. Ontology-based knowledge engineering approaches add auxiliary knowledge to the architecture and help in making the result generalize to real-world problems. Knowledge-based approaches are well suited for text-based data than traditional NLP. Ontology- and knowledge-based approaches have been proposed in [6–14], in support of the proposed approach. Park et al. [15] have used the interaction concept to find the influential user in social networking sites. Since this focuses on referencing relationships, it does not care about real-world knowledge.

3 Proposed System Architecture

Figure 1 shows the proposed OntoBlogDis influencing blogger predictor architecture. This OntoBlogDis architecture is divided into 5 phases.

The first phase comprises extracting data from Social Web and loading blogger profiles and the linked semantic blogger data is loaded, extracted, and converted into

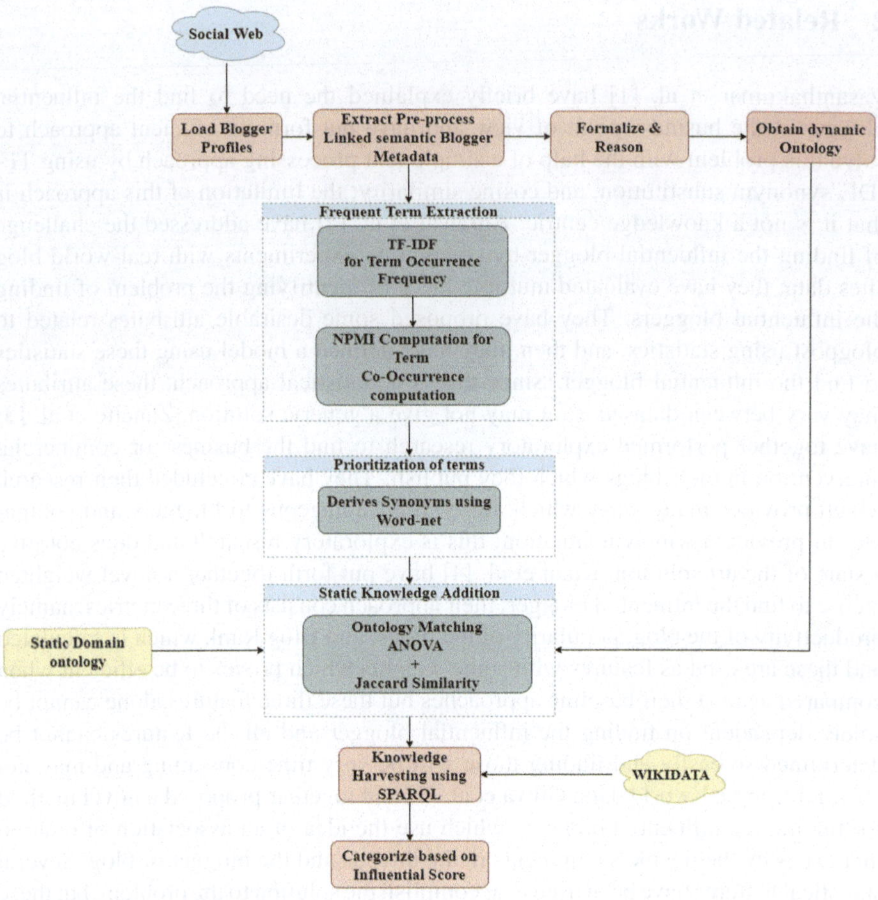

Fig. 1 Proposed OntoBlogDis architecture diagram

a data frame and is pre-processed. This pre-processing step consists of converting data into lowercase, then the article's text is tokenized, stop words are removed, then duplicate words and punctuations are removed, then word stemming and lemmatization is done, and then, words are vectorized. Then, formalizing and reasoning is done on the dataset and domain ontology is created and they are run in reasoners to verify ontology for which Pellet reasoner is used.

The second phase consists of frequent term extraction where term frequency–inverse frequency (TF-IDF) [16] is calculated. TF-IDF is generally used for information retrieval and is used to calculate the impact of every word in the document and rank them based on it so that the significance of each word in every individual document is found out. The expression of TF-IDF is shown in Eq. (1).

$$\mathrm{TF} - \mathrm{IDF}(t, d) = \mathrm{TF}(t, d) \times \log\left(\frac{N}{\mathrm{DF} + 1}\right) \quad (1)$$

On successful incorporation of the TF-IDF, the normalized pointwise mutual information (NPMI) [17] score is computed. PMI is a measure of relationships used in the field of information analysis and statistics. PMI score is calculated using the formula as shown in Eq. (2). Normalized pointwise mutual information score can be got by normalizing the PMI score between {[− 1, + 1]}, giving − 1 for events that do not come together, 0 for independent events, and + 1 for complete co-occurrence. NPMI can be calculated using the formula as shown in Eq. (3), where events X and Y co-occur more than if they were independent.

$$\mathrm{PMI}(X, Y) = \log \frac{P(x, y)}{P(x)P(y)} \tag{2}$$

$$\mathrm{NPMI} = \frac{\mathrm{PMI}}{\log P(x, y)} \tag{3}$$

The NPMI score is calculated for each word using this score; terms that occur frequently can be found out. The third phase involves prioritization of the frequent terms extracted from the previous phase, and these terms are synonymized using WordNet 3.0. WordNet is a lexico-syntactic knowledge base comprising of words and their grammatical constructs linked with each other by semantic relations like synonyms, hyponyms, etc. It enhances the probability of the knowledge and the grammatical structure by which the depth of knowledge linked to semantic blogger metadata is increased. The fourth phase of the proposed framework is the static knowledge addition phase. The main purpose of this phase is to perform ontology matching for which an analysis of variance (ANOVA) integrated Jaccard similarity has been encompassed. ANOVA test can be defined as the set of statistical models and related techniques of estimation used to determine variations between the means of the group in the sample. The p value of the similarity of the terms and their respective domain ontology is found out. Similarly, the Jaccard similarity score is also calculated for the ontology and the instances of the dataset. The Jaccard similarity [18] is a similarity function that is used for computing the term similarity between pair of terms or sentences or two documents based on sample sets, which is based on intersection over the union. Equation (4) represents Jaccard similarity where A and B are two sets containing text.

$$\text{Jaccard Similarity} = \frac{A \cap B}{A \cup B} \tag{4}$$

The Jaccard similarity and the p value are added and the average is obtained by averaging the two entities which act as the semantic similarity measure. If the similarity score is greater than 0.7, then the pair of terms are considered similar. In this phase, there are three inputs; one is the static domain ontology and the other is the dynamic domain ontology obtained from formalizing and reasoning, and further, the metadata is got from prioritization of the terms; all these three are compared for similarity and static knowledge is added. The next phase is the static knowledge addition phase where the knowledge harvesting is achieved using SPARQL. Knowledge harvesting is realized to contextualize the local data with that of the real-world global knowledge by adding background information and details to expand the knowledge visibility with that of the contents in the existing world, for which Wikidata has been employed. Wikidata is a free open-source knowledge base that acts as central storage for Wikipedia, Wikivoyage, Wikisource, and other projects. Once the knowledge has been harvested, it is categorized based on the influential score that is used for predicting the top most influential blogger.

4 Implementation and Performance Evaluation

The proposed system has been successfully implemented using Windows 10 Operating System. For this implementation, Intel Core i5 8th Gen Processor and 8 GB RAM were employed. NLTK, Re, num2words are the python NLP tool libraries that have been imbibed for pre-processing data and performing TF-IDF. Math library was used to perform NPMI computation. Sklearn is a python library that has prebuilt statistics and machine learning models; this library was used to calculate ANOVA scores. To model domain ontologies, Web Protégé which is an open-source online ontology editor has been used to create and visualize ontologies for all the events. For reasoning and verification, Pellet reasoner has been instilled into the approach; Pellet is an open-source ontology reasoner which has been used with OWL API libraries. It includes nominal optimization and gradual reasoning. A SPARQL-based agent that runs on JAVA was used for querying auxiliary knowledge, which has a state and behavior. The state of the agent is to aggregate similar knowledge from the knowledge base, while the behavior is to furnish the informative knowledge. The knowledge base that was used for this is the Wikidata knowledge base.

The experimentations were performed on Kaggle's blog authorship corpus dataset. The Blog Authorship Corpus dataset consists of a compilation of nineteen thousand three hundred and twenty bloggers compiled from blogger.com in the year 2004. This dataset comprises a collection of six lakhs twenty-eight thousand three and eighty-eight posts and more than 140 million words—approximately thirty posts and seven thousand two-fifty words per person. Every one of these blogs is given as an individual document, the name of which is indicated by the ID#, blogger, and the self-provided gender, age, industry, and the astrological sign of the blogger. To this dataset, some custom-made data is crawled explicitly and combined with the Kaggle dataset to improve the quality and reliability of the dataset.

The precision, recall, accuracy, F-measure, and false negative rate are the potential metrics and are used as performance evaluation metrics. The accuracy, F-measures, and false negative rate are computed and depicted in Figs. 2 and 3 using Eqs. (5)–(9), respectively.

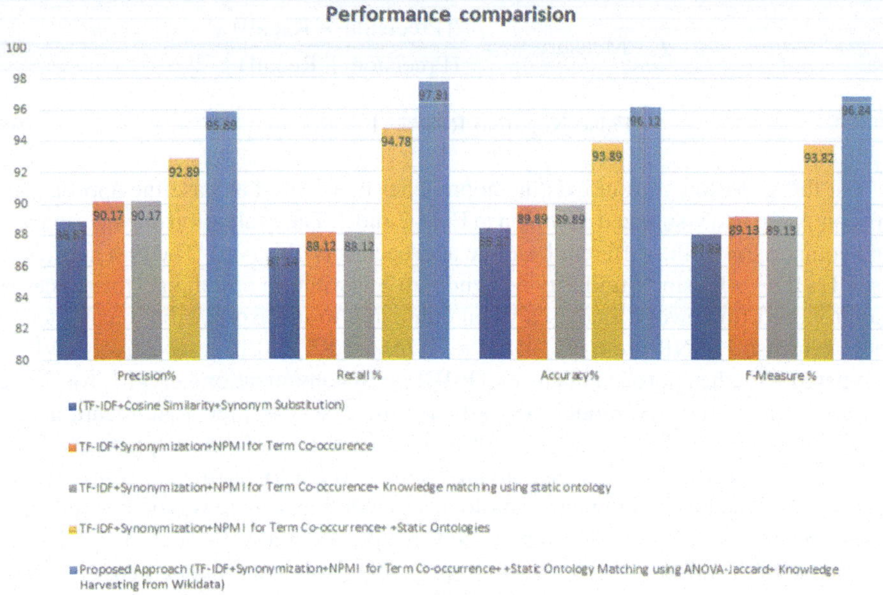

Fig. 2 Performance comparison of the proposed approach with baseline approaches

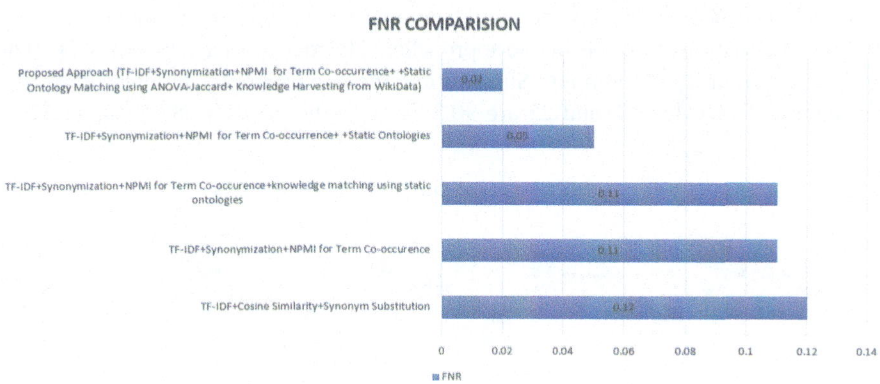

Fig. 3 FNR comparison with the baseline approaches

$$\text{Precision }\% = \frac{\text{True number of Positives}}{\text{True number of Positives} + \text{False number of Positives}} \tag{5}$$

$$\text{Recall }\% = \frac{\text{True number of Positives}}{\text{True number of Positives} + \text{False number of Negatives}} \tag{6}$$

$$\text{Accuracy }\% = \frac{\text{Proportion corrects that has passed ground truth tests}}{\text{Total No. of independent terms extracted from Blogger profiles}} \tag{7}$$

$$F\text{ - Measure }\% = \frac{2(\text{Precision} \times \text{Recall})}{(\text{Precision} + \text{Recall})} \tag{8}$$

$$\text{False Negative Rate} = 1 - \text{Recall} \tag{9}$$

For this experiment, four existing approaches considered as baseline approaches are compared and visualized as shown in Figs. 2 and 3. The evaluation of performance and comparison of the different baseline approaches is observed. The first approach uses TF-IDF + Cosine Similarity + Synonym Substitution and the performance of the proposed framework increases when compared to it for precision, recall, accuracy, *F*-measure, FNR are 6.3%, 10.67%, 7.75%, 8.85%, 0.10, respectively. When compared with the approach that uses TF-IDF + Synonymization + NPMI for Term Co-occurrence proposed model, performance increases for precision, recall, accuracy, F-measure, FNR are 5.72%, 9.69%, 7.75%, 7.707%, 0.96, respectively. The third approach uses TF-IDF + Synonymization + NPMI for Term Co-occurrence + Knowledge Matching using Static Ontologies, performance increases of the proposed model when compared to it for precision, recall, accuracy, *F*-measure, FNR are 4.71%, 3.92%, 4.06%, 4.32%, 0.04, respectively. The fourth approach that uses TF-IDF + Synonymization + NPMI for Term Co-occurrence + Static Ontologies, performance increases of the proposed model when compared to it for precision, recall, accuracy, *F*-measure, FNR are 3, 3.03, 2.23, 3.02, 0.03%.

Figure 4 shows the visualization of the comparison of the *F*-measure and number of recommendations for the baseline approaches. The first approach that uses TF-IDF + Cosine Similarity + Synonym Substitution; the *F*-measure for number of recommendations 5, 10, 15, 20, and 25 are 90.32%, 89.18%, 88.81%, 87.42%, 85.32%,

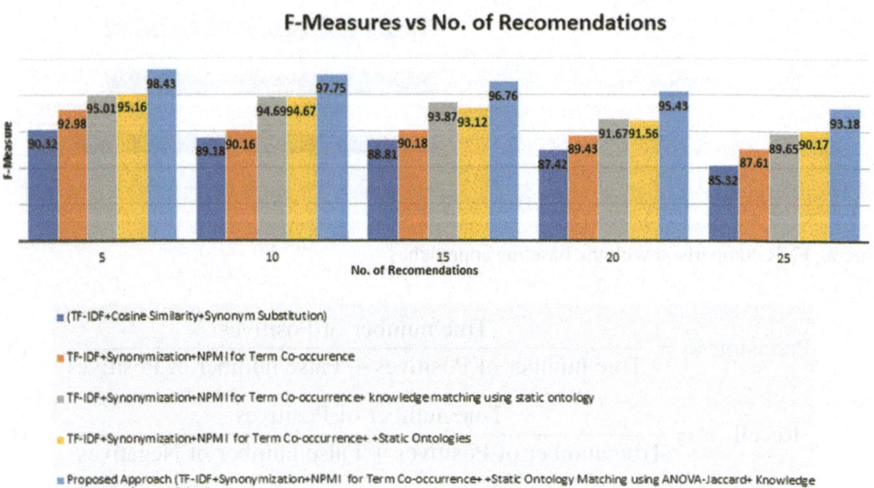

Fig. 4 Comparison for *F*-measure versus no. of recommendations for all the approaches

Table 1 Comparison of performance of the proposed model with its benchmark models

Benchmark models	Precision (%)	Recall (%)	Accuracy (%)	*F*-measure (%)
FIUSIC [15]	84.31	86.34	85.03	85.31
PTIMBR [1]	88.87	87.14	88.37	87.99
OntoBlogDis	95.89	97.83	96.12	96.84

respectively. The second approach that uses TF-IDF + Synonymization + NPMI for Term Co-occurrence, the *F*-measure for number of recommendations 5, 10, 15, 20, and 25 are 92.98%, 91.16%, 90.18%, 89.43%, 87.61%, respectively. The third approach that uses TF-IDF + Synonymization + NPMI for Term Co-occurrence + Knowledge Matching using Static Ontologies, the *F*-measure for number of Recommendations 5,10,15,20 and 25 are 95.01%, 94.69%, 93.87%, 91.67%, 89.65%, respectively. The fourth approach that uses TF-IDF + Synonymization + NPMI for Term Co-occurrence + Static Ontologies, the *F*-measure for number of recommendations 5, 10, 15, 20, and 25 are 95.16, 94.67, 93.12, 91.56, 90.17, respectively. For the proposed approach, F-measure for number of recommendations 5, 10, 15, 20, and 25 are 98.43, 97.75, 96.76, 95.43, 93.18, respectively. From this, it can be observed again that the proposed model's performance is superior than that of the baseline approaches.

The proposed approach is also compared with 2 other benchmark approaches as shown in Table 1. It can be seen that the performance percentage increase of the OntoBlogDis in terms of precision, recall, accuracy, F-measure when compared with the FIUSIC [15] is 11.58%, 11.49%, 11.09%, 11.52%, respectively, and when compared with the PTIMBR [1] the percentage increase is 7.02%, 10.69%, 7.75%, 8.85%, respectively.

From the performance comparison as shown in Figs. 2 and 3 and the analysis of the number of recommendations as in Fig. 4, it is observed that the proposed model's performance is higher when compared to the baseline approaches. The superiority of the proposed model in terms of performance when compared to the other model is because of the use of TF-IDF + Synonymization + NPMI for Term Co-occurrence + Static Ontology Matching using ANOVA-Jaccard + Knowledge Harvesting from Wikidata which makes it a user-driven model and also, it is a knowledge-centric approach where the dataset is linked with the external real-world knowledge using knowledge harvesting. Whereas the baseline approaches are not bothered about the real-world knowledge and it is solely dependent on the dataset that they use so there is no cognitive knowledge addition. Similarly when looking at the benchmark approaches as shown in Table 1, the FIUSIC approach uses continuous referencing relationship but does not care about the semantic relationship and is not the knowledge-centric approach, whereas the PTIMBR uses TF-IDF and cosine similarity and synonym substitution which is also one of our baseline approaches, performs better than the FIUSIC because of the knowledge engineering treatment given to the problem; at the same time, the proposed approach overtakes both of the approaches in terms of performance because of the auxiliary knowledge addition.

As we have used cognitive knowledge addition, the gap between the current dataset and the real world is filled. Since we have static ontology and synonymization as well as Wikidata is used as the knowledge base, the OntoBlogDis becomes a full cover approach, that is complete knowledge addition where background knowledge, contextualization, all of this are taken into care. The stated baseline approaches were mainly focusing on synonymization and were using an obsolete score, whereas the proposed approach uses ANOVA combined with Jaccard similarity which adds to the proposed approach's efficiency.

5 Conclusion

Finding the credibility and authenticity of every blog content on that topic is tough so if we can find the influential blogger then we can use that as a key to solve these problems and also to authenticate the upcoming blog on this content whether it has some novelty from the blogs having same topics or not, and based on it can be accepted for publishing. It can be noted that using ontologies and knowledge harvesting techniques adds some auxiliary knowledge to the problem and this knowledge-centric approach adds to the superiority of proposed approaches performance. Further, it can be concluded that than using traditional NLP approaches, knowledge engineering approaches that use statistical models and ontologies can be very effective to address this type of similar problem. The proposed OntoBlogDis yields an overall F-measure of 96.84%.

References

1. Vasanthakumar, G.U., Priyanka, R., Vanitha, C., Bhavani, S., Asha, B.R., Shenoy, P., Venugopal, K.R.: PTMIBSS: profiling top most influential blogger using synonym substitution approach. ICTACT J. Soft Comput. **7**(2), 1408–1420 (2017)
2. Agarwal, N., Liu, H., Tang, L., Yu, P.: Identifying the influential bloggers in a community WSDM '08. In: Proceedings of the 2008 International Conference on Web Search and Data Mining, pp. 207–218. ACM, California, USA (2008)
3. Zanette, M., Brito, E., Coutinho, M.: New influentials: an exploratory study on blogs. J. Direct Data Digit. Mark. Pract. **15**, 36–46 (2013)
4. Khan, H.U., Daud, A., Malik, T.A.: MIIB: a metric to identify top influential bloggers in a community. PLoS ONE **10**(9), 1–15 (2015)
5. Gliwa, B., Zygmunt, A.: Finding influential bloggers. Int. J. Mach. Learn. Comput. **5**(2), 127–131 (2015)
6. Wongthontham, P., Abu-Salih, B.: Ontology-based approach for identifying the credibility domain in social big data. J. Organ. Comput. Electron. Commer. **28**(4), 354–357 (2018)
7. Taieb, M.A.H., Aouicha, M.B., Hamadou, A.B.: Ontology-based approach for measuring semantic similarity. Eng. Appl. Artif. Intel. **36**, 238–261 (2014)
8. Todorović, M.L., Petrović, D.Č., Mihić, M.M., Obradović, V.L., Bushuyev, S.D.: Project success analysis framework: a knowledge-based approach in project management. Int. J. Proj. Manag. **33**(4), 772–783 (2015)

9. Kumar, A., Deepak, G., Santhanavijayan, A.: HeTOnto: a novel approach for conceptualization, modeling, visualization, and formalization of domain centric ontologies for heat transfer. In: 2020 IEEE International Conference on Electronics, Computing and Communication Technologies (CONECCT), pp. 1–6. IEEE, Bangalore, India (2020)
10. Deepak, G., Kasaraneni, D.: OntoCommerce: an ontology focused semantic framework for personalised product recommendation for user targeted e-commerce. Int. J. Comput. Aided Eng. Technol. **11**(4/5):449–466 (2019)
11. Gulzar, Z., Anny Leema, A., Deepak, G.: PCRS: personalized course recommender system based on hybrid approach. Procedia Comput. Sci. **125**, 518–524 (2018)
12. Deepak, G., Teja, V., Santhanavijayan, A.: A novel firefly driven scheme for resume parsing and matching based on entity linking paradigm. J. Discrete Math. Sci. Cryptogr. **23**(1), 157–165 (2020)
13. Haribabu, S., Sai Kumar, P.S., Padhy, S., Deepak, G., Santhanavijayan, A., Kumar, N.D.: A novel approach for ontology focused inter- domain personalized search based on semantic set expansion. In: 2019 Fifteenth International Conference on Information Processing (ICINPRO), 2019, pp. 1–5. IEEE, Bengaluru, India (2019)
14. Deepak, G., Naresh Kumar, G., Sai Yashaswea Bharadwaj, V.S.N., Santhanavijayan, A.: OntoQuest: an ontological strategy for automatic question generation for e-assessment using static and dynamic knowledge. In: 2019 Fifteenth International Conference on Information Processing (ICINPRO), pp. 1–6. IEEE, Bengaluru, India (2019)
15. Park, H.-J., Rho, S.: A study on the selection of SNS influenced users by interactivity: focusing on the blogosphere with continuous reference relationships. J. Korea Electron. Commerce Assoc. 17(4), 69–93 (2012)
16. Shahzad, Q., Ali, R.: Text mining: use of TF-IDF to examine the relevance of words to documents. Int. J. Comput. Appl. **181**(1), 25–29 (2018)
17. Bouma, G.: Normalized (pointwise) mutual information in collocation extraction. In: Proceedings of the Biennial GSCL Conference, pp. 31–40. Potsdam, Germany (2009)
18. Niwattanakul, S., Singthongchai, J., Naenudorn, E., Wanapu, S.: Using of Jaccard coefficient for keywords similarity. Proc. Int. Multiconf. Eng. Comput. Sci. **1**(6), 380–384 (2013)

An Ontology-Based Semantic Approach for First Aid Prediction in Aviation Services Incorporating RBFNN Over a Cloud Server

Gerard Deepak, D. Naresh Kumar, Deepak Surya, Khizer Razak, and K. R. Venugopal

Abstract Mid-air medical emergencies can prove to be fatal if appropriate infrastructure is not established in aircrafts. The alarmingly increasing rates of heart attacks around the world pose a huge threat of life endangerment anywhere and anytime. Especially when the passengers are at high altitudes, the airline crew despite being trained may not always be prepared to deal with the worst of unexpected situations. Since such cases need to be dealt in a very prompt and appropriate manner, an arrangement has to be made to receive swift external medical help whenever required. This paper draws attention upon telemedicine in the aviation sector, and how emergencies are handled mid-air. Internet of things plays a major role in virtually connecting the ailing patient with a specialist doctor, who would in turn suggest life-saving measures through remote diagnosis. In order to lessen the diagnosis time and speed up the process of providing appropriate first aid, semantic web can simultaneously help the healthcare professional to better understand and interpret the vague symptoms or conditions told, thereby suggesting some accurate remedies. The concept of hypernymy would enable effective facilitation the same. As instances as such can happen in massive numbers (sometimes even simultaneously in different locations), a centralized content management system is always required and should be continuously maintained up to date with every relevant data. All of this is stored and processed in the cloud. As a result, this paper focuses on how sensory inputs coupled with stored data in the cloud, alongside the hypernymy detection concept of the semantic web, would substantially increase the accuracy of medical diagnosis. This would in turn help save more lives in time of crisis. Also, the present challenges, along with the scope for improvement in the model, are discussed.

G. Deepak (✉) · D. N. Kumar · D. Surya
Department of Computer Science and Engineering, National Institute of Technology, Tiruchirappalli, India

K. Razak
Department of Radiodiagnosis, All India Institute of Medical Sciences (AIIMS), Mangalagiri, India

K. R. Venugopal
Vice Chancellor, Bangalore University, Bangalore, India

A. Noor et al. (eds.), *Proceedings of Emerging Trends and Technologies on Intelligent Systems*, Advances in Intelligent Systems and Computing 1371,
https://doi.org/10.1007/978-981-16-3097-2_5

Keywords IoT · Sense graph · Cloud platform · Cloud computing

1 Introduction

Over past few decades, rapid technological advancement has taken a giant leap, thus transforming lives in every sphere. Even the medical domain has developed expeditiously over the years. While a number of the regular medical institutions are already providing world-class healthcare facilities, the same is not properly reached out to every nook and corner. One such place is inside the aircrafts that are flying mid-air in high altitudes. It would take time to land the aircraft and treat the patient, especially in the case of emergencies (such as cardiac arrests and unexpected labour of pregnant women). This may even cost lives of these victims. As a result, telemedicine in the aviation arena has turned out to become a mandatory requirement, which would prove to be a game-changing way towards mid-air service. Especially with the reduction of cost of in-flight Wi-Fi, this provides long-term benefits to the passengers.

Telemedicine is basically an extension of information communication technology to the healthcare sector. It axes the tremendous barrier of high altitudes to improve first-air treatment access to airline passenger's mid-air, which is usually not available otherwise. During times of critical situations, it becomes a medium to save lives. The two terms 'telemedicine' and 'telehealth' are often confused with one another due to sounding similar in meaning. However, there exists a thin blurred line that distinguishes them. While telehealth refers to non-clinical remote services, telemedicine specifically refers to remote clinical services. Services such as administrative meetings and provider trainings are come under the umbrella of telehealth. In simpler words, while telemedicine is used for providing treatment, telehealth is about the work done to set up the required appropriate infrastructure and train personnel to efficiently carry out the first aid.

Providing emergency services that of telemedical to victims of cardiac arrests is the main focus of this paper, while also taking into account how telehealth services play a significant role in facilitating telemedicine in the long run. All this is aided by the internet of things (IoT), which provides a medium to contact any of the readily available qualified medical practitioners. IoT sends the live data generated by the medical sensors to the doctor, through a smart-phone application or the web. If in case, the patient's past medical history is required, there is provision to access it from the cloud.

When any airline passenger experiences cardiac arrest while flying, the person has to be kept under stringent observation by medical practitioners. Since the cabin crews are already trained to handle situations to some extent, guiding them properly during emergencies would only help them better in assisting the patients. In the time of heart attacks, continuous monitoring of victim is essential, so that respective first aid reaches the victim at appropriate time. This precarious condition can be subdued under the presence of a qualified healthcare professional (if present in the vicinity). However, this is not always the case. In addition to that, neglecting such situations can prove fatal to victims, and hence, immediate healthcare attention becomes quintessential. Internet comes into play in these dire conditions by providing a medium to connect victims with doctors across the globe. This forms the main motivation for the entire paper.

The outcome of this entire procedure is to connect the patients' sensory data with the professional medical staff through a proper telemedicine content management system that would give suggestions regarding on-time treatment measures. Just to make this entire procedure smoother and easier for the doctor (to analyse) and faster for the patient (to receive quick treatment), the concept of semantic web is also incorporated into this procedure. This paper suggests the use of hypernymy to help the doctor better understand the symptoms being conveyed, whilst looking through the past medical records. With the aid of this information, it would be much easier to conclude what should be done in a lesser span of time. The paper is concluded by discussing the future prospects of this methodology, along with coeval gaps (in terms of methodology and technology) that need to be addressed.

The rest of the paper contains: related works are explained in Sect. 2, while proposed architecture is explained in Sect. 3; Sect. 4 depicts a detailed idea of the Implementation and results of the proposed model. The paper is concluded in Sect. 5.

2 Related Works

Kathryn et al. [1] have analysed the casualty reports of over 7000 patients who have been presented for first aid over a period of 63 days and have realized that the correlation of the first aid assisted with the factors such as crowd size, maximum daily temperature, and humidity was significantly higher. This work has concluded that subsequent analysis of several such events could be helpful in the support of accurate predictor models. Flabouris et al. [2] have proposed a work to analyse, for any public event, the importance of primary medical care by identifying initial symptoms of patients. This system is further used to improve the efficiency of medical care. Tannvik et al. [3] have proposed a review on the first aid that is provided to trauma victims. This study also sheds some light on the frequency of the first aid provided by providing a systematic review according to the PRISMA guidelines. Donaldson et al. [4] have submitted a review of all consecutive incidents in-flight based on the reports for QANTAS airplanes. The review included all the incidents that required medical attention from doctors. This study has presented a ranked order of several health issues that can potentially take place in an aircraft and the first aid that is to be administered. Joseph et al. [5] have reviewed a kit used in an airline and the care provided to the passengers on the plane. This study presents analytical information about the number of times the kit has been used and the symptoms that the passengers have experienced. This study explains the most common ill symptoms that are shown by passengers while travelling via an aircraft. Anjali et al. [6] have put forward a survey about the use of internet of things and its applications for the medical equipment. This study proposes that the incorporation of IoT can not only increase the efficiency of the equipment due to increasing data but also paves way for improved public health. The work also explains the use of IoT devices to monitor health conditions of toddlers to operation theatres. Tyagi et al. [7] have presented the applications of both IoT and cloud architecture to address important parameters for efficient diagnosis based on symptoms in the field of health care. This work also proposes a cloud-based framework for beneficial implementations of healthcare solutions. Colin Puri et al. [8] have proposed the need and advantages of integration of ontologies to solve the problems in the field of health care. This paper explains the need for the use of multiple ontologies for the continuously increasing streaming data through several sources, which requires advanced data analysis. In [9–18], several works explaining the use of ontologies as dynamic knowledge bases to solve several problems in various domains including the fields of engineering, health care and so on are discussed.

3 Proposed Architecture

The proposed architecture is depicted in Fig. 1. Smart IoT devices are prevalent in the present day and age. These devices are essential for low-latency data transfer as they enable continuous streaming of data. The data that flow through these devices can also be incorporated into several cloud services now available which include Google Cloud Platform, Amazon Web Services and Microsoft Azure. These platforms provide users with tools that help extract various insights from the data, preprocess and help formulate machine learning models and facility to deploy these models at scale.

Consider Google Cloud Platform for a given streaming sensory data. Google Cloud Dataflow helps user in processing streaming and batch data; these data are then ingested using Cloud Pub/Sub. Cloud IoT core manages device connections and monitors the functioning of IoT devices. Big Query is where all the queries are run to extract reliable insights from the data. Big Query enables faster cloud querying with the help of VM compute instances. For training, deploying, and tuning the hyperparameters and thereby running ML models, users are provided with Cloud ML engine. All these cloud services run in conjunction with each other so as to enable proper workflow (Fig. 2).

To provide appropriate treatment without hurdles, there should be a systematic arrangement made to manage the entire process. First and foremost, heart-rate sensors should be set up within the airlines to monitor the pulse when necessary. This sensor would be capable of receiving mass volumes of sensor data. The input data are in analog form, so it is then processed using an analog-to-digital converter (ADC) for being readable by the system. The same is then sent to online a web-based server specially allocated for this purpose. It can be assumed in this case that the airline premise is the client. From the continuous medical data that is collected, diagnosis can be done, and appropriate treatment measures can be suggested.

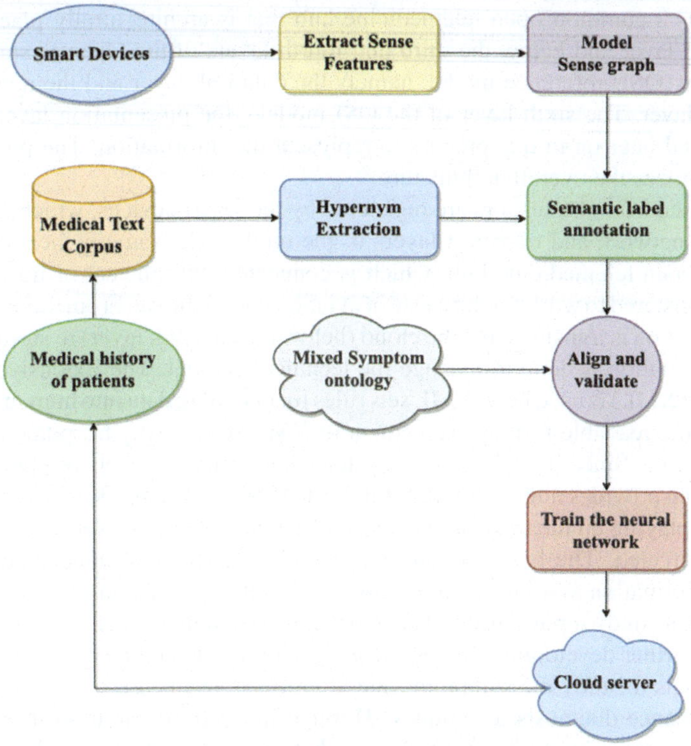

Fig. 1 Proposed architecture

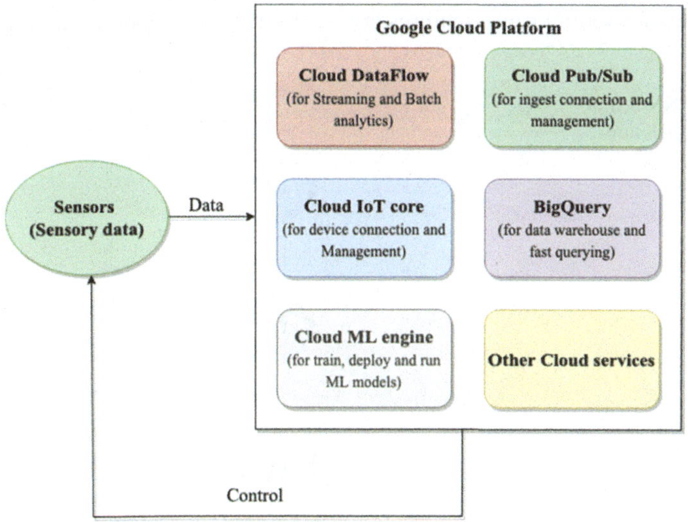

Fig. 2 Google Cloud Platform workflow

There is a common open telemedicine hub that is architecturally placed above the second layer and below the third and fourth layers of the Open Systems Interconnection (OSI) reference model, namely the data link layer and the network and transport layer. The sixth layer of the OSI model—the presentation layer—would also be used later on in this process to represent the information. The presentation layer optimizes the overall architecture.

Input data from the sensors are registered by the smart devices, which are placed above the network and transport layers of the model. The same is then sent to the common open telemedicine hub, which is connected with the set of hospitals and doctors registered to work for the cause of aviation telemedicine. From there, a copy of the information is transferred to the cloud (below the data link layer) to store the data. Data presentation is done to manage the technical content. The Extensive Markup Language (XML) is used here. XML sets rules for encoding data into human-readable and machine-readable format. Agreement of XML schemes by hospitals facilitates data exchange. Since the telemedical systems have varying levels of complexity, a solution that provides interoperability for the distributed systems is provided (Fig. 3).

For displaying all the medical data, a customized webpage or smart-phone application is created. The hospitals and airlines use this for the management of the overall information system. In some cases, a cognitive algorithm is also developed in parallel, to study input sensory data and the treatment that is being suggested each time. On further developing the cognitive algorithm, it would be easier for medical professionals to study the symptoms and aftereffects from case to case, that would further enhance diagnosis techniques. Through in-depth research on the cognitive algorithm, a stage could be reached such that the system itself can diagnose and provide treatment measures.

After the data are streamed from sensors, various important features of the data are extracted, and essential sense features are used for further modelling. A knowledge

Fig. 3 RBFNN neural network architecture

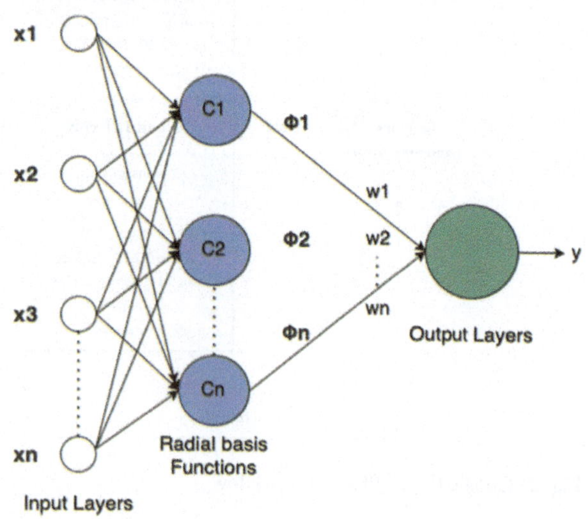

graph is vital for the model to have proper understanding on the input features and also predict required outputs properly. For this purpose, a sense graph is modelled incorporating several symptoms that lead to a certain disease. These symptoms and the graph help the crew in the flight to quickly understand the probable health issue that the victim might have. This will in turn help in understanding the gravity of the issue.

Semantic label annotation makes use of hypernym extraction from the open-source medical Text corpus. Hypernym extraction helps understand the nuances in the symptoms of several diseases. This helps the model to properly classify a certain disease based on symptoms from that of another. However, the success of this overall system does not merely depend on the efficiency of the system architecture and design, but also correctness of inputs provided by the airline's cabin crew. The more accurate is the data retrieval, the better the treatment. Therefore, incomplete data entry should strictly be checked for, as it would disrupt the completeness and consistency of data. Within continuous increase in complexity of the total architecture of information technology, there arises the dire need for system integration. Bit-to-bit data interchange for machine-to-machine interaction needs to use a standard protocol. Here, the sender's and receiver's sides apply the exact same sizes for fields and characters. The whole process is carried out through a secure channel to avoid intrusion of patient data, thus protecting their privacy.

4 Implementation and Results

The entire model is implemented in Python 3 using the open-source Google Colaboratory environment. Semantic model using radial basis function neural network or RBFNN as the feedforward layer was used for the implementation. RBFNN is one of the extremely fast, yet unusual algorithms and is generally used to solve many classification and regression problems. A framework with incorporating web PMI and cosine similarity was employed for the model. Threshold of 0.5 was passed for web PMI and that of cosine similarity.

A dataset was manually formulated with consultation from several medical experts and was verified by a co-author who is a medical practitioner and a researcher. Several conditions with respect to anxiety, cardio-thoracic problems, cardiovascular problems, hypoxia, respiratory disorders, claustrophobia, phobia and anxiety-based issues predominantly exhibited by flyers, food and beverage allergies, panic attacks, attacks due to variations in blood pressure, hormonal and adrenaline disorders, disorders due to blood sugar fluctuations were some of the key conditions for which the symptoms along with their manifestations were collected and used in the approach.

The mixed-symptom Ontology comprises 247 primary symptoms and 704 auxiliary symptoms such that the symptoms are hierarchically modelled as a concept, sub-concept, and individuals. The relations are expressed between the concept, sub-concept pairs along with axioms which are capable of dynamic alignment with specific conditions based on reasoning. The Ontologies are modelled by extracting standard text from textbooks of General Medicine, Pharmacology, Radio-diagnosis, Pulmonary Medicine, Community Medicine, and other associated medical reference books and journals under the supervision of a medical practitioner. Web Protégé is used for static Ontology modelling, and OntoColab is used to dynamically generate Ontologies from these texts. The inconsistencies of the Ontology are evaluated using Pellet reasoner. However, a medical practitioner has checked, evaluated, and adhered to the modelled Ontologies. The Ontology modelled is a detailed 7-level implementation of the core classes, which are highly specialized and granular in nature.

Classification model using RBFNN learns recursively based on expert's opinion and medical data of several patients. A dynamic knowledge graph is formulated with ground truth. Expert's opinions from medical textbooks were taken to improve the performance of the model while retaining its reliability. Currently, this architecture can be employed for immediate first aid, which can be administered by a non-medical practitioner based on the first-aid kit available on the aircraft. Around 247 variations of symptoms connected to hypoxia, fear, and anxiety. Usage of advanced 3-level semantic approach integrated with a neural network approach (RBFNN) has paved way for improved performance of the proposed architecture. The model has achieved improved accuracy without the need for a deep neural network. The model is trained for more than 11,000 cases and tested around 3400 cases. Sensors are integrated using the cloud architecture. Accuracy, precision, and recall were based on the first aid recommended.

The performance is evaluated by considering precision, recall, accuracy, *F*-measure, false discovery rate (FDR) as potential metrics. Precision, recall, and accuracy are depicted in Eqs. (1)–(3), respectively. The *F*-measure and FNR are computed as illustrated in Eq. (4), respectively, and Fig. 4. compares the accuracy distribution of the baseline models with the proposed architecture. Accuracy is computed as the average proportion corrects of each query that is highly similar to the ground truth.

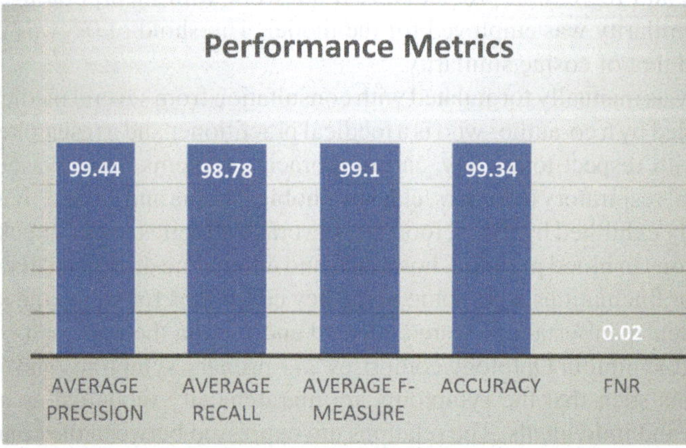

Fig. 4 Performance metrics

$$\text{Precision} = \frac{\text{First-aid Retreived} \cap \text{First-aid Relevant}}{\text{First-aid Retreived}} \quad (1)$$

$$\text{Recall} = \frac{\text{First-aid Retreived} \cap \text{First-aid Relevant}}{\text{First-aid Relevant}} \quad (2)$$

$$\text{Accuracy} = \frac{\text{Proportion corrects of each query passed ground truth test}}{\text{Total queries}} \quad (3)$$

$$F\text{-measure} = \frac{2 * \text{Precision} * \text{Recall}}{\text{Precision} + \text{Recall}} \quad (4)$$

$$\text{FNR} = 1 - \text{Recall} \quad (5)$$

Algorithm 1. First Aid Recommendation Algorithm

Input: Query with symptoms

Output: Recommended first aid for the patient
Begin
Step 1:
Take query q as input
Step 2:
Preprocessing of query:
Import General_Stop_words, Numbers, Punctuations
#Tokenization:
Tokeknized_q = Splitting the terms in the query q
#Normalization:
Normalized_temp = Tokenized_q – term
Normalized_q = Noramlized_temp (remove Punctuations)
#Synonymization:
For each term in Normalized_q:
Search for synonyms of term using WordNet 3.0
Synonymized = Normalized_q + synonyms
User_query = Synonymized_q
Step 3:
Train RBFNN model using input data
Summing and first level scoring
Medical history from cloud server
Cosine Similarity
Second level scoring
Pass all the data to RBFNN to get the final scoring
Step 4:
Predict first aid using weights of the model
Output top 3 scored precautions to be administered
End

From Fig. 4, it is inferable that the proposed approach yields an average precision of 99.44 and average accuracy of 99.34. This model performs significantly better when compared to current methods. This increase in performance is due to the

Table 1 Performance comparison of the proposed approach with its variations

Metric	Proposed model for prediction of first aid	Absence of mixed symptoms Ontology in the proposed model	Absence of RBFNN	Absence of sense graph and annotation
Precision (%)	99.44	84.53	87.12	89.44
Recall (%)	98.78	82.14	85.64	86.32
Accuracy (%)	99.34	83.46	83.27	87.18
F-Measure (%)	99.1	83.31	86.37	87.85
FNR	0.02	0.18	0.15	0.14

incorporation of sense graph and the ontologies that provide domain and background knowledge based on the medical history of the patient in the cloud server. This datum is completely protected and is not circulated to any third-party vendor due to privacy concerns. Symptoms and the first aid along with the condition are perfectly mapped between sense graph and background ontologies. Incorporation of RBFNNs adds increase in performance to the model in synergy with cosine similarity. RBFNNs present many advantages like that of ease of design, flexible control systems, adaptability, and input noise tolerance. All these features make it best for the problem at hand and is employed.

The performance metrics of the proposed model are depicted in Table 1. As there are no current models present that incorporate various techniques presented in the proposed architecture, comparison of the method with those of existing methods is tough. Table 1 explains the effect of incorporating various additions to the model while demonstrating the performance of the model in the absence of such mentioned techniques. It is seen that the model yields an accuracy of 99.34, while the absence of symptom ontology has reduced the accuracy significantly to 83.46. The accuracy of the model in the absence of RBFNN for classification has found to be 83.27, while in the absence of sense graph the accuracy was reported to be 87.18. The reason for improved accuracy can be explained by the incorporation of symptom Ontology to using RBFNN for classification. Incorporation of sense graph and annotation has only increased the accuracy of the model to a greater extent.

5 Conclusions

Analysing patients' sensory data and telemedicine content management is the primary objective of the work presented in this paper. As per current works done in the field of health care, it is observed that there is a significant delay in the diagnosis and treatment of a symptom while travelling in aircraft. By incorporating the concept of semantic web, this entire process can be accelerated. This acceleration can also be increased significantly by the integration of cloud environment and IoT architecture. Hypernymy technique encompassing RBFNN and cosine similarity is proposed for improved efficiency for the model. By the incorporation of such techniques, the delay in the process of medical administration can be reduced and hence help save many lives. An average *F*-measure is 99.1.

References

1. Zeitz, K.M., Schneider, D.P., Jarrett, D., Zeitz, C.J.: Mass gathering events: retrospective analysis of patient presentations over seven years. Prehosp. Disaster Med. **17**(3), 147–150 (2002)
2. Flabouris, A., Bridgewater, F.: An analysis of demand for first-aid care at a major public event. Prehosp. Disaster Med. **11**(1), 48–54 (1996)
3. Tannvik, T.D., Bakke, H.K., Wisborg, T.: A systematic literature review on first aid provided by laypeople to trauma victims. Acta Anaesthesiol. Scand. **56**(10), 1222–1227 (2012)
4. Donaldson, E., Pearnt, J.: First aid in the air. Aust. N. Z. J. Surg. **66**(7), 431–434 (1996)
5. Cottrell, J.J., Callaghan, J.T., Kohn, G.M., Hensler, E.C., Rogers, R.M.: In-flight medical emergencies: one year of experience with the enhanced medical kit. JAMA **262**(12), 1653–1656 (1989)
6. Yeole, A.S., Kalbande, D.R.: Use of internet of things (IoT) in healthcare: a survey. In: Proceedings of the ACM Symposium on Women in Research, pp. 71–76 (2016)
7. Tyagi, S., Agarwal, A., Maheshwari, P.: A conceptual framework for IoT-based healthcare system using cloud computing. In: 6th International Conference-Cloud System and Big Data Engineering (Confluence), pp. 503–507. IEEE (2016)
8. Puri, C.A., Gomadam, K., Jain, P., Yeh, P.Z., Verma, K.: Multiple ontologies in healthcare information technology: motivations and recommendation for ontology mapping and alignment. In: ICBO (2011)
9. Deepak, G., Santhanavijayan, A.: OntoBestFit: a best-fit occurrence estimation strategy for RDF driven faceted semantic search. Comput. Commun. **160**, 284–298 (2020)
10. Pushpa, C.N., Deepak, G., Kumar, A., Thriveni, J., Venugopal, K.R.: OntoDisco: Improving web service discovery by hybridization of ontology focused concept clustering and interface semantics. In: IEEE International Conference on Electronics, Computing and Communication Technologies, pp. 1–5 (2020)
11. Deepak, G., Kasaraneni, D.: OntoCommerce: an ontology focused semantic framework for personalised product recommendation for user targeted e-commerce. Int. J. Comput. Aided Eng. Technol. **11**(4–5), 449–466 (2019)
12. Gulzar, Z., Anny Leema, A., Deepak, G.: Pcrs: personalized course recommender system based on hybrid approach. Procedia Comput. Sci. 125, 518–524 (2018)
13. Deepak, G., Teja, V., Santhanavijayan, A.: A novel firefly driven scheme for resume parsing and matching based on entity linking paradigm. J. Disc. Math. Sci. Cryptogr. **23**(1), 157–165 (2020)
14. Haribabu, S., Sai Kumar, P.S., Padhy, S., Deepak, G., Santhanavijayan, A., Kumar, N.D.: A novel approach for ontology focused inter-domain personalized search based on semantic set expansion. In: 2019 Fifteenth International Conference on Information Processing (2019)
15. Deepak, G., Naresh Kumar, G., Sai Yashaswea Bharadwaj, V.S.N., Santhanavijayan, A.: OntoQuest: an ontological strategy for automatic question generation for e-assessment using static and dynamic knowledge. In: Fifteenth International Conference on Information Processing (ICINPRO), pp. 1–6 (2019)
16. Santhanavijayan, A., Naresh Kumar, D., Deepak, G.: A semantic-aware strategy for automatic speech recognition incorporating deep learning models. In: Intelligent System Design, pp. 247–254. Springer, Berlin

17. Deepak, G., et al.: Design and evaluation of conceptual ontologies for electrochemistry as a domain. In: 2019 IEEE International WIE Conference on Electrical and Computer Engineering (2019)
18. Deepak, G., Priyadarshini, J.S.: Personalized and enhanced hybridized semantic algorithm for web image retrieval incorporating ontology classification, strategic query expansion, and content-based analysis. Comput. Electr. Eng. **72**, 14–25 (2018)

Benchmarking Analysis of CNN Architectures for Artificial Intelligence Platforms

Nishi Jha, Pooja Rawat, and Abhishek Tiwari

Abstract The prompt innovations in Digital Technologies with the availability of credible data have led to the emergence of an era of Artificial Intelligence and Deep Learning. This demonstrates their effectiveness in solving complex problems, particularly in image classification and object recognition applications, with the assistance of Convolutional Neural Networks (CNNs). However, such algorithms need to be executed within a certain time frame specifically applications like High Performance Computing (HPC), Autonomous vehicles, Gaming etc. The requirement of time certainty brings hardware accelerators into the picture. These hardware accelerators not only accelerate time-critical tasks but have also been proven effective in enhancing the throughput of CNNs. In this research, we have tried to tune in performances of different CNN models i.e. Alexnet, SqueezeNet1.1, GoogleNet-v1, and VGG-16 of the Caffe framework which have been pre-trained and converged, using the Bench-Marking and Cross Check Tools of OpenVINO Toolkit. The tool define the performance of CNN Models based on latency, throughput, absolute and relative differences in each layer of these models. The CNN models have been simulated on platforms like Intel i5-8265 CPU, 1.60 Ghz and Integrated GPU UHD graphics 620 using OpenVINO Toolkit, which helped in running the simulations on Windows 10. Furthermore, this study is expected to direct the future development of an efficient accelerator on specialized hardware accelerators and also be useful for deep learning researchers.

Keywords Convolutional Neural Network (CNN) · Cross-Check Tool · Benchmarking · OpenVINO · Alexnet · SqueezeNet · GoogleNet · VGG-16 · Artificial Intelligence (AI)

N. Jha (✉) · P. Rawat
Guru Gobind Singh Indraprastha University, Delhi, India

N. Jha · P. Rawat · A. Tiwari
Centre for Development of Advanced Computing (C-DAC), Noida, Uttar Pradesh, India

A. Noor et al. (eds.), *Proceedings of Emerging Trends and Technologies on Intelligent Systems*, Advances in Intelligent Systems and Computing 1371,
https://doi.org/10.1007/978-981-16-3097-2_6

1 Introduction

With the technological developments and availability of tenable data, Deep Learning and Artificial Intelligence have become popular. They have shown their effectiveness and advance abilities in solving complex problems that were not possible earlier. The offloading of common Artificial Intelligence and Deep Learning tasks to hardware such as CPU and GPUs to obtain a considerable amount of acceleration on them is one of the major need for such AI-enabled systems.

Now, with the technologies getting more and more complex its implementation on Low-power devices have become difficult. Thus, such a situation demands an increase in processing power which can be achieved with the usage of AI Hardware Accelerators [12].

With this paper, we are trying to understand which accelerator can provide an efficient result with all these complexities and to do so we have organized the rest of the paper as follows: Sect. 2 briefly explains Convolution Neural Networks [1] and its architecture; Sect. 3 evaluates trained datasets and use them as an input for Benchmarking [2, 3]; in Sect. 4, we discuss the different parameters that had helped to create a comparison table of each model and its layers; in Sect. 5, we present the experimental results of Cross-Check tool [14] and Benchmarking application [15] and Sect. 6 indicates the conclusion of the results and related Future work.

2 Convolutional Neural Networks

CNN's are trained off-line using the back-propagation process which is then used to perform recognition tasks using the feed-forward process [1]. They are widely used in modern Artificial Intelligence (AI) systems and have shown remarkable performance in speech recognition, computer vision, and image recognition. In a simple word what CNN does is, it extracts the feature of the image and convert it into a lower dimension without losing its characteristics.

2.1 Architecture of CNN

The architecture of CNNs [1] consists of layers which include convolution, activation functions, normalization, pooling, and characteristics of fully connected layers.

Convolution (CONV): A convolution operation [8] is the production of a matrix smaller in size than the original image matrix, which is representing pixels, by sliding a small window (called filter, feature identifier, or kernel) of size $k \wedge k$ over the image (called input feature map (FM)), to produce an output feature neuron value.

Activation Function: The activation function of the neuron defines the output of the neuron given a set of inputs [8]. Activation functions introduce nonlinear properties to the network. This helps the network learn any complex relationship between input and output. Sigmoid and ReLu has been used in the experiment for this research work.

Normalization: When training a neural n/w, we want to normalize our data in some way ahead of time as part of the pre-processing step [11]. At this step, data is prepared to get ready for training. Normalization has the objective of transforming the data to put all the data points on the same scale.

Pooling: It is used to rapidly reduce the spatial size of the representation, thereby reducing the number of parameters and computation in the network [11].

Fully Connected Layer: A common form of a convolutional neural network architecture comprises stacks of a few convolutional and ReLU layers, followed by layers for pooling and this pattern is repeated until the image has merged spatially to a small size [8].

3 Datasets

In our research study, ImageNet dataset have been used for analyzing the performance of CNN Models on different devices. ImageNet is an image database in which each node of the hierarchy is depicted by hundreds and thousands of images [8]. Here we have 14 million images, each 224 by 224 pixels. To use this database, we download the datasets which will come with two folders i.e. "train" and "val" that contains training and validation set respectively. Inside these folders there are separate folders for each class, which has over 1000 images. Now we put these classes as an input test set for the tools.

4 Simulation Results and Observations

Here we analyze and accelerate the convolutional neural network models on Intel i5-8265 CPU, 1.60 GHz and Integrated GPU UHD graphics 620 using the Intel OpenVINO toolkit [5], and compare the results obtained. The simulation results of all four convolution neural network models were performed on the Windows 10 Version.

Full Inference Flow [13] as indicated in Fig. 1, defines the description of the workflow as below:

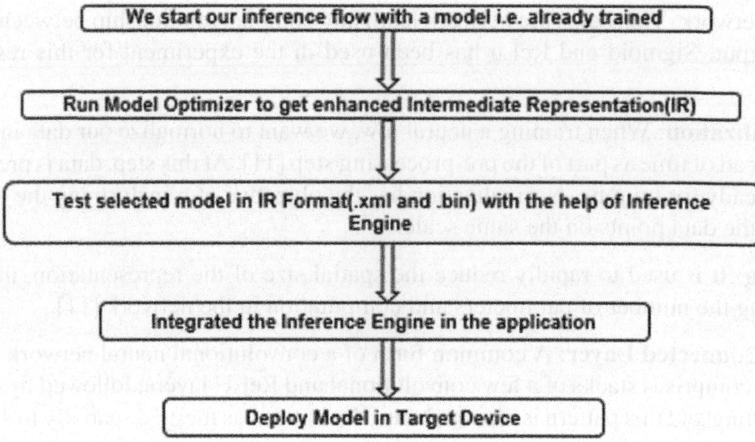

Fig. 1 Full Inference Flow

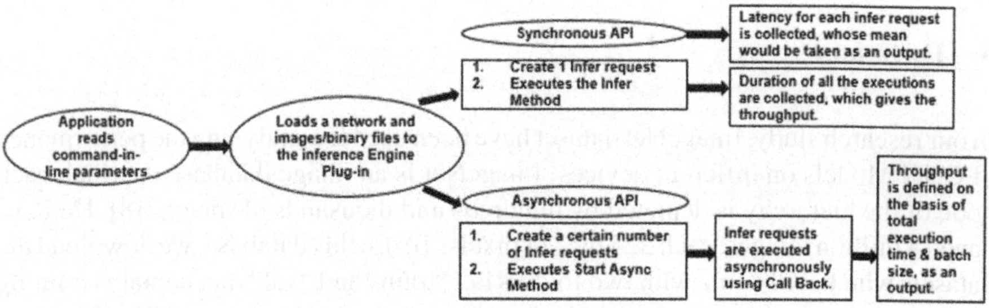

Fig. 2 Adopted Flow for Benchmarking Application

4.1 Benchmarking Application and Flow

The benchmarking experiment performed here demonstrates high-performance gains on our Neural networks when deployed on CPU and integrated GPU [5]. The results can help one to determine which hardware is best suited for their application. We have mainly used two inference modes (i.e., Asynchronous API and Synchronous API, as shown in Fig. 2 [15] to draw a comparison table for the four CNN Models.

The application also collects per-layer Performance Measurement (PM) for each executed infer request.

4.2 *Cross-Check Tool*

The Cross-Check Tool is an application that enables the comparison of Accuracy and Performance metrics for two model inferences which are performed on two different devices with different precision [14].

In this paper, we have drawn comparison results by simulating each layer as well as the overall model using the tool. The tool uses some performance metrics like accuracy, precision, sensitivity, and specificity to understand how the models are working on different platforms with different precision and give results in form of following parameters:

1. **Absolute Difference**: It defines the difference between the accuracy results shown by the CNN model on GPU and CPU respectively with FP32 precision when given the same datasheet as an input. The tool defines this value for each layer and also for the entire model.
2. **Relative Difference**: It defines the difference between the speedup time while recognizing the features of the given image datasets on both CPU and GPU. It can be calculated using the total execution time taken by the model.

In the following sections, the result obtained by both the methodologies i.e. benchmarking application and Cross-Check Tool will be discussed.

5 Results and Discussions

In the asynchronous case, the performance of an individual infer request is usually of less concern. Instead, one typically execute multiple requests asynchronously and measure the throughput [6] in images per second by dividing the number of images that were processed by the processing time. In contrast, for the synchronous case, the time to a single frame is more important.

The comparison of AlexNet [4], SqueezeNet1.1 [9], VGG 16 [8], and GoogleNet-v1 [10], which were performed on CPU and GPU with data format FP32 and FP16 respectively shown in Table 1). FPS defines the throughput i.e. time taken by the model to process the datasets, obtained through the simulations and Latency [6] is defined in "ms" i.e. millisecond unit.

Table 1 Comparison of four models with batch size =1 and batch size =32

Model name	CPU(B.S=1)		CPU(B.S=32)		GPU(B.S=1)		GPU(B.S=32)	
	FPS	Latency	FPS	Latency	FPS	Latency	FPS	Latency
AlexNet	48.06	85.08	132.58	960.45	110.82	33.77	195.29	646.64
SqueezeNet 1.1	230.06	17.11	180.48	659.61	125.87	31.63	228	556.57
VGG 16	6.11	712.4	2.81	39426.47	24.89	158.47	19.57	6250.56
GoogleNet v1	57.02	76.62	54.45	2413.71	24.09	160.05	120.98	1044.86

5.1 FPS Versus Batch Size

Figures 3, 4, and 5 shows the frame per second (FPS) versus Batch size comparison of Alexnet, Squeezenet1.1, Googlenet-v1 on CPU(FP32), and GPU(FP16). The graphs illustrate the throughput performance of the models on increasing batch size.

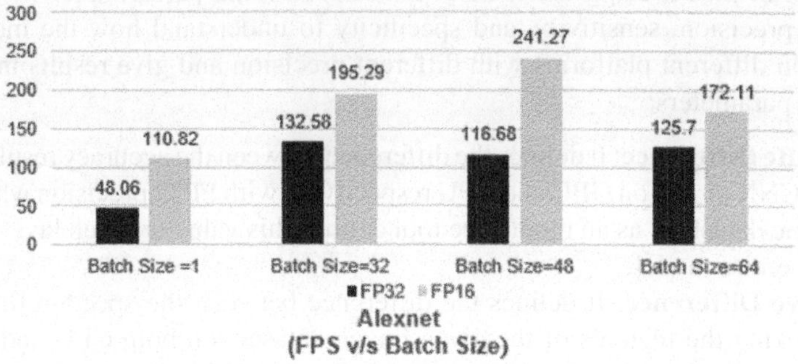

Fig. 3 FPS v/s Batch size for AlexNet

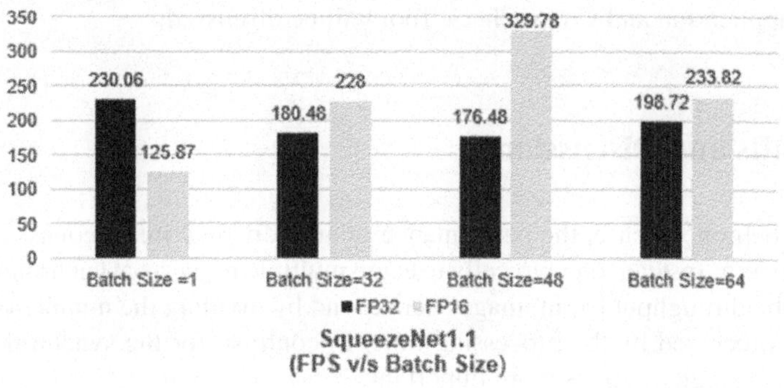

Fig. 4 FPS v/s Batch size for SqueezeNet 1.1

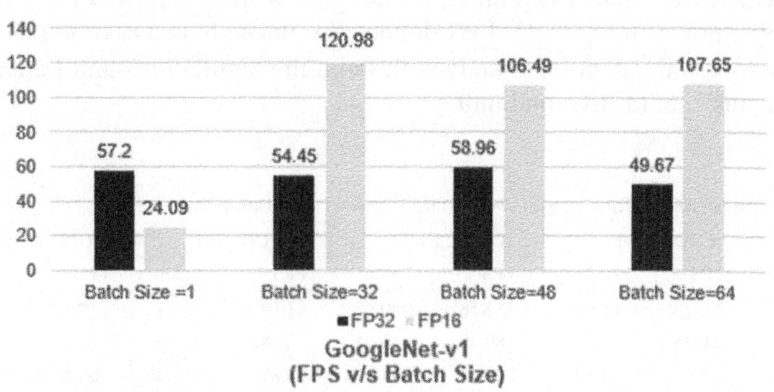

Fig. 5 FPS v/s Batch size for GoogleNet-v1

Results obtained from the simulation of the CNN models on the Cross-Check Tool are shown in Tables 2, 3, 4, 5, and 6. Here, the parameters like Absolute Difference, Relative Difference, Actual Value, Absolute Actual Value. Reference Value, Absolute Reference Value for each layer of the CNN model have been defined in tabular form.

5.2 AlexNet

AlexNet has 22 layers, out of which 5 layers are of Convolution, 9 layers are of activation(ReLu), 3 of Pooling Layers, 2 Normalization Layers, 3 Fully Connected Layer and 1 layer of Softmax [4]. When we simulate the model on the Cross-Check Tool, the tool executes the model on both CPU and GPU, calculates accuracy and precision metrics and provides an output for each layer separately. Below mentioned are the tested results.

The below table (see Table 2), indicates the maximum and minimum values obtained for mentioned parameters and its experimental obtained results of each 22 layers.

Here are some conclusions we can draw from tested results.

1. Relative Difference = Absolute Difference $\times$ 10^{20}.
2. For all the layers of AlexNet Minimum value of Absolute Difference and Relative Difference is Zero.
3. Except for Convolution and Fully Connected Layer, minimum values for all the parameters are Zero.
4. In all the cases, Reference Value=Actual Value except for Convolution 2 Layer.
5. In the case of Norm Layer, Activation Layer, and Pooling Layer both minimum and maximum values of Reference Value=Actual Value=Absolute Reference Value=Absolute Actual Value respectively.

5.3 SqueezeNet 1.1

SqueezeNet has 66 layers, out of which 5 layers are of Convolution, 9 layers are of Activation(ReLu), 3 of Pooling Layers, 2 Normalization Layers, 3 Fully Connected Layer and 1 layer of Softmax [9]. In the architecture of SqueezeNet, Fire module acts as a building block for CNN.

(Table 3 and 4) comprises

1. A squeeze convolution layer (which has only 1 $\times$ 1 filters), and
2. Feeding into an expand layer that has a mix of 1 $\times$ 1 and 3 $\times$ 3 convolution filters.

We expose three tunable dimensions (hyperparameters) in a fire module: $s_{1\times1}$, $e_{1\times1}$, and $e_{3\times3}$. When we use fire modules, we set, $\mathbf{s}_{1\times1} < \mathbf{e}_{1\times1}+e_{3\times3}$, so the squeeze layer helps to limit the number of input channels to the 3 $\times$ 3 filters.

Table 2 Cross-Check Tool Values for AlexNet

	Activation Layer		Pooling Layer		Normalization Layer		Fully Connected		Convolution Layer	
	Max	Min	Max	Min	Max	Min	Max	Min	Max	Min
Abs. Diff(max)	1.4×10^{-3}	1.2×10^{-5}	1.2×10^{-3}	1.06×10^{-4}	1.3×10^{-3}	4.72×10^{-4}	1.70×10^{-4}	9.50×10^{-6}	1.6×10^{-3}	1.6×10^{-4}
Actual(max)	2.1×10^{3}	1.3×10^{1}	1.8×10^{2}	1.38×10^{2}	1.3×10^{2}	1.3×10^{2}	6.70×10^{1}	1.30×10^{1}	2.1×10^{3}	1.81×10^{2}
Actual(min)	0	0	0	0	0	0	−5.8	-8.50×10^{1}	-1.5×10^{2}	-2.3×10^{3}
Ref value(max)	2.1×10^{3}	1.3×10^{1}	1.8×10^{2}	1.38×10^{2}	1.3×10^{2}	1.3×10^{2}	6.7×10^{1}	1.30×10^{1}	2.1×10^{3}	1.8×10^{2}
Ref value(min)	0	0	0	0	0	0	−5.8	-8.5×10^{1}	-1.5×10^{2}	-2.3×10^{3}
Abs. actual(max)	2.1×10^{3}	1.3×10^{1}	1.8×10^{2}	1.38×10^{2}	1.3×10^{2}	1.3×10^{2}	8.50×10^{1}	1.80×10^{1}	2.3×10^{3}	1.81×10^{2}
Abs. actual(min)	0	0	0	0	0	0	0	0	7.9×10^{-3}	1.50×10^{-5}
Abs. Ref.(max)	2.1×10^{3}	1.3×10^{1}	1.8×10^{2}	1.38×10^{2}	1.3×10^{2}	1.3×10^{2}	8.5×10^{1}	1.80×10^{1}	2.3×10^{3}	1.81×10^{2}
Abs. Ref.(min)	0	0	0	0	0	0	3.2×10^{-2}	2.5×10^{-4}	7.9×10^{-3}	2.0×10^{-5}

Table 3 Cross-Check Tool Values for SqueezeNet 1.1

	Convolution-Expand (1 × 1)		Convolution-Expand (3 × 3)		Convolution-Squeeze		Activation-Squeeze	
	Max	Min	Max	Min	Max	Min	Max	Min
Abs. Diff(max)	4.27×10^{-4}	0	1.46×10^{-3}	4.37×10^{-4}	6.70×10^{-4}	0	6.71×10^{-4}	0
Actual(max)	9.30×10^{2}	2.03×10^{2}	1.6×10^{3}	4.13×10^{2}	1.85×10^{3}	5.93×10^{2}	1.85×10^{3}	5.93×10^{2}
Actual(min)	-1.04×10^{3}	-7.70×10^{2}	-2.30×10^{3}	-8.12×10^{2}	-8.82×10^{2}	-1.8×10^{3}	0	0
Ref Value(max)	9.30×10^{2}	2.03×10^{2}	1.6×10^{3}	4.13×10^{2}	1.85×10^{3}	5.93×10^{2}	1.85×10^{3}	5.93×10^{2}
Ref Value(min)	-1.04×10^{3}	-7.70×10^{2}	-2.30×10^{3}	-8.12×10^{2}	-8.82×10^{2}	-1.8×10^{3}	0	0
Abs. actual(max)	7.08×10^{3}	2.79×10^{2}	2.34×10^{3}	8.26×10^{2}	1.8×10^{3}	8.3×10^{-4}	1.85×10^{3}	5.93×10^{2}
Abs. actual(min)	1.12×10^{-2}	2.20×10^{-4}	3.73×10^{-2}	1.05×10^{-6}	1.65×10^{5}	8.5×10^{-4}	0	0
Abs. Ref.(max)	7.08×10^{3}	2.79×10^{2}	2.34×10^{3}	8.26×10^{2}	1.8×10^{3}	8.3×10^{-4}	1.85×10^{3}	5.93×10^{2}
Abs. Ref.(min)	1.12×10^{-2}	2.20×10^{-4}	3.73×10^{-2}	1.05×10^{-6}	1.65×10^{5}	8.5×10^{-4}	0	0

Table 4 Cross-Check Tool Values for SqueezeNet 1.1

	Activation-Expand (1 × 1)		Activation-Expand (3 × 3)		Concat Layer		Pooling Layer	
	Max	Min	Max	Min	Max	Min	Max	Min
Abs. Diff(max)	4.27×10^{-4}	0	8.54×10^{-4}	3.43×10^{-4}	8.5×10^{-4}	3.4×10^{-4}	6.1×10^{-4}	0
Actual(max)	9.3×10^{2}	2.3×10^{2}	1.6×10^{3}	4.13×10^{2}	1.6×10^{3}	4.13×10^{2}	1.27×10^{3}	3.72×10^{1}
Actual(min)	0	0	0	0	0	0	1.71	0
Ref Value(max)	9.3×10^{2}	2.3×10^{2}	1.6×10^{3}	4.13×10^{2}	1.6×10^{3}	4.13×10^{2}	1.27×10^{3}	3.72×10^{1}
Ref Value(min)	0	0	0	0	0	0	1.71	0
Abs. Actual(max)	9.3×10^{2}	2.3×10^{2}	1.6×10^{3}	4.13×10^{2}	1.6×10^{3}	4.13×10^{2}	1.27×10^{3}	3.72×10^{1}
Abs. Actual(min)	0	0	0	0	0	0	1.71	0
Abs. Ref.(max)	9.3×10^{2}	2.3×10^{2}	1.6×10^{3}	4.13×10^{2}	1.6×10^{3}	4.13×10^{2}	1.27×10^{3}	3.72×10^{1}
Abs. Ref.(min)	0	0	0	0	0	0	1.71	0

Table 5 Cross-Check Tool Values for VGG 16

	Activation Layer		Pooling Layer		Fully Connected		Convolution Layer	
	Max	Min	Max	Min	Max	Min	Max	Min
Abs Diff(max)	7.85×10^{-3}	0	7.81×10^{-3}	1.98×10^{-4}	1.70×10^{-4}	1.04×10^{-5}	8.3×10^{-3}	0
Actual(max)	1.26×10^{4}	1.01×10^{1}	1.24×10^{4}	2.49×10^{2}	3.7×10^{1}	1.0×10^{1}	1.26×10^{4}	2.49×10^{2}
Actual(min)	0	0	0	0	−5.34	-6.75×10^{1}	-5.76×10^{2}	-1.20×10^{4}
Ref Value(max)	1.26×10^{4}	1.01×10^{1}	1.24×10^{4}	2.49×10^{2}	3.7×10^{1}	1.0×10^{1}	1.26×10^{4}	2.49×10^{2}
Ref Value(min)	0	0	0	0	−5.34	-6.75×10^{1}	-5.76×10^{2}	-1.20×10^{4}
Abs. Actual(max)	1.26×10^{4}	1.01×10^{1}	1.24×10^{4}	2.49×10^{2}	6.75×10^{1}	1.15×10^{1}	1.26×10^{4}	5.76×10^{2}
Abs. Actual(min)	0	0	0	0	2.79×10^{-3}	6.52×10^{-4}	2.2×10^{-3}	1.86×10^{-7}
Abs. Ref(max)	1.26×10^{4}	1.01×10^{1}	1.24×10^{4}	2.49×10^{2}	6.75×10^{1}	1.15×10^{1}	1.26×10^{4}	5.76×10^{2}
Abs. Ref(min)	0	0	0	0	2.79×10^{-3}	6.52×10^{-4}	2.2×10^{-3}	1.86×10^{-7}

Table 6 Cross-Check Tool Values for GoogleNet

	Activation Layer		Pooling Layer		Normalization Layer		Concat Layer		Convolution Layer	
	Max	Min	Max	Min	Max	Min	Max	Min	Max	Min
Abs Diff(max)	2.9×10^{-3}	1.62×10^{-5}	2.86×10^{-3}	3.80×10^{-6}	1.55×10^{-3}	5.06×10^{-7}	8.50×10^{-4}	6.38×10^{-5}	2.4×10^{-2}	1.71×10^{-5}
Actual(max)	6.2×10^{2}	1.6×10^{1}	3.51×10^{3}	8.7	1.38×10^{2}	3.57×10^{-1}	6.02×10^{2}	3.87×10^{1}	3.5×10^{3}	1.60×10^{1}
Actual(min)	0	0	0	0	1.23×10^{-7}	0	0	0	-2.04×10^{1}	-8.20×10^{2}
Ref Value(max)	6.2×10^{2}	1.6×10^{1}	3.51×10^{3}	8.7	1.38×10^{2}	3.57×10^{-1}	6.02×10^{2}	3.87×10^{1}	3.5×10^{3}	1.60×10^{1}
Ref Value(min)	0	0	0	0	1.23×10^{-7}	0	0	0	-2.04×10^{1}	-8.20×10^{2}
Abs. actual(max)	6.2×10^{2}	1.6×10^{1}	3.51×10^{3}	8.7	1.38×10^{2}	3.57×10^{-1}	6.02×10^{2}	3.87×10^{1}	8.22×10^{2}	2.4×10^{1}
Abs. actual(min)	0	0	0	0	1.23×10^{-7}	0	0	0	-5.21×10^{-2}	9.89×10^{-6}
Abs. Ref.Value(max)	6.2×10^{2}	1.6×10^{1}	3.51×10^{3}	8.7	1.38×10^{2}	3.57×10^{-1}	6.02×10^{2}	3.87×10^{1}	8.22×10^{2}	2.4×10^{1}

We can draw important conclusions based on the results obtained by running on cross-check tool for SqueezeNet 1.1.

1. Relative Difference = 10^{20} × Absolute Difference.
2. Except for the convolution layer (squeeze and expand), the minimum values of all parameters are zero.
3. The minimum value of absolute difference and the relative difference is zero.
4. For the convolution layer,
    (a) Absolute actual value(max)=Absolute reference value(max).
    (b) Reference value(max)=Actual value(max).
    (c) Reference value(min)=Actual value(min).
    (d) But the minimum value of absolute actual value and absolute reference value are not the same.
5. Except for the convolution layer, the maximum values of actual value, absolute actual value, reference value, and the absolute reference value of the normalization layer, pooling layer, concatenation layer, and activation layer are the same.

5.4 VGG 16

VGG has 38 layers, out of which 5 layers are of Convolution each having 2 or 3 sub-layers making it 13 Layers for Convolution. Similarly, it has 7 layers are of Activation(ReLu), which again have 2 or 3 sub-layers making in total 15 layers. Then it has 5 Pooling Layers, 3 Fully Connected Layers and 2 layers of Softmax.

The above table (see Table 5), defines the maximum and minimum values obtained for the mentioned parameters w.r.t experimental values we got of these parameters for each of 38 layers. Here are major conclusions which can be drawn from tested results.

1. Relative difference = 10^{20} × Absolute difference.
2. Except for the fully-connected and convolution layer, the minimum values of all parameters are zero.
3. In the case of the activation layer and pooling layer, the maximum values of actual value, absolute actual value, reference value, and absolute reference value are the same.
4. For all layers in VGG-16 the minimum value of absolute difference and the relative difference is zero.
5. For convolution and fully-connected layer,
    (a) Absolute actual value(max)=Absolute reference value(max).
    (b) Reference value(max)=Actual value(max).
    (c) Reference value(min)=Actual value(min).

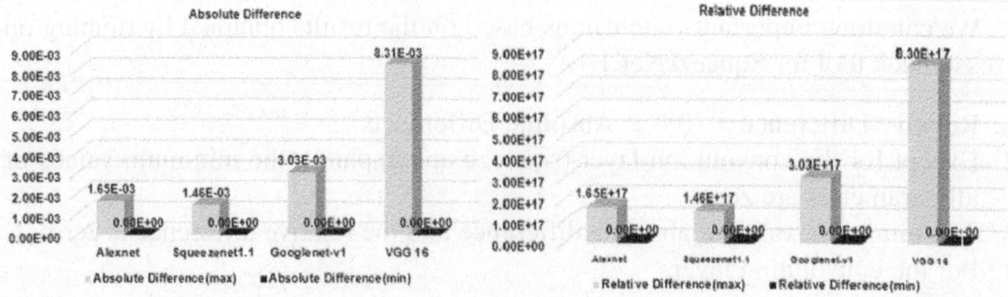

Fig. 6 Absolute and Relative Differences for all the CNN Models

6. In the case of the fully-connected layer, the minimum value of absolute actual value and the absolute reference value is the same, but not the same in the case of the convolution layer.

5.5 GoogleNet/Inception

GoogLeNet [10] has 142 layers, when it is simulated on Cross-Check Tool on both the devices i.e. CPU and GPU, it calculates accuracy and precision metrics and provide an output for each layer separately. Below are the mentioned tested results.

The above table (see Table 6), defines the maximum and minimum values obtained for the above mentioned parameters and experimental values we got of these parameters for each of 142 layers. Here are the conclusions we can draw from tested results.

1. Relative Difference = 10^{20} × Absolute difference.
2. Except for the normalization and convolution layer, the minimum values of all parameters are zero.
3. For the convolution layer,
   (a) Absolute actual value(max)=Absolute reference value(max).
   (b) Reference value(max)=Actual value(max).
   (c) Reference value(min)=Actual value(min).
   (d) But the minimum value of absolute actual value and absolute reference value are not the same.
4. For all layers in Googlenet-v1, the minimum value of absolute difference and the relative difference is zero.
5. Except for the convolution layer, the maximum values of actual value, absolute actual value, reference value, and the absolute reference value of the normalization layer, pooling layer, concatenation layer, and activation layer are the same.
6. In the case of the normalization layer, the minimum values of actual value, absolute actual value, reference value, and absolute reference value are the same.

5.6 *Overall Observed Results for CNN Models*

Above shown (Fig. 6) is a graphical representation (drawn using Excel sheet) of the parameter i.e. absolute difference and relative difference (mentioned in Sect. 4.2) for all the four CNN models.

6 Conclusion and Future Work

From the results of the Benchmarking application, it is observed that GPU gives better acceleration compared to CPU. The results clearly indicate high throughput and low latency on a GPU platform (Intel integrated graphics). It is observed that if batch size increases, the performance of Alexnet, Squeezenet1.1, and Googlenet-v1 also increases. We also observed that the throughput of Alexnet, Squeezenet1.1, and Googlenet-v1 is better in the FP16 data format when the batch size increases. The performance of Googlenet-v1 (Table 1) is almost double when the batch size increases.

The Cross-Check Tool gives maximum and minimum values of parameters like absolute difference relative difference, and actual value for each layer of the CNN model. If we see Fig. 6, we can conclude that the maximum difference for the parameters is seen in VGG 16. Similarly in Benchmarking application also CNN Model i.e., VGG 16 showed the worst latency and throughput results. Further, this knowledge can help us to develop more efficient CNN Architectures.

For future work, the Performance and other Acceleration parameters of all the above models will be analysed on FPGA and will be compared accordingly. During the analysis, we also saw that in GoogleNet, most of the delays were caused due to the convolution layer and to improve the metric i.e. latency, we can use Winograd's Minimal Filtering Algorithms [7].

References

1. Albawi, S., Mohammed, T.A., Al-Zawi, S.: Understanding of a convolutional neural network. In: International Conference on Engineering and Technology. IEEE Press, Turkey (2017)
2. Lin, Z., Yih, M., Ota, J.M., Owens, J.D., Muyan-Ozcelik, P.: Benchmarking deep learning frameworks and investigating FPGA deployment for traffic sign classification and detection. IEEE Trans. Intell. Veh. 4(3), 385–395 (2019)
3. Ashwin, P.R.: Acceleration of deep learning image classification model. Int. Res. J. Eng. Technol. 7(6), 421–424 (2020)
4. Jmour, N., Zayen, S., Abdelkrim, A.: Convolutional neural networks for image classification. In: International Conference on Advanced Systems and Electric Technologies, pp. 397–402. IEEE Press, Tunisian (2018)
5. Jin, Z., Finkel, H.: Analyzing deep learning model inferences for image classification using OpenVINO. In: International Parallel and Distributed Processing Symposium Workshops, pp. 908–911. IEEE Press, USA (2020)

6. Demidovskij, A., Gorbachev, Y., Fedorov, M., Slavutin, I., Tugarev, A., Fatekhov, M., Tarkan, Y.: OpenVINO Deep Learning Workbench-Comprehensive Analysis and Tuning of Neural Networks Inference. In: International Conference on Computer Vision Workshop, pp. 783–787. IEEE Press, South Korea (2019)
7. Wang, D., Liu, L., Liu, L., Fan, H., Tang, Z., Bai, Z.: An OpenCL-based FPGA accelerator with the Winograd's minimal filtering algorithm for convolution neuron networks. In: 5th International Conference on Computer and Communications, pp. 277–282. IEEE Press, China (2019)
8. Shawahna, A., Sait, S.M., El-Maleh, A.: FPGA-based accelerators of deep learning networks for learning and classification–a review. IEEE Access **7**, 7823–7859 (2018)
9. Mohanraj, V., Guda, R., Rao, J.V.K.: Hybrid approach for pixel-wise semantic segmentation using SegNet and SqueezeNet for embedded platforms. In: Bapi, R.S., Rao, K.S., Prasad, M.V.N.K. (eds.) First International Conference on Artificial Intelligence and cognitive Computing, AICC, vol. 815. Springer, Heidelberg (2018), pp. 155–163
10. Szegedy, C., Liu, W., Jia, Y., Sermanet, P., Reed, S., Anguelov, D., Erhan, D., Vanhoucke, V., Rabinovich, A.: Going deeper with convolutions. In: IEEE Conference on Computer Vision and Pattern Recognition, pp. 1–9. IEEE Press, USA (2015)
11. Kyriakos, A., Kitsakis, V., Louropoulos, A., Papatheofanous, E.A., Patronas, I., Reisis, D.: High-performance accelerator for CNN applications. In: 29th International Symposium on Power and Timing Modeling. Optimization and Simulation, pp. 135–140. IEEE Press, Greece (2019)
12. Wang, Y., Wang, Q., Shi, S., He, X., Tang, Z., Zhao, K., Chu, X.: Benchmarking the performance and energy efficiency of AI accelerators for AI training. In: 20th ACM International Symposium on Cluster. Cloud and Internet Computing, pp. 744–751. IEEE Press, Australia (2020)
13. OpenVINO Toolkit Overview. http://docs.openvinotoolkit.org/latest/index.html
14. Cross Check Tool-OpenVINO Toolkit, http://docs.openvinotoolkit.org/2019R2/docs_IE_DG_Cross_Check_Tool.html
15. BenchmarkPython*Tool-OpenVINO™ Toolkit, http://docs.openvinotoolkit.org/latest/openvino_inference_engine_tools_benchmark_tool_README.html

Welding Defect Inspection Using Deep Learning

Hasan Asif and Shailendra Kumar

Abstract To supplant human shortcomings in Welding Inspection System performed manually by humans, an intelligent computer vision inspection leveraging deep learning technologies method is proposed. In the proposed model, the top layer of the VGG16 model was replaced by a fully connected layer with a softmax activation function. Rather than providing an extensive comparison between different transfer learning models (Inception V3, Xception, VGG16, VGG19, ResNet50), we analyzed the VGG16 model in-depth, as it performed well in initial comparison on accuracy and training time as compared with other models. All models were trained using a welding image dataset to identify and detect various welding defects, including Cracks, Over-Roll, Under-Fill, Porosity, and Mechanical Damage. The proposed VGG16 model was evaluated to verify how accurately, faster, and precisely it performs multiple-class classification. Furthermore, the proposed model was optimized by hyper-parameter tuning of Learning rate and Batch size, and an overall 8.5% increase in the accuracy was observed. The experimental results show that in several aspects, VGG16 had performed quite accurately in classifying defects and achieved the highest accuracy of 85%.

Keywords Deep learning · Welding defect detection · Transfer learning

1 Introduction

In the manufacturing industry, there are inherent challenges in ensuring consistency in product quality, both from an esthetic point of view, as well as functional performance. Defects in the metal can be very hazardous in railroads, gas pipes, wheels, etc. [1]. Fortunately, progress in computer science, control theory, robotics, and artificial intelligence [2] has opened a road to automatic welding defect detection that can detect welding defects without any or very less human dependency. Designing a deep learning-based detection system has been widely applied in many fields

H. Asif (✉) · S. Kumar
Mechanical Engineering Department, Meerut Institute of Engineering and Technology, Meerut, UP, India

A. Noor et al. (eds.), *Proceedings of Emerging Trends and Technologies on Intelligent Systems*, Advances in Intelligent Systems and Computing 1371,
https://doi.org/10.1007/978-981-16-3097-2_7

namely medical, security, and sports. Presently, the methods available to supervise welding quality are lacking in real-time capability. Traditionally, it is done manually or through nondestructive testing (NDT), which are expensive and time-consuming techniques [3]. Previous research explored the capability of deep learning methods to solve the problem of anomaly detection. The results are impressive but require a complex implementation and significant computing power.

In 2017, Hou et al. proposed an automatic detection method based on a deep neural network, using the sliding-window approach to detect on the whole X-ray image, images were collected from an openly available dataset called GDXay, later trained a deep neural net (DNN) model having three hidden layers of 576-196-100 neurons and maximum accuracy achieved is 91%. The author used Otsu's method to reduce the inter-class variance of thresholded black and white regions [4]. Zhang et al. experimented with a novel approach to detect the welding defect, by dividing the original image into sub-images of 32 * 32 pixels and passed through the model and used Inception and MobileNet as the base of the multi-model ensemble for feature extraction and detection. Accuracy of model classifying four different classes was always above 94% after trained on 7606 images. Which indicates the potential application in the industry [5]. In the work of Wu et al. an end-to-end learning method for defect detection in turbine blades was presented. They proposed a framework for generic defect detection by employing the ResNet as the basic structure for feature extraction. The model can effectively detect the defect area of as small as 1.3% of the image [6]. Similarly, Xia et al. trained the ResNet model on a customized dataset, having five different classes of defect for keyhole TIG welding. Modeling results with a detailed feature map of ResNet. It shows that the model can effectively identify the defect [7]. Zhu et al. proposed a deep learning framework, based on a convolution neural network (CNN) and a Random Forest, to classify weld surface defects. The combined network approach improved the overall performance [8]. Pan et al. researched automatic surface defect detection based on deep transfer learning. MobileNet model was reconfigured by adding another Full Connection layer (FC-128) and a Softmax classifier result in smaller model size, less calculation time, and better accuracy compared with other traditional neural network and transfer learning models [9]. Ferguson et al. proposed a system for casting defect detection, by using X-ray images for training the ResNet-101 model. The model outperformed the state-of-the-art performance [10]. In 2010, Kasban et al. proposed a method to detect and identify welding defects in radiography images using a novel feature extraction technique. The model was tested on 60 radiography images containing 73 defects. Results show reliability in the proposed system for autonomous defect identification from radiography images in the presence of noise and blurring [11].

Nowadays, a lot of research is being carried out to strive for an automatic weld defect identification method. It is clear from the above literature review that the automatic detection of weld defects in images can be effectively deployed in the industry.

This paper proposed a system to detect and classify welding defects. Image data to train the deep learning models were collected from publicly available data. The use of transfer learning reduces the requirement of a large training data for learning without compromising the precise classification of a welding defect. These images were processed and then used as the input for feature extraction using transfer learning. The feasibility of the proposed technique is strengthened by the evaluation results.

1.1 Motive for Work

To ensure customer satisfaction in esthetics and performance, each welded product needs to be checked thoroughly at the end of the production line, before dispatch to the customer for any visual defects. A typical checking process requires an operator to verify and validate defect-free weld. The issue gets more compounded with the fact that humans in this traditional welding system performed all information tasks, including sensing, analysis, decision-making, operation, control, cognition, and learning. Needless to say, this process causes high fatigue and has a high chance of human error. Hence, the motivation is to have an autonomous system that can monitor and detect defects in near real-time to eliminate human fatigue, as well as improve production efficiencies. The most important goal is to have defect-free weld in products reaching the market which improves the Brand value and provides a better customer experience.

2 Transfer Learning in Image Classification

In recent years, convolutional neural networks (CNNs) have achieved remarkable performance in a variety of defect image classification applications [12–14]. In deep CNN, deeper layer tries to learn more specific feature related to the task, on the other hand, the initial layer of deep CNN model learns more low-level feature like edges. As compared to deeper layers, initial layers are tougher to train, due to vanishing gradient issues.

A deep neural network model will not be able to perform the classification task accurately until the model's requirements of diverse and large labeled data were satisfied and stringent supervision that the model is learning well. A promising alternative to this drawback is to use transfer learning. When it comes to image classification, the ImageNet challenge is the perfect benchmark for computer vision classification algorithms—and the leaderboard for this challenge has been dominated by convolutional neural networks and deep learning techniques since 2012. With the commence of AlexNet [15] in the Imagnet Competition in 2012, essential breakthrough in computer vision was witnessed. Since then, transfer learning has achieved great success in computer vision tasks. From then on, a wide range of transfer learning models has been proposed such as VGG16 [16], Inception V3 [17], Xception [18], and ResNet [19] models for extensive image classification. Juxtapose with the established feature extraction methods, transfer learning leveraging knowledge gained in one or more base tasks is transferred and applied to improve the learning of a target task. These methods save time and give better performance. This transfer learning approach can contribute to our image classifier in many ways. Firstly, it can unfold the issue of low classification accuracy, due to the fewer data samples. As it leverages the benefit of already trained weight on a huge ImageNet dataset and in our proposed model, we used VGG16, which was used to win the ILSVR (ImageNet) competition in 2014, and aimed to improve the capability of the model to draw as many features possible from the image by increasing the complexity of the network. The deep architecture of transfer learning CNN models results in a network that is more coherent and positively impacts the prediction efficacy.

2.1 VGG16 Model Architecture

The whole structure of VGG16 [16] is given in Fig. 1. VGG16 architecture is as follows: Firstly, input an image of size 224 * 224 * 3, where 224 * 224 represents the length and width of an image, and 3 represents three channels of an RGB image. The input size can be changed as per the quality of the image. In VGG16, all the convolution layers have the same filter size of 3 * 3, the stride is equaled to 1 and padding is the same. Similarly, in the max-pooling layer, filter size is 2 * 2 and the stride is 2. The first and second convolution layers extract information from the image with 64 kernels, followed by a max-pooling layer. The third and fourth convolution layer includes 128 kernels of the same size as in previous and then max-pooling layer. Then, there are three sets of layers, and in each set, there are three convolution layers

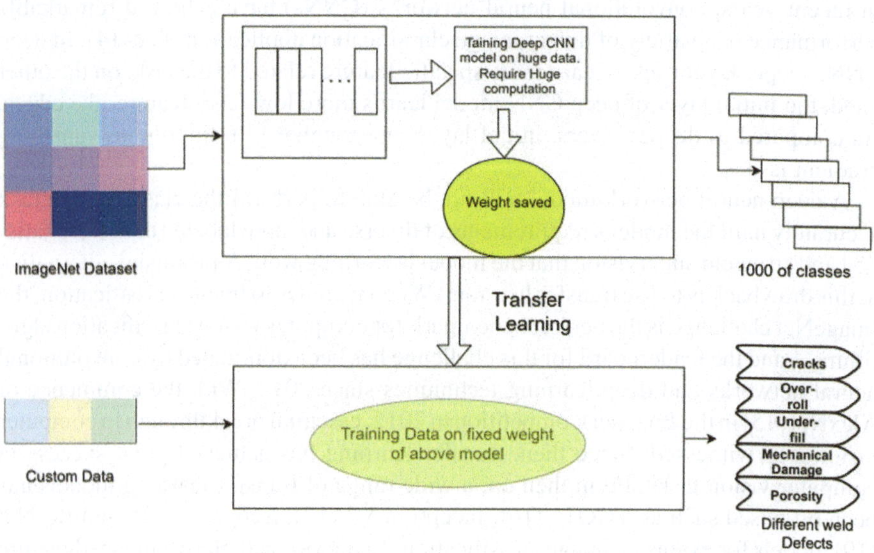

Fig. 1 Idea of transfer learning

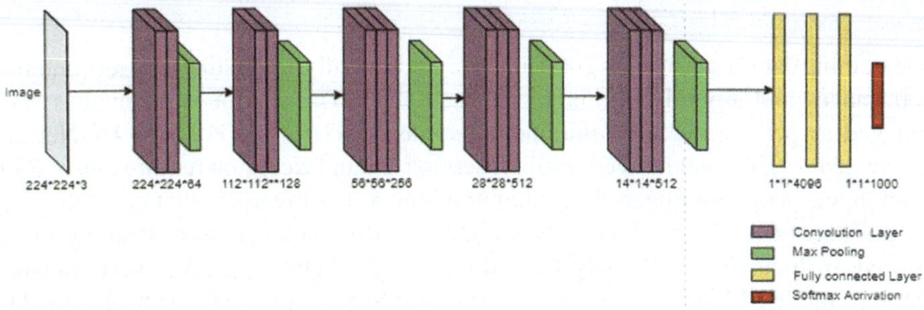

Fig. 2 Skeleton of VGG16 model

and one max-pooling layer. The number of kernels is 256, 512 and 512, respectively, in every set, followed by three fully connected layers connected one after other, the first two fully connected layers have 4096 and the last one has 1000 neurons. Softmax layer is linked at the end of the classification model (Fig. 2).

2.2 *Dataset*

Dataset used in our study was downloaded directly from Kaggle. The diversity of images is like the fuel for the model in classification. The dataset comprises 500 images of five different types of welding defects. For the training of the model, 80% of the images were used and left 20% of images were used in testing the model accuracy. Each image in the dataset is 650 * 500 in size. Proceeding to the preprocessing stage, the number of channels remained the same as 3 and size to 224 * 224 or 299 * 299 as per the model requirement. Eventually, this helped us sap the computational load. Followed the preprocessing stage was the data augmentation stage to add heterogeneity in the dataset without adding the new data such that a better deep learning model can be built using them. Applied techniques like Geometric transformation, Rotation, and flipping over the images increase the data size. Data augmentation prevents over-fitting by modifying limited data. To reduce over-fitting during model training, data augmentation processes such as rotation, flipping, and zooming are taken on before the training stage.

Hence, significant improvement in the performance of the algorithm was achieved [20]. We then normalize both training and testing dataset and configured as the required input to the neural network.

2.3 Training Different Models

The training dataset comprises 2000 images. Additionally, the testing dataset contains 500 images and around 100 images of each class. The size of input images was changed as per the model requirement such as VGG16, VGG19, and ResNet all accept 224 × 224 input images, while Inception V3 and Xception require 299 × 299 pixel inputs. Training images helped in grabbing a set of features that can delineate the attribute of welding defects. The weights of different layers were kept the same as were in ImageNets. We only trained the last prediction layer of every transfer learning model. Models were trained using the Keras API with a TensorFlow 2.0 backend running in GoogleColab. Adam optimizer was used to reach the global minima while training our model.

To find the optimum value of the learning rate and batch size, hyper-parameter tuning was performed. The data are shared in Figs. 4 and 5. Parameters were selected by monitoring the convergence. In the proposed model, the top layer was replaced by adding a fully connected layer having five classes that followed by the softmax activation function. The skeleton of the proposed system is given in Fig. 3.

2.4 Evaluation Metric

Following the training stage, the performance of different models was examined under various performance metrics in terms of accuracy, recall, precision, and F1-score using Eqs. 1, 2, 3, and 4. Here, TP means the actual image is defective and the prediction by the model is correct. Similarly, TN means the actual image is non-defective and the prediction by the model is correct. Whereas FP means the actual image is defective and the prediction by the model is not correct and FN means the actual image is non-defective and the prediction by the model is not correct.

$$\text{Accuracy} = \frac{\text{TTN}}{\text{TP} + \text{FP} + \text{TN} + \text{FN}} \tag{1}$$

$$\text{Precision} = \frac{\text{TP}}{\text{TP} + \text{FP}} \tag{2}$$

$$\text{Recall} = \frac{\text{TP}}{\text{TP} + \text{FN}} \tag{3}$$

$$\text{F1} - \text{score} = 2 \times \frac{\text{Precision} * \text{Recall}}{\text{Precision} + \text{Recall}} \tag{4}$$

F1-score value is very important because it considers both recall and precision. A high F1-score value would be only possible when both precision and recall are

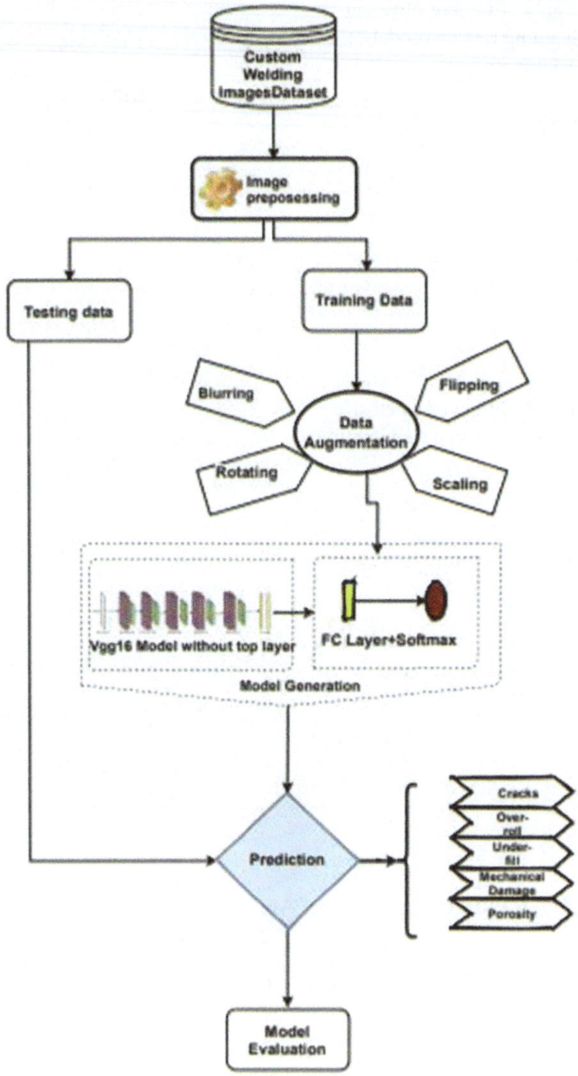

Fig. 3 Structure of the proposed system

high. F1-score close to 1 is the indication of the model performing well in effectively identifying the correct classes.

To accomplish any deep learning task, a good amount of annotated data is always needed to properly train the model. We had assessed the consequence of different batch sizes and learning rates on how accurately the model predicts. In addition to this, every model was evaluated on how much time it needed to train on given welding image and storage.

Fig. 4 Effect of different learning rate on model accuracy

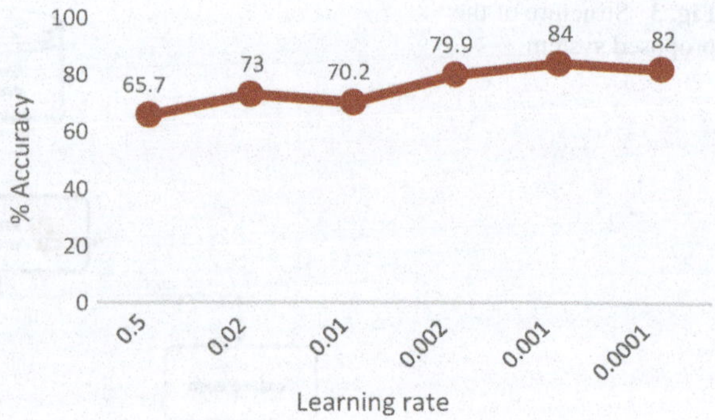

Fig. 5 Effect of batch size on accuracy and training time

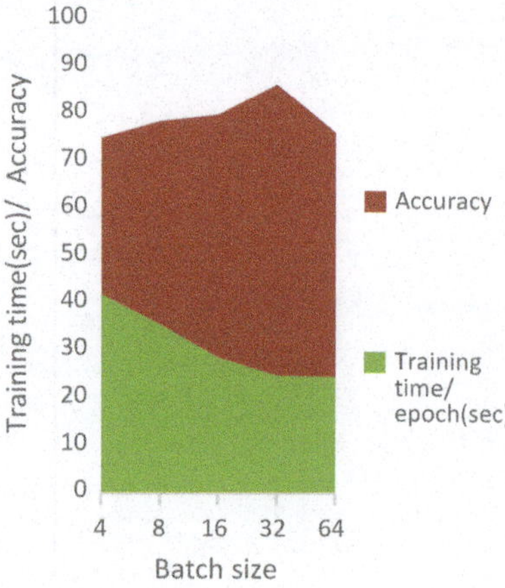

3 Results

Table 1 shows the accuracy of different models in identifying welding defect. The highest accuracy of 76.5% was achieved, using VGG16 model. Among them, the accuracy of VGG19 and Inception V3 was relatively low, accounting for 69.4% and 68%, respectively. In comparison with other models, ResNet50 and Xception performed averagely on correctly classifying defects in the test image. In overall,

Table 1 Classification accuracy of a different model for initial comparison

Model	Accuracy (%)
ResNet50	71.2
VGG19	69.4
VGG16	76.5
Inception V3	68
Xception	70.8

Table. 2 Training time and size of different model

Model	Training time (sec per epoch)	Size (MB)
ResNet50	30	104
VGG19	25	574
VGG16	22	533
Inception V3	26	96
Xception	29	91

the weld defect detection accuracy is consider in the case of VGG16, ResNet50, and Xception. Suggesting their feasibility in real-time usage.

The experiment results shown in Table 2 illustrate the average time required to complete one epoch of training and the size of the models. Table 2 shows that the training time of VGG16 was less than other models, however, the size was considerably large. Xception holds the lowest size, but its training time was higher than VGG16; ResNet50 has the highest training time of an average of 30s per epoch with the model size of 104 MB; Vgg19 and Inception V3 have almost the same training time but there is a huge difference in model size, VGG19 has the biggest size among all, whereas Inception V3 estimated around 96 MB. In comparison with other models, the VGG16 model has the least training time although it is larger in size than many other networks. VGG16 has the best accuracy to identify weld defects and the least training time, owing to this, it was chosen for further evaluation.

Table 4 shows the output from the model when tested. It explains the number of times the proposed VGG16 model correctly or incorrectly identified the defect in test images. Precision, Recall, and F1-score are presented in Table 3. The model achieved a precision of 87% when the image contains Cracks, probably model can be characterized the feature well in this case, similarly, Recall and F1-score were 86.13 and 86.56% respectively. For image having Porosity model had the highest F1-score and recall. However, the Under-Fill defect can be detected easily by humans, but the model had the lowest precision and F1-score of 80.76 and 81.92%, whereas recall was 83.16%. The identification of the image having Mechanical Damage achieved a precision of 83.33%, recall of 86.95%, and F1-score of 85.10%. The model achieved the highest precision was in detecting Over-Roll images and achieved a precision of 87.36%, but recall of 81.37% which is the lowest among others, and an F1-score of 84.09%.

Table. 3 Evaluation of identification of different defects by proposed VGG 16 model

Model	F1-score (%)	Precision (%)	Recall (%)
Cracks	86.56	87	86.13
Over-roll	84.09	87.36	81.37
Under-fill	81.92	80.76	83.16
Porosity	87.07	86.66	87.5
Mechanical damage	85.10	83.33	86.95

Table. 4 Scatter plot of VGG16 model

Target ↓	Predicted →				
	Cracks	Over-roll	Under-fill	Porosity	Mechanical damage
Cracks	87	3	0	8	3
Over-roll	11	83	6	0	2
Under-fill	0	8	84	4	5
Porosity	2	1	4	91	6
Mechanical damage	0	0	10	2	80

On hyper-parameter tuning, the learning rate of the Adam optimizer was tuned using the values range from 0.5 to 0.0001. Figure 4 depicts the effect of different learning rates on the accuracy of the proposed VGG16 model. It is noted that the learning rate had a striking effect on the model accuracy. Highest accuracy was achieved with a 0.001 learning rate helping to reach the minimum of the loss function efficiently. The model was poorly converged at the learning rate of 0.05.

This paper has also studied different batch sizes to optimize the model. The effect of batch size on model accuracy with the training time per epoch is displayed in Fig. 5. The figure illustrates that there was no significant change in accuracy as batch size increases from 16. In overall, accuracy hovers between 75.1 and 86.4%. The batch size was chosen between 4 and 64, larger batch sizes were not considered because of the small data size. At low and high batch sizes, model struggles with the over-fitting or is stuck at local minima and cannot converge optimally. In contrast, the average time to train each epoch took a considerable toll on computing power, as the batch size reduced below 32 training time was increased steeply.

4 Conclusion

The paper proposed a method of defect identification in welding images, leveraging the transfer learning for the task. By incorporating transfer learning, data requirement to train the model dropped significantly. Firstly, five different transfer learning models were deployed to classify defects and the best performance was achieved by a VGG16 model in the initial evaluation, achieving the highest accuracy of 76.5% and least training time. Secondly, in the proposed model, top layers were altered by adding a fully connected softmax function in VGG16 model. To optimize, the proposed VGG16 model was tuned for Learning rate and Batch size. Results showed that learning rate had a more prominent impact on accuracy than the batch size, an overall 8.5% increase in the accuracy was observed. Finally, we evaluated the proposed model by calculating precision, recall, and F1-score after fine-tuning. Results showed that our model was classifying the defect quite satisfactorily, the mean F1-score was as high as 87.07% for different defect identification, and the model can detect defect1 and defect4 efficiently with the precision of 87 and 86.66%.

The future scope is to investigate the application of the proposed technique on the large dataset of images and improve the classification performance by exploring other algorithms.

References

1. Mahmoudi, A., Regragui, F.: Welding defect detection by segmentation of radiographic images. In: 2009 World Congr. Comput. Sci. Inf. Eng. (2009). https://doi.org/10.1109/CSIE.2009.501
2. Wanga, B., Hub, S.J., Suna, L., Freihei, T.: Intelligent welding system technologies: State-of-the-art review and perspectives. J. Manuf. Syst. **56**, 373–391 (2020). https://doi.org/10.1016/j.jmsy.2020.06.020
3. Mirapeix, J., Garcı´a-Allende, P.B., Cobo, A., Conde, O.M., López-Higuera, J.M.: Real-time arc-welding defect detection and classification with principal component analysis and artificial neural networks. NDT&E Int. **40**, 315–323 (2007). https://doi.org/10.1016/j.ndteint.2006.12.001
4. Hou, W., Wei, Y., Guo J., Jin, Y., Zhu, C.: Automatic detection of welding defects using deep neural network. J. Phys. Conf. Ser. **933** (2017). https://doi.org/10.1088/1742-6596/933/1/012006
5. Zhang, H., Chen, Z., Zhang, C., Xi, J., Le, X.: Weld defect detection based on deep learning method. In: IEEE 15th Int. Conf. Automation Sci. Eng, pp. 1574–1579 (2019)
6. Yupei, Wu., Guo, Di., Liu, H., Huang, Y.: An end-to-end learning method for industrial defect detection. Assem. Autom. **40**, 31–39 (2020). https://doi.org/10.1108/AA-08-2018-114
7. Xia, C., Pan, Z., Fei, Z., Zhang, S., Li, H.: Vision based defects detection for Keyhole TIG welding using deep learning with visual explanation. J. Manuf. Process. **56**, 845–855 (2020). https://doi.org/10.1016/j.jmapro.2020.05.033
8. Zhu, I., Ge, W., Liu, Z.: Deep learning-based classification of weld surface defects. Appl. Sci. **9**, 3312 (2019). https://doi.org/10.3390/app9163312
9. Pan, H., Pang, Z., Wang, Y., Wang, Y., Chen, L.: A new image recognition and classification method combining transfer learning algorithm and mobilenet model for welding defects. IEEE Access, **8**, pp. 119951–119960, (2020). doi: https://doi.org/10.1109/ACCESS.2020.3005450
10. Ferguson, M.K., Ronay, A., Tina Lee, Y.T., Law, K.H.: Detection and segmentation of manufacturing defects with convolutional neural networks and transfer learning. Smart Sustain Manuf. Syst. (2018). https://doi.org/10.1520/SSMS20180033
11. Kasban, H., et al.: Welding defect detection from radiography images with a cepstral approach. NDT&E Int. **44**, 226–231 (2011). https://doi.org/10.1016/j.ndteint.2010.10.005
12. Wen, Z., Zhao, Q., Tong, L., CNN-based minor fabric defects detection. Int. J. Clothing Sci. Technol. **32** (2020). https://doi.org/10.1108/IJCST-11-2019-0177
13. Soukup, D., Huber-Mork, R.: Convolutional neural networks for steel surface defect detection from photometric stereo images. ISVC (2014). https://doi.org/10.1007/978-3-319-14249-4_64
14. Chen, H., Pang, Y., Qidi, Hu., Liu, K.: Solar cell surface defect inspection based on multispectral convolutional neural network. J. Intell. Manuf. (2018). https://doi.org/10.1007/s10845-018-1458-z
15. Krizhevsky, A., Sutskever, I., Hinton, G.E.: Imagenet classification with deep convolutional neural networks. pp. 1097–110 (2012). https://papers.nips.cc/paper/2012/file/c399862d3b9d6b76c8436e924a68c45b-Paper.pdf
16. Simonyan, K., Zisserman, A.: Very deep convolutional networks for large-scale image recognition. Int. Conf. Learn. Representations (2015). https://arxiv.org/pdf/1409.1556.pdf
17. Szegedy, C., Vanhoucke, V., Ioffe, S., Shlens, J., Wojna, Z.: Rethinking the inception architecture for computer vision. IEEE Conf. Computer Vision Pattern Recogn. (2016). https://doi.org/10.1109/CVPR.2016.308

18. Chollet, F., Xception: Deep learning with depth wise separable convolutions. In: 7 IEEE Conf. Computer Vision Pattern Recogni (2017). https://doi.org/10.1109/CVPR.2017.195
19. He, K., Zhang, X., Ren, S., Sun, J., Deep residual learning for image recognition. In: The IEEE Conf. Computer Vision Pattern Recogn. (CVPR), pp. 770–778 (2016). https://arxiv.org/pdf/1512.03385.pdf
20. Shorten, C., Khoshgoftaar, T.M.: A survey on image data augmentation for deep learning. J. Big Data **60** (2019). https://doi.org/10.1186/s40537-019-0197-0

Image Generation Using GAN and Its Classification Using SVM and CNN

Aadarsh Singh, Aashutosh Bansal, Nishant Chauhan, Satya Prakash Sahu, and Deepak Kumar Dewangan

Abstract In the computer vision world, generative adversarial networks have acquired huge recognition because of their data generation potential without modeling the function of probability density. Generative adversarial networks (GANs) have capability to generate new samples similar to data they were trained on. A smart means of integrating samples without label into preparation and applying higher-order accuracy is adversarial loss generated by discriminator. In this paper, GAN is implemented on MNIST dataset and recognizes latent representation of the feature for digit generation. The model also consists of support vector machine (SVM) which is initially trained on the same MNIST dataset that is being used by GAN for generating new data similar to that of MNIST. The data generated by GAN are then passed through the pre-trained SVM classifier for predicting its equivalent label. The support vector machine model which was used with the MNIST dataset obtains the accuracy of 96.67%. Moreover, the proposed model with convolution neural network (CNN) on the same MNIST dataset produces an accuracy of 99.22%.

Keywords Generative adversarial network · Support vector machine · Convolutional neural networks · MNIST

1 Introduction

Feature extraction has been utilized in various image processing-based domains [1–6], and in recent years, computer vision-based image processing [7, 8] and supervised learning with convolutional networks (CNNs) has seen huge adoption in intelligent applications [9] and [10]. In this direction, the problem of securing the hand-written digital documents and their processing is being employed in recent deep learning techniques including generative adversarial network (GAN). Generative adversarial networks [11] have recently gained tremendous interest within the machine learning ecosystem since it offers a learning process which is fresh and simple for generative models. In applications of GAN, such as computer vision and NLP, it has produced

A. Singh (✉) · A. Bansal · N. Chauhan · S. P. Sahu · D. K. Dewangan
Department of Information Technology, National Institute of Technology, Raipur, India

A. Noor et al. (eds.), *Proceedings of Emerging Trends and Technologies on Intelligent Systems*, Advances in Intelligent Systems and Computing 1371,
https://doi.org/10.1007/978-981-16-3097-2_8

promising progress. The idea is to generate sample images, and comparing them with the subsequent module ensures the authenticity of the application domain. Similarly, support vector machine (SVM) can be defined by a separating hyperplane and is a discriminative classifier. The SVM algorithm produces an optimal hyperplane that categorizes new instances, given labeled training data (supervised). Convolutional neural networks (CNNs) as a supervised learning scheme have seen tremendous acceptance in computer vision tasks in recent times. Relatively, less focus was paid to unsupervised learning with CNNs [14]. CNNs are somewhat similar to regular neural networks, composed of neurons that have weights and biases that can be trained. Each neuron obtains any input, and then, the dot product is done and is optionally accompanied with a nonlinearity [14, 15]. A single distinct score feature is now reflected in the entire network. On the one hand, from raw image pixels to class scores on the other one. And on the final fully connected layer, they still have a loss function and the info we built for studying standard neural networks holds good even now [16]. However, the potential of GAN has not been experimented enough using some classifiers including SVM and feature learning capabilities of CNN. In the proposed approach, GANs have been utilized to generate MNIST dataset like images and its potential has been experimented with SVM and CNN and can be visualized from Fig. 1.

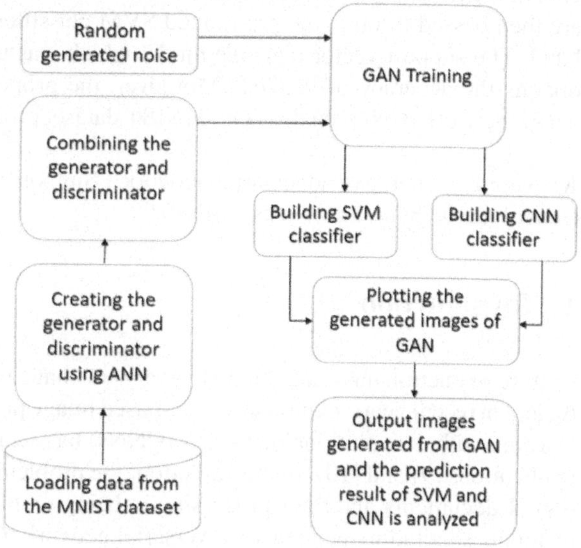

Fig. 1 Process flow of the proposed approach

2 Literature Review

The authors suggest deep convolutional generative adversarial networks also known as DCGAN in [11], and have definite architecture-related limitations and are efficient for unsupervised machine learning applications. Inside a Laplacian pyramid system, the authors in [14] use a cascade of convolutional models to produce images without using any paired data, which learn from image-to-image conversion [14]. Similarly, Alex et al. implemented GANs as generative units alongside RNNs to construct sequential data prevailing in the domain of speech and Language processing [20]. Generative adversarial networks have also been used recently in the sense of subject modeling [21]. InfoGAN [22] is an information-theoretical extension of the GANs capable of unmonitored learning of disentangled representations. The authors introduce different architecture-related features and training protocols in [22], which need to be adapted to enhance GAN training. The recently suggested Wasserstein GAN (WGAN) [23] increases stability problems and to some degree mitigates the problem of mode-collapse. It utilizes the earth-distance mover's as the criterion to get the distance present in between the real, induced distribution, in contrast to the traditional Jensen-Shannon divergence. The authors implement reinforcement learning ideas, then measure theory and implement a loss of discrimination, which is termed as essential loss. Thus, findings were similar, with less hyper-parameters and better theoretical guarantees, to the state-of-the-art. Going through the study [24] shows an upgrade over Wasserstein GANs [25] which further lightens the issue of tuning hyper-parameters, maintaining the stability intact. In such conditions, which are usually applied due to critical weight clipping in WGAN, gradient penalty is utilized in addition to WGAN to remove the volatility.

3 Material and Methods

To identify the performance of GAN with SVM and CNN, the proposed approach consists of the implementation of the following stages:

3.1 The Generative Adversarial Networks Model

It is a deep learning technique which is unsupervised. A GAN has a generator balanced with a discriminator, which is an adversarial network. Multilayer perceptrons will be both generators and discriminators. The task of the generator would be to produce data really close to the data which are being utilized for training. Generator-generated data should be identical to the actual data. As shown in Fig. 2, the discriminator uses two input sets: one input is from the dataset of training (actual data), while additional input is generator-generated dataset [19].

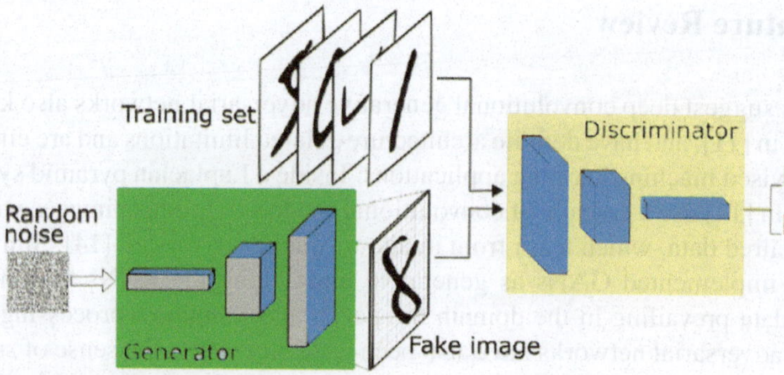

Fig. 2 Approach of simple Generative Adversarial Network (GAN) model

Let us take into consideration pairs of *x* and *y*, where *x* is the input resembling image data or text data or voice data and *y* denotes targets for the corresponding input. Taking into account some random distribution noise, feed the noise to generator G to generate the false pair of input-label *x* with label *y* as 0 for pair *x* and *y* and then this false pair with actual pair *x* with label *y* as 1 and feed this as input to discriminator. The discriminator is a neural network binary assignment and thus measures the loss for each false *x* and true *x* and then averages it as loss at D. Each network has targets in order to put these two networks during the training against each other. In producing the real form of results, the generator G becomes stronger and stronger, and the discriminator network *D* also becomes smarter and smarter in recognizing real versus fake data [17, 18].

The actual discriminator is a double classifier, so when we feed the genuine information, the model will deliver a high likelihood of genuine information and a low likelihood of bogus information (generator yield). A score somewhere in the range of 0 and 1 is certainly given by $D(x)$, $D(G(z))$.

Let us consider some random noise as *z*. We know the generator take its input as random noise, i.e., *z*, and produces output fake images i.e. Xfake. Let us consider *x* as real image of training dataset, i.e., Xreal. The discriminator output would be a probability which it gives by taking input both real and fake images.

(a) The task of discriminator is to maximize the probability for real images and minimize the probability for fake images.
(b) The task of Generator is to produce image so that discriminator should provide high probability value for its generated image.

3.2 *Creating the Generator (G) Neural Network*

Commitment of the generator neural organization in the GAN takes after a progression arbitrarily created digits which are known as latent samples. After it is prepared, the generator is able to deliver pictures from dormant samples. The generator is just a completely associated network which takes an inactive example and produces 784 information focuses that may be resized to a 28 × 28 digit picture that is the resolution utilized by the MNIST images. The generator neural organization comprise one information layer, 3 concealed layers and one yield layer. All 3 shrouded layers utilized "defective ReLU" while the yield layer utilizes 'tanh' initiation work [13, 15].

3.3 Creating the Discriminator (D) Neural Network

A classifier which is prepared utilizing administered learning procedure is the discriminator. It decides when given picture is a phony picture or genuine picture. We train the discriminator for both generator-generated and MNIST images. Discriminator will take the contribution from genuine information which is of size 784 and furthermore pictures created from generator. This neural organization likewise comprises one info layer, 3 concealed layers and one yield layer. After each shrouded layer, a dropout of 30% of perceptrons is done [15]. While significant neural association approaches have starting late displayed critical results similar to mix quality yet it really accompanies huge computational time [13]. While the preparation of GAN, the loads of the discriminator is fixed and we implement that by setting a teachable banner of discriminator as false. The enactment is sigmoid to disclose to us likelihood of whether the information picture is genuine or not (gives esteem between 0 and 1).

3.4 Creating SVM Classifier

A SVM denotes a distinctive classifier which is characterized by its separating hyperplanes. Here, we have prepared an SVM classifier on MNIST dataset in order to foresee the new models produced by GAN. At the end of the day, our SVM classifier is pre-prepared and its information is pictures produced by GAN. In this manner, it needs to just order the pictures created by GAN [17]. One by one 'rbf' and 'linear' kernels for SVM classifier have been evaluated based on the accuracy obtained in both cases.

3.5 Creating a CNN Classifier

We utilize a CNN model to classify on the pictures created by the GAN. Here, we have prepared a CNN classifier on MNIST dataset in order to make forecast on the new information being produced by the GAN. We can foresee the class for new information cases utilizing our settled grouping model in Keras utilizing the

Layer (type)	Output Shape	Param #
dense_1 (Dense)	(None, 256)	25856
leaky_re_lu_1 (LeakyReLU)	(None, 256)	0
dense_2 (Dense)	(None, 512)	131584
leaky_re_lu_2 (LeakyReLU)	(None, 512)	0
dense_3 (Dense)	(None, 1024)	525312
leaky_re_lu_3 (LeakyReLU)	(None, 1024)	0
dense_4 (Dense)	(None, 784)	803600

Total params: 1,486,352
Trainable params: 1,486,352
Non-trainable params: 0

Layer (type)	Output Shape	Param #
dense_9 (Dense)	(None, 1024)	803840
leaky_re_lu_7 (LeakyReLU)	(None, 1024)	0
dropout_3 (Dropout)	(None, 1024)	0
dense_10 (Dense)	(None, 512)	524800
leaky_re_lu_8 (LeakyReLU)	(None, 512)	0
dropout_4 (Dropout)	(None, 512)	0
dense_11 (Dense)	(None, 256)	131328
leaky_re_lu_9 (LeakyReLU)	(None, 256)	0
dense_12 (Dense)	(None, 1)	257

Total params: 1,460,225
Trainable params: 1,460,225
Non-trainable params: 0

Layer (type)	Output Shape	Param #
input_1 (InputLayer)	(None, 100)	0
sequential_1 (Sequential)	(None, 784)	1486352
sequential_2 (Sequential)	(None, 1)	1460225

Total params: 2,946,577
Trainable params: 1,486,352
Non-trainable params: 1,460,225

Fig. 3 Layered wise configuration details of the proposed model

'anticipate classes()' function. The initiation work utilized is 'ReLU' for shrouded layer and 'softmax' for the yield layer as we have ten classes as yield [16].

3.6 Creating GAN

We move toward creating GAN (configuration details given in Fig. 3) by combining generator and discriminator neural network. The input of GAN is noise data of size 100 units that goes into the generator and the output generated is of size 784 which is being fed into the discriminator which provides the output of GAN [17].

3.7 Training the GAN

We feed inactive guides to GAN by setting the ordinary outcome (label) to 1 (real) as we foresee that the generator ought to make a sensible picture, and we envision that discriminator should state it is certifiable or close to real. The generator from the start conveys garbage pictures and incident regard is enormous. Consequently, back-multiplication revives the generators heaps for make more sensible pictures with getting ready continues. This is the manner in which we train the generator by methods for getting ready GAN. While discriminator is describing pictures created as

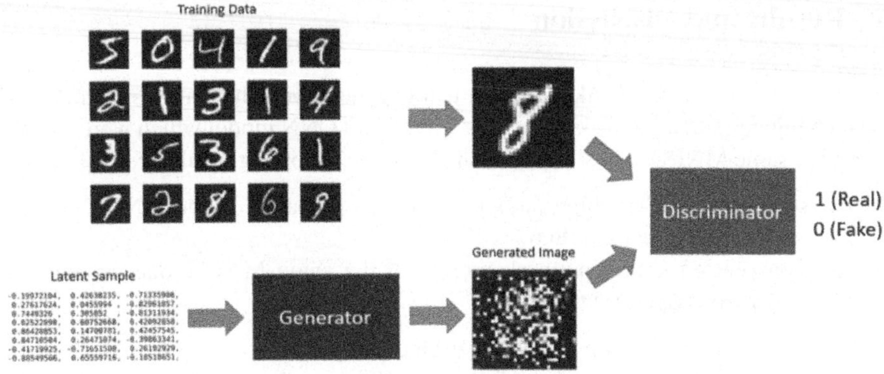

Fig. 4 The training of discriminator network (D)

certified or fake, we need not bother with it to get impacted. Thus, we set the discriminator network as non-workable during the preparation of the generator network [12] followed by the training loop [13]:

(1) Make the discriminator network (D) as teachable/trainable.
(2) Do the preparation of the discriminator organization (D) with the certifiable MNIST pictures and the photographs made by the generator organization to arrange the variable, fake pictures.
(3) Make the discriminator network (D) as non-teachable/non-training mode.
(4) Train the generator as a component of the GAN. The dormant models are taken care of into the GAN and let the generator network produce digit pictures, and use the discriminator network portray picture. At that point, the hover should ideally continue until they are both arranged well and cannot be improved any further as shown in Fig. 4.

4 Experimental Setup

In our essential venture, we have utilized the 'MNIST' information base huge dataset with manually written numbers (0–9) which ordinarily utilized as the purpose of preparing various picture handling systems. The MNIST information base contains 10,000 testing pictures and 60,000 training pictures. We have utilized 'Google Colab' for the advancement of our project. Colaboratory is a Google research project made to help scatter AI instruction and exploration. It is a Jupyter note pad climate that requires no arrangement to utilize and runs completely in the cloud. This implies that as long as you have a Google account, you can unreservedly prepare your models on a 'K80 GPU'.

5 Results and Discussion

After generating MNIST like images using generative adversarial network and performing classification task with SVM model and CNN model which were trained using the same MNIST dataset as used in GAN, the following results were obtained:

- The support vector machine model which was prepared on the MNIST dataset gave an exactness level of 96.67.
- The convolution neural network model on the same MNIST dataset gave an accuracy percentage of 99.22.

Figure 5 shows confusion matrix for each classifier.

Table 1 gives the quantitative performance of our model as shown below.

Few number of maps (shown in Fig. 6) depicting the CNN model's implementation are as follows.

(a) The very first chart illustrates the differences as one line chart in training accuracy versus epoch and as another line chart in evaluation accuracy versus epoch.

(b) Other chart illustrates the differences as one line chart of training loss versus epoch, and as another line chart of evaluation loss versus epoch.

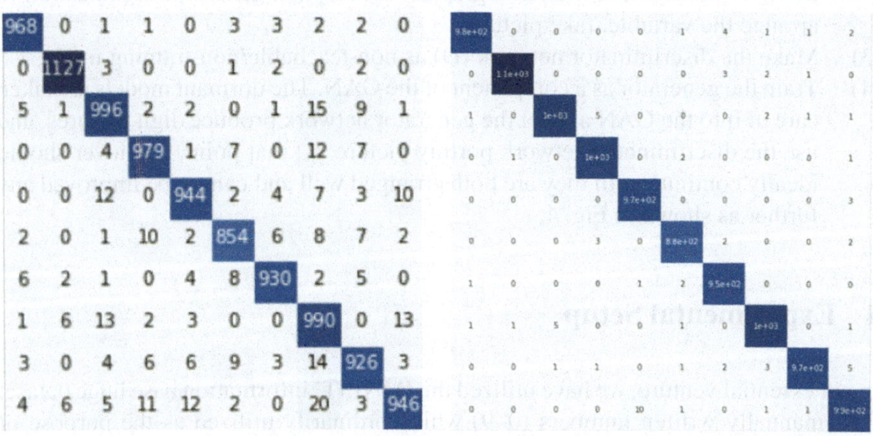

(a) (b)

Fig. 5 Performance of trained model using confusion matrix **a** SVM, **b** CNN

Table 1 Quantitative classification performance of GAN using SVM and CNN

GAN classifier	Accuracy	F1	Precision	Recall
SVM	0.9667	0.9657	0.9660	0.9665
CNN	0.9920	0.9921	0.9922	0.9921

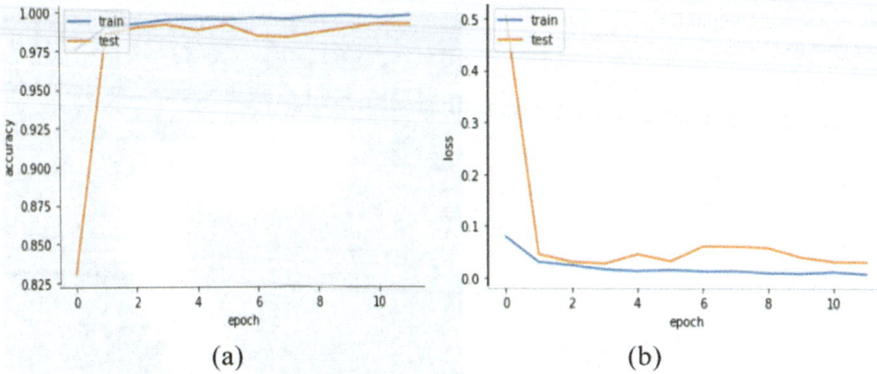

Fig. 6 Plot depicting the performance of CNN model **a** accuracy versus epoch, **b** loss versus epoch

Now the output of the generated images from GAN was passed on to the two classifiers predict functions and predicted output was generated next to the images generated by GAN. It is to be noted that the GAN was trained up to 400 epochs.

Figures 7 and 8 show a general output of our model. The heat map image is generated using GAN. The number written in small just above the GAN generated output is the prediction from our classifiers, namely SVM and CNN. The images generated initially are vague and hard to differentiate, whereas the pictures obtained in later phases are quite clear and can be easily classified using a well-trained classifier.

Previous work on MNIST dataset that includes Dupont et al. [26] has obtained an accuracy of 98.2% for classification of given MNIST like digit image. Further the method by Ravi et al. [27] was able to classify the MNIST images with an accuracy of 95%. Compared to this, our model GAN + SVM obtained an accuracy of 96.6% and GAN + CNN obtained an accuracy of 99.2%. Following Table 2 shows the

Fig. 7 Sample output for initial epochs

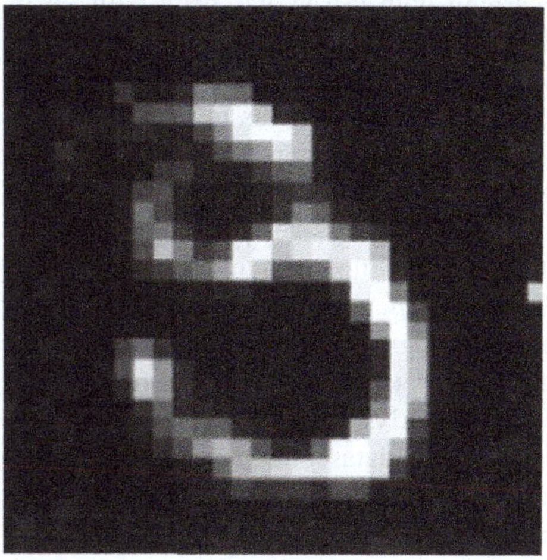

Fig. 8 General output for later epochs

Table 2 Comparison of proposed method with state-of-the-art techniques

Performance measure	Dupont et al. [26]	Ravi et al. [27]	GAN + SVM (Proposed)	GAN + CNN (Proposed)
Accuracy	98.2%	95%	96.6%	99.2%

comparison with the state-of-the-art techniques.

6 Conclusion and Future Scope

Notwithstanding creating excellent pictures, a methodology for semi-directed learning with GANs has been built up that includes the discriminator delivering an extra yield demonstrating the mark of the info using SVM and CNN. This methodology empowers forefront results on datasets with not many marked models. Given the advancements in individual picture super-resolution precision and speed employing quicker and deeper convolution neural networks, one core issue stays mostly unaddressed: how would they restore the finer texture information while we super-resolve massive upscaling parameters. The action of super-resolution methods focused on refinement is motivated mainly through the preference of the objective function, on MNIST, for instance, 99.22% accuracy with completely regulated methodologies utilizing each of the 60,000 named samples. GANs further can be used for various other purposes like Image DE blurring and generating super-resolution images. GANs can not only be used with text or image dataset but can also be used with speech dataset for generating fresh speech data which is very much realistic and convincing on its own. Many more such applications of GANs can tremendously affect our current technologies.

References

1. Bhattacharya, N., Dewangan, D.K., Dewangan, K.K.: An efficacious matching of finger knuckle print images using Gabor feature. In: ICT Based Innovations, pp. 153–162. Springer, Singapore (2018)
2. Pandey, P., Dewangan, K.K., Dewangan, D.K.: Enhancing the quality of satellite images using fuzzy inference system. In: 2017 International Conference on Energy, Communication, Data Analytics and Soft Computing (ICECDS), pp. 3087–3092. IEEE (2017)
3. Pandey, P., Dewangan, K.K., Dewangan, D.K.: Satellite image enhancement techniques—a comparative study. In: 2017 International Conference on Energy, Communication, Data Analytics and Soft Computing (ICECDS), pp. 597–602. IEEE (2017)
4. Pandey, P., Dewangan, K.K., Dewangan, D.K.: Enhancing the quality of satellite images by preprocessing and contrast enhancement. In: 2017 International Conference on Communication and Signal Processing (ICCSP), pp. 0056–0060. IEEE (2017)
5. Dewangan, D.K., Rathore,Y.: Image quality estimation of images using full reference and no reference method. Int. J. Adv. Res. Comput. Sci. **2**(5) (2011)
6. Dewangan, D.K., Rathore, Y.: Image quality costing of compressed image using full reference method. Int. J. Tech. **1**(2), 68–71 (2011)
7. Dewangan, D.K., Sahu, S.P.: Driving behaviour analysis of intelligent vehicle system for lane detection using vision-sensor. IEEE Sens. J. (2020)
8. Dewangan, D.K., Sahu, S.P.: Real time object tracking for intelligent vehicle. In: 2020 First International Conference on Power, Control and Computing Technologies (ICPC2T), pp. 134–138. IEEE (2020)
9. Dewangan, D.K., Sahu, S.P.: Deep learning-based speed bump detection model for intelligent vehicle system using raspberry pi. IEEE Sens. J. **21**(3), 3570–3578 (2020)
10. Dewangan, D.K., Sahu, S.P.: PotNet: Pothole detection for autonomous vehicle system using convolutional neural network. Electron. Lett. **57**(2), 53–56 (2021)
11. Isola, P, Zhu, J.-Y., Zhou, T., Efros, A.A.: Image-to-image translation with conditional adversarial networks. In: Proceedings of the IEEE Conference on Computer Vision and Pattern Recognition, pp. 1125–1134 (2017)
12. Ledig, C., Theis, L., Huszár, F., Caballero, J., Cunningham, A., Acosta, A., Aitken, A. et al.: Photo-realistic single image super-resolution using a generative adversarial network. In: Proceedings of the IEEE Conference on Computer Vision and Pattern Recognition, pp. 4681–4690 (2017)
13. Li, C., Wand, M.: Precomputed real-time texture synthesis with Markovian generative adversarial networks. In: European Conference on Computer Vision, pp. 702–716. Springer, Cham (2016)
14. Radford, A., Metz, L., Chintala, S.: Unsupervised representation learning with deep convolutional generative adversarial networks. arXiv preprint arXiv:1511.06434 (2015)
15. Bastien, F., Lamblin, P., Pascanu, R., Bergstra, J., Goodfellow, I., Bergeron, A., Bouchard, N., Warde-Farley, D., Bengio, Y.: Theano: new features and speed improvements. arXiv preprint arXiv:1211.5590 (2012)
16. Bengio, Y.: Learning deep architectures for AI. Now Publishers Inc, (2009)
17. Bengio, Y., Mesnil, G., Dauphin, Y., Rifai, S.: Better mixing via deep representations. In: International Conference on Machine Learning, pp. 552–560. PMLR (2013)
18. Bengio, Y., Yao, L., Alain, G., Vincent, P.: Generalized denoising auto-encoders as generative models. arXiv preprint arXiv:1305.6663 (2013)
19. Goodfellow, I., Pouget-Abadie, J., Mirza, M., Xu, B., Warde-Farley, D., Ozair, S., Courville, A., Bengio, Y.: Generative adversarial networks. arXiv preprint arXiv:1406.2661 (2014)
20. Zhu, J.-Y., Park, T., Isola, P., Efros, A.A.: Unpaired image-to-image translation using cycle-consistent adversarial networks. In: Proceedings of the IEEE International Conference on Computer Vision, pp. 2223–2232 (2017)
21. Lamb, A., Goyal, A., Zhang, Y., Zhang, S., Courville, A., Bengio, Y.: Professor forcing: A new algorithm for training recurrent networks. arXiv preprint arXiv:1610.09038 (2016)

22. Glover, J.: Modeling documents with generative adversarial networks. arXiv preprint arXiv: 1612.09122 (2016)
23. Chen, X., Duan, Y., Houthooft, R., Schulman, J., Sutskever, I., Abbeel, P.: Infogan: Interpretable representation learning by information maximizing generative adversarial nets. arXiv preprint arXiv:1606.03657 (2016)
24. Arjovsky, M., Chintala, S., Bottou, L.: Wasserstein generative adversarial networks. In: International Conference on Machine Learning, pp. 214–223. PMLR (2017)
25. Gulrajani, I., Ahmed, F., Arjovsky, M., Dumoulin, V., Courville, A.: Improved training of Wasserstein gans. arXiv preprint arXiv:1704.00028 (2017)
26. Dupont, E., Doucet, A., The, Y.W.: Augmented neural odes. arXiv preprint arXiv:1904.01681 (2019)
27. Ravi, S.: Projectionnet: Learning efficient on-device deep networks using neural projections. arXiv preprint arXiv:1708.00630 (2017)

VDNet: Vehicle Detection Network Using Computer Vision and Deep Learning Mechanism for Intelligent Vehicle System

Apoorva Ojha, Satya Prakash Sahu, and Deepak Kumar Dewangan

Abstract Computer vision using deep learning has revolutionized the detection system and vehicle detection is no exception to it. The importance of vehicle classification and detection is in Intelligent Vehicle and Transportation systems which require making critical decisions based on this information; therefore it is a prominent area of work. This paper presents a vehicle detection model predicated on convolution neural network using bounding box annotations for marking the region of interest. The model is tuned to give best performance by evaluating the different parameter configurations. The implementation is done using Python and OpenCV and is trained upon Google Colab's free GPU access. The paper presents an efficient stepwise explanation of the process flow in detecting cars in various real life scenes. The proposed model is trained on the combination of two benchmark dataset and attains 94.66% accuracy and 95.13% precision.

Keywords Intelligent system · Vehicle detection · Object detection · CNN · Deep learning

1 Introduction

The rapid advancements in the automobile industry over the past decade has been a direct result of innovations in computer vision field. Deep Learning techniques for computer vision like Convolution Neural Networks (CNN) [1] have proved to be highly efficient in solving the problems in various domains. Since long, various deep CNNs are used for image classification, segmentation, object detection and tracking, as they excel in extracting features from the image pixels by processing through various layers of the network. However, feature extraction has been utilized in various image processing-based domains [2–7] and in recent years, computer vision-based image processing [8, 9] and supervised learning with convolutional networks (CNNs) has seen huge adoption in intelligent applications [10, 11].

A. Ojha (✉) · S. P. Sahu · D. K. Dewangan
Department of Information Technology, National Institute of Technology, Raipur, India

A. Noor et al. (eds.), *Proceedings of Emerging Trends and Technologies on Intelligent Systems*, Advances in Intelligent Systems and Computing 1371,
https://doi.org/10.1007/978-981-16-3097-2_9

The growing density of vehicles on roads has made an intelligent system a necessity which will take advantage of technologies to create "more intelligent" roads, vehicles and users [12]. The Intelligent Transportation Systems (ITS) consisting of autonomous cars and intelligent traffic management system are flourishing with the aid of vision-based detection techniques. For autonomous cars to be effective it requires an efficient vision-based detection method along with the sensors, to perceive its surrounding entities like lane, other vehicles, potholes, and traffic signals. Vehicle detection is one of the crucial building blocks of these intelligent systems, which enables us to identify and localize the car in images, videos or in real time. It can assist a wide range of tasks like autonomous driving, vehicle counting to evaluate the traffic density, finding a particular stolen vehicle, traffic monitoring and generating e-challan. It can also be employed in real-time critical situations like detecting an ambulance or locating an accident, so that quick action can be taken. The application of vehicle detection in such important areas and crucial conditions has attracted a lot of attention of the researchers. Thus these areas require an error free and robust system for detection. Vehicle detection is a challenging task as it involves immense variations with respect to weather conditions, lighting, background scenes and occlusion. Therefore, the demand of a system that handles the dynamic environment condition is vital. Any model which can be applied in real time, first needs to be proficient on stored images acquired from real scenarios.

This study is particularly centered on the car detection owing to the huge variation in categories in them. The system trains convolution neural network on a mixture of benchmark KITTI and LISA datasets to classify the object as a car or non-car, and further test it to detect the cars on KITTI test dataset images and actual scenario images. Figure 1 illustrates the steps of the process flow for the study.

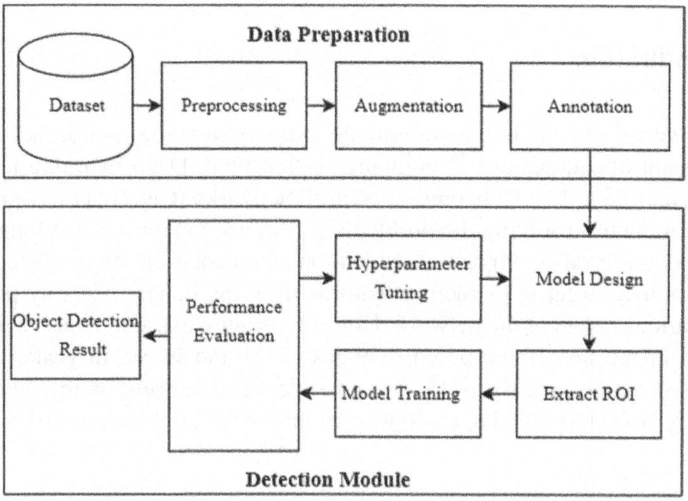

Fig. 1 Process flow for vehicle detection

2 Related Works

The vehicle detection is studied widely and different approaches have been used by various researchers to achieve the target. These can be grouped under two categories:

2.1 Based on Image Processing

Traditionally many vehicle detection methods were used with the application of the feature extractors. HAAR feature-based algorithm is used in [13], which then applies machine learning classifier to detect the cars and display their count. HAAR and Triangle features are used in the paper [14] for cascade of classifiers and monolithic classifier, respectively, and finally using Kalaman filter for detection. The authors in [15] have implemented appearance-based approach by using HAAR like feature to collect color intensity characteristics along with SVM. Some authors have used HOG (Histogram of Oriented Gradients) [16] and πHOG [17] to extract the feature map then fed it to classifiers like SVM and Adaboost, respectively. In [18], SIFT features is employed along with SVM and then Gaussian pyramid with sliding window is used to to narrow down the correct detections. In [19], a night time vehicle detection system detects headlamps and rear lamps using thresholding and cross correlation to pair the lamps. The study in [20] proposed an emergency vehicle's detection using Hue Saturation and Value (HSV) color segmentation and SVM with focus on detecting the siren on the vehicle.

2.2 Based on Deep Learning

Convolution Neural Network in object recognition got the spotlight when a CNN named AlexNet [21], composed of 8 layers—5 convolution and 3 fully connected layers, won ImageNet visual recognition competition by giving lowest error of 15.3%. Since then many state of the art algorithms have been developed to constantly improve the object detection precision. In the study [22], the authors combined the region proposal with CNN, called it R-CNN, which improved the mean average precision (mAP) on PASCAL VOC 2010 data from 35.1 to 53.7%. Limitation of the R-CNN was that it was a very slow algorithm. In [23], the authors introduced an architecture YOLO which had an excellent speed of detection and achieved 57.9% mAP on PASCAL VOC 2012 dataset. Limitation of this was the detection accuracy of the small targets is relatively low. In [24], the authors provided an incremental improvement to YOLO as version 3 with darknet-53 architecture. Other techniques like R-FCN [25] and SSD [26] networks have found prevalent use in vehicle detection owing to their better results.

The authors in the study [27] proposed an improvement to Faster R-CNN which uses region proposals and compares it with the Faster RCNN. In [28], the authors used a combination of passive (camera images) and active (LiDAR) sensors' information to build a model composed of Caffe CNN, integrated with LiDAR for autonomous driving and achieved 80% precision. Limitation was the small size and less variety in the dataset, consequently not considering the different weather conditions, lighting and different camera angles. In study [29], vehicle detection system for aerial images was proposed with improved YOLO V3 architecture and 95.7% precision was achieved, but this architecture was limited only for infrared aerial images. The aerial images have focus on the top viewpoint of the objects and have low resolution and low contrast in comparison to the normal ground view images. The aerial images also have limited set of features as they have a very small object size and images may be even occluded by tree branches, whereas the normal images are sharper, with the different features of the car like color and boundaries being distinctly visible. We have used normal camera images for training this model taken from different viewpoints. The authors in [30] used a hybrid deep convolution neural network composed of two parts- feature extractor and MLP classifier, to facilitate multi scaling in satellite vehicle images. The study [31], presents the color and depth analysis using multi-scale deep convolution neural network for vehicle detection.

3 Materials and Method

3.1 Dataset Preparation

Deep Learning works effectively in the presence of big and diverse datasets. For this experiment we used a combination of two benchmark dataset namely KITTI's [32] 2D object detection vision benchmark suite and LISA (Laboratory for Intelligent and Safe Automobiles) [33]. In the pre-processing module, the video sequences of LISA Sunny and Urban scenes were broken into frames and saved in the desired image format. First 1500 images of KITTI's training set and LISA's Sunny and Urban video sequences together contributed to 2100 images for the dataset. A snapshot of the dataset is shown in Fig. 2. Thus the input images to the model include diverse scenes. Further, data augmentation has been used to generalize our model and fit it more properly. In order to further include variations in the dataset minor alterations were applied like shifting, horizontal flipping, mirroring, rotation of 30° and translation, to the images. Factors like brightness and zoom were also adjusted. In the final stage of this phase, annotation of the images has been carried out using an open source graphical image annotation tool called LabelImg [34], which is a program written in Python and have a user friendly interface, and saves annotations in XML format (see Fig. 3). Though annotations may contain many classes, the main focus for this study was using binary classes i.e. car (encoded as 1) and non-car (encoded as 0).

Fig. 2 Snapshot of dataset

Fig. 3 Snapshot of annotation file


```
<?xml version="1.0"?>
- <annotation>
    <filename>000201.png</filename>
    <segmented>0</segmented>
  - <size>
      <depth>3</depth>
      <width>1242</width>
      <height>375</height>
    </size>
  - <source>
      <annotation>Dummy</annotation>
      <database>Dummy</database>
      <image>Dummy</image>
    </source>
  - <object>
      <name>car</name>
      <difficult>0</difficult>
      <occluded>0</occluded>
      <truncated>0</truncated>
      <pose>Unspecified</pose>
    - <bndbox>
        <xmin>1113.43</xmin>
        <xmax>1242.0</xmax>
        <ymin>128.41</ymin>
        <ymax>375.0</ymax>
      </bndbox>
    </object>
  - <object>
      <name>car</name>
      <difficult>0</difficult>
      <occluded>0</occluded>
      <truncated>0</truncated>
      <pose>Unspecified</pose>
    - <bndbox>
        <xmin>806.19</xmin>
        <xmax>1242.0</xmax>
        <ymin>176.71</ymin>
        <ymax>375.0</ymax>
      </bndbox>
    </object>
```

3.2 Detection Module

The prominent deep learning model—Convolution Neural Network (CNN) or convnets are popular for their use with images in classification, detection, recognition, and recommender system. It contains multiple hidden layers to learn the features and generate a feature map. Convolution Layer applies the convolution operation with selected filter size on the image pixels. The filter is placed over the image pixel values, each pixel is multiplied by the corresponding filter value and at the end all the values of this filter position are added to get one output pixel, and then filter slides to the next position. Rectified Linear Unit (ReLU) is an activation function that normalizes the negative output of the convolution layers to zero value. It is applied element wise on each pixel value. Pooling is the next most important layer, whose main function is to shrink the size of the input matrix. We have chosen max pool operation which uses a defined window size, and maximum value from each window position on the image pixels is selected for output. Pooling operation is applied individually on each channel. Before input is fed to the next layer—Fully Connected Layer (FC); we need to flatten the 3 dimensional output of previous layer into a one dimension. In this layer, each input unit is connected to all the neurons, thus the name fully connected. The last layer is generally FC layer having neurons equal to the number of output classes. Softmax function is used as the last layer's activation and provides the final output. It gives the probabilistic values for the input vector, so that they add up to give 1. Table 1 presents the expression associated with each layer.Where 'nh × nw' image with 'nc' channels is input and filter of size $f \times f$ is used with 'p' padding of input, 's' strides (sliding the window by 's' pixel), 'nf' number of filters, and input vector 'z' have k values. For our CNN model we have

Table 1 Expression for functional layers

S. No.	Functional layer	Expression
1	Convolution layer	$[nhXnwXnc] * [fXfXnf] = \frac{nh + 2p - f}{s} + 1X\frac{nw + 2p - f}{s} + 1Xnf$
2	ReLU	$g(z) = \begin{cases} z, & \text{if} z > 0 \\ 0, & \text{if} z \leq 0 \end{cases} = \max(0, z)$
3	Max pooling	$[nhXnwXnc] * [fXf] = \frac{nh - f}{s} + 1X\frac{nw - f}{s} + 1Xnc$
4	Softmax	$e(z_i) = \frac{e^{z_i}}{\sum_{j=1}^{k} e^{z_j}}$

Table 2 Configuration of the proposed model

Layer	Operational layer	Output	Filter size
0	Input	64,64,3	–
1	Convolution	64,64,20	3 × 3
2	ReLU	64,64,20	–
3	Max pooling	32,32,20	2 × 2
4	Convolution	32,32,50	3 × 3
5	ReLU	32,32,50	–
6	Max pooling	16,16,50	2 × 2
7	Flatten	12,800	–
8	Dense	900	–
9	ReLU	900	–
10	Dense	2	–
11	Softmax	2	–

used two sets of Convolution + ReLU and max pooling layer, and one Fully Convolution layer. The 20 and 50 filters of size 3 × 3 are used for convolution operation and 2 × 2 filters for pooling layers.

Further, extraction of region of interest (ROI) has been carried out using the annotation file in XML format, which is parsed to detect and extract the coordinates of the bounding boxes. These regions are then resized to the same shape, accumulated together and then provided as an input to the model. Afterwards, model training has been done using Google Colaboratory, which facilitates us to train our model using cloud and free GPU access. Training is executed using Tesla T4 GPU with RAM of 12.72 GB. Foremost task is to zip and upload our data to the Google Drive, so that we can have the liberty to mount the drive and use the dataset at any time, instead of uploading it for each session. We started the experiment using a basic model design and evaluated the training results, then tuned the hyper-parameters and model design to finally get the most suitable configuration for our objective. The final model was trained for 1000 epochs and it took total 2 h 27 min and 35 s to be completed. The following Table 2 represents the model configuration.

4 Results and Discussion

4.1 Qualitative Analysis

The behavior of the system while learning can be visualized by the two crucial graphical plots of the process, i.e. accuracy graph and the loss graph. Overviewing these plots can help us understand whether the model is trained only for the given dataset or it can be extended to new test data, i.e. we can recognize the situation such overfitting or underfitting and tune the model and hyper parameters accordingly. The accuracy graph is a plot of training and validation accuracy for each epoch, as exhibited in the Fig. 4, both the accuracies are increasing at each step and then gradually saturates at 0.95. Also there is almost no gap in both the accuracy plots, meaning the system is well fitted. The Fig. 5 is the loss plot illustrating the training and validation loss decreasing from 0.6 to 0.1. We can infer that the model has comparable performance for both training and validation sets. The Loss function

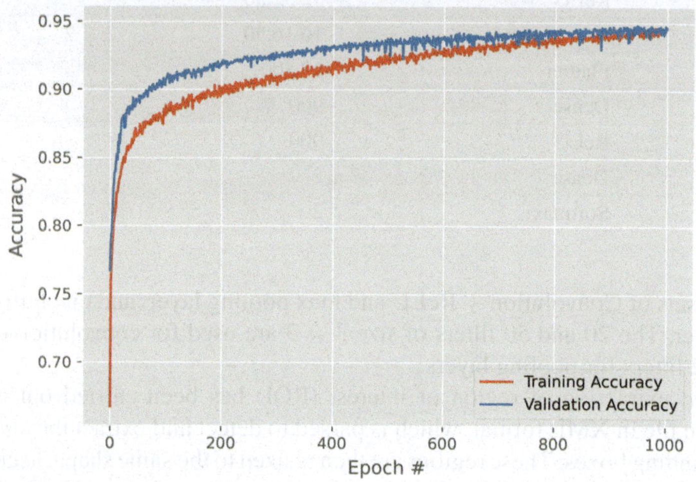

Fig. 4 Accuracy graph

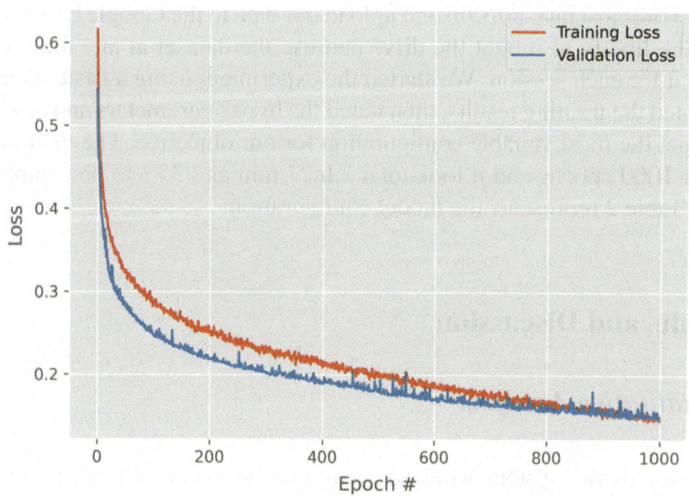

Fig. 5 Loss graph

Table 3 Expression for performance evaluation

S. No.	Performance metric	Expression
1	Accuracy	$\frac{TP+TN}{TP+FP+TN+FN}$
2	Precision	$\frac{TP}{TP+FP}$
3	Recall	$\frac{TP}{TP+FN}$
4	F1 Score	$2 * \frac{Precision*Recall}{Precision+Recall}$

Table 4 Quantitative performance evaluation for our experiment

Accuracy	Precision	Recall	F1 score
94.66%	95.13%	96.75%	95.93%

used is Binary Cross Entropy (also known as log loss) as there are two classes i.e. car and non-car.

4.2 Quantitative Analysis

The performance of the model is quantified in terms of the standard object detection metrics like accuracy, precision, recall and F1 Score [15] using **true positive (TP), false positive (FP), true negative (TN) and false negative** (**FN**). The following Table 3 gives the expressions used for performance evaluation.

The estimations of above metrics for our system are tabulated below in Table 4:

4.3 Detection Results and Comparison of Performance

The proposed model's performance is tested on the images of KITTI test data and images captured by us in different scenes and illumination conditions. The model was able to identify the maximum number of cars on the images. Figure 6 displays the result of the model tested on the images of KITTI dataset and Fig. 7 exhibit the detection results for various images taken by us from different angles and different times of the day.

A better performance irrespective of the environment conditions is an advantage that our approach provides. Mamun and Deb have used HAAR like feature to evaluate the color characteristics along with SVM [15]. They have achieved 86.75% precision that doesn't perform well for complex and night scenes. At the same time, the study [19] gave vehicle detection model specifically for night time, using rear and head lamps with cross correlation for detection and achieved 93.56% average accuracy. Their model will not be able to work well in illuminated areas where light intensity

Fig. 6 Detection result of KITTI's test images

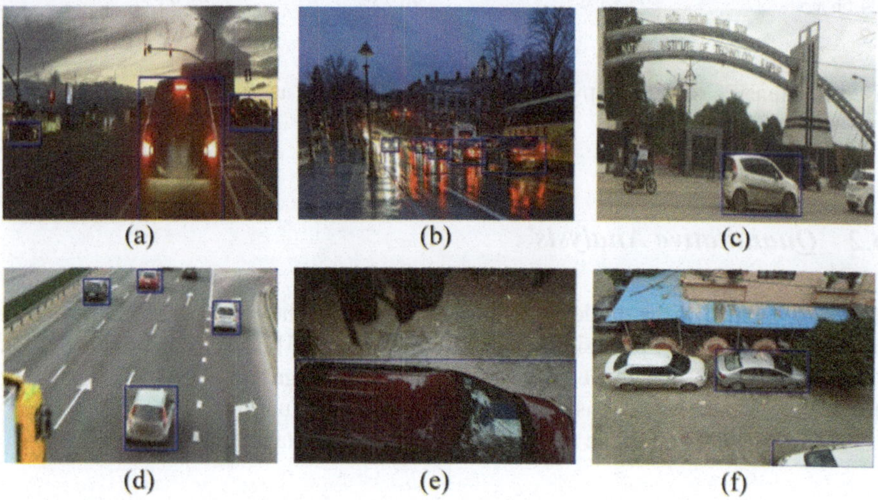

Fig. 7 Detection results in different scenes and angles in real time scenarios

of the lamps is less. Mitral et al. in [27] gave an improved version for Faster RCNN with increased performance of 85.7% F1 Score for heavy vehicles and 75% for light vehicles eliminates the limitation of duplicate label detection for the same object.

By comparing our result with the other methods, we can clearly observe from Table 5 that our performance is better than the other models. Even though our dataset is

Table 5 Performance comparison of the proposed model with state-of-the-art methods

	Accuracy (%)	Precision (%)	F1 score (%)
Mamun and Deb [15]	–	86.75	87.03
Zaarane et al. [19]	93.56	–	–
Mittal et al. [27]	–	–	85.7
Proposed model	94.66	95.13	95.93

balanced, and F1 score is used to evaluate performance mainly for unbalanced dataset, our F1 Score is also quite appreciable. We believe that our model has attained superior performance and is fit to implement.

5 Conclusion

This paper presents a CNN model for Vision-based Car detection which was successfully implemented in Tensorflow Keras GPU. This experiment was carried on the benchmark dataset and can be extended to other custom datasets. During this study we examined various model configurations for CNN with focus on obtaining the best precision and recall, along with the adequately fitted model. We found that our CNN model with specified tuned hyper-parameters provides the best result for our dataset. We successfully applied this model to 2100 images and achieved 94.66% accuracy and 95.13% precision.

Our future work will be focused on the various challenges like occlusion and multi scaling problem in the detection of the subject of interest. We also aim to implement the system with real-time data and optimize accordingly. Though there are wide range of state of the art techniques available for vehicle classification, detection and tracking, there is still a wide scope for enhancement to deal with the challenges related to obstacles and live movement, leading the vehicle detection research to remain a progressive topic for further exploration.

References

1. Albawi, S., Mohammed, T.A., Al-Zawi, S.: Understanding of a convolutional neural network. In: 2017 International Conference on Engineering and Technology (ICET), Jan 2018, pp. 1–6 (2018). https://doi.org/10.1109/ICEngTechnol.2017.8308186
2. Bhattacharya, N., Dewangan, D.K., Dewangan, K.K.: An efficacious matching of finger knuckle print images using Gabor feature. Adv. Intell. Syst. Comput. **653**, 153–162 (2018). https://doi.org/10.1007/978-981-10-6602-3_15
3. Pandey, P., Dewangan, K.K., Dewangan, D.K.: Enhancing the quality of satellite images using fuzzy inference system. In: 2017 International Conference on Energy, Communication, Data Analytics and Soft Computing (ICECDS), pp. 3087–3092 (2017). https://doi.org/10.1109/ICECDS.2017.8390024
4. Pandey, P., Dewangan, K.K., Dewangan, D.K.: Satellite image enhancement techniques—a comparative study. In: 2017 International Conference on Energy, Communication, Data Analytics and Soft Computing (ICECDS), pp. 597–602 (2017). https://doi.org/10.1109/ICECDS.2017.8389506
5. Pandey, P., Dewangan, K.K., Dewangan, D.K.: Enhancing the quality of satellite images by preprocessing and contrast enhancement. In: 2017 International Conference on Communication and Signal Processing (ICCSP), ICCSP 2017, Jan 2018, pp. 56–60 (2018). https://doi.org/10.1109/ICCSP.2017.8286525
6. Dewangan, D.K., Rathore, Y.: Image quality estimation of images using full reference and no reference method. Int. J. Adv. Res. Comput. Sci. **2** (2011)

7. Dewangan, D., Rathore, Y.K.: Image Quality Costing of Compressed Image Using Full Reference Method (2016)
8. Dewangan, D.K., Sahu, S.P.: Driving behaviour analysis of intelligent vehicle system for lane detection using vision-sensor. IEEE Sens. J. **21**, 6367–6375 (2020). https://doi.org/10.1109/jsen.2020.3037340
9. Dewangan, D.K., Sahu, S.P.: Real time object tracking for intelligent vehicle. In: 2020 First International Conference on Power, Control and Computing Technologies (ICPC2T), ICPC2T 2020, pp. 134–138 (2020). https://doi.org/10.1109/ICPC2T48082.2020.9071478
10. Dewangan, D.K., Sahu, S.P.: Deep learning-based speed bump detection model for intelligent vehicle system using Raspberry Pi. IEEE Sens. J. **21**, 3570–3578 (2021)
11. Dewangan, D.K., Sahu, S.P.: PotNet: Pothole detection for autonomous vehicle system using convolutional neural network. Electron. Lett. 2–5 (2020). https://doi.org/10.1049/ell2.12062
12. Yang, Z., Pun-Cheng, L.S.C.: Vehicle detection in intelligent transportation systems and its applications under varying environments: A review. Image Vis. Comput. **69**, 143–154 (2018). https://doi.org/10.1016/j.imavis.2017.09.008
13. Choudhury, S., Chattopadhyay, S.P., Hazra, T.K.: Vehicle detection and counting using haar feature-based classifier. In: 2017 8th Annual Industrial Automation and Electromechanical Engineering Conference (IEMECON), pp. 106–109 (2017). https://doi.org/10.1109/IEMECON.2017.8079571
14. Haselhoff, A., Kummert, A.: A vehicle detection system based on haar and triangle features. IEEE Intell. Veh. Symp. Proc. 261–266 (2009). https://doi.org/10.1109/IVS.2009.5164288
15. Mamun, M.A. Al, Deb, K.: An approach for recognizing vehicle based on appearance. In: 2019 International Conference on Computer, Communication, Chemical, Materials and Electronic Engineering (IC4ME2), pp. 11–12 (2019). https://doi.org/10.1109/IC4ME247184.2019.9036598
16. Tuermer, S., Kurz, F., Reinartz, P., Stilla, U.: Airborne vehicle detection in dense urban areas using HoG features and disparity maps. IEEE J. Sel. Top. Appl. Earth Obs. Remote Sens. **6**, 2327–2337 (2013). https://doi.org/10.1109/JSTARS.2013.2242846
17. Kim, J., Baek, J., Kim, E.: A novel on-road vehicle detection method using π HOG. IEEE Trans. Intell. Transp. Syst. **16**, 3414–3429 (2015). https://doi.org/10.1109/TITS.2015.2465296
18. Yawen, T., Jinxu, G.: Research on vehicle detection technology based on SIFT feature. In: 2018 8th International Conference on Electronics Information and Emergency Communication (ICEIEC), ICEIEC 2018, pp. 274–278 (2018). https://doi.org/10.1109/ICEIEC.2018.8473575
19. Zaarane, A., Slimani, I., Al Okaishi, W., Atouf, I., Hamdoun, A.: An automated night-time vehicle detection system for driving assistance based on cross-correlation. In: 2019 International Conference on Systems of Collaboration Big Data, Internet of Things & Security (SysCoBIoTS), SysCoBIoTS 2019, pp. 1–5 (2019). https://doi.org/10.1109/SysCoBIoTS48768.2019.9028038
20. Razalli, H., Ramli, R., Alkawaz, M.H.: Emergency vehicle recognition and classification method using HSV color segmentation. In: 2020 16th IEEE International Colloquium on Signal Processing & Its Applications (CSPA), CSPA 2020, pp. 284–289 (2020). https://doi.org/10.1109/CSPA48992.2020.9068695
21. Krizhevsky, A., Sutskever, I., Hinton, G.E.: ImageNet classification with deep convolutional neural networks. Commun. ACM. **60**, 84–90 (2012)
22. Girshick, R., Donahue, J., Darrell, T., Malik, J., Berkeley, U.C., Malik, J.: Rich feature hierarchies for accurate object detection and semantic segmentation. In: Proceedings of the IEEE Conference on Computer Vision and Pattern Recognition (CVPR), 1, 5000 (2014). https://doi.org/10.1109/CVPR.2014.81
23. Redmon, J., Divvala, S., Girshick, R., Farhadi, A.: You only look once: Unified, real-time object detection. In: Proceedings of the IEEE Conference on Computer Vision and Pattern Recognition (CVPR), Dec 2016, pp. 779–788 (2016). https://doi.org/10.1109/CVPR.2016.91
24. Redmon, J., Farhadi, A.: YOLOv3: An incremental improvement. arXiv. (2018)
25. Li, Y., He, K., Sun, J., others: R-fcn: Object detection via region-based fully convolutional networks. Adv. Neural Inf. Process. Syst. 379–387 (2016)

26. Liu, W., Anguelov, D., Erhan, D., Szegedy, C., Reed, S., Fu, C.Y., Berg, A.C.: SSD: Single shot multibox detector. Lect. Notes Comput. Sci. (including Subser. Lect. Notes Artif. Intell. Lect. Notes Bioinformatics). 9905 LNCS, 21–37 (2016). https://doi.org/10.1007/978-3-319-46448-0_2
27. Mittal, U., Potnuru, R., Chawla, P.: Vehicle detection and classification using improved faster region based convolution neural network. In: 2020 8th International Conference on Reliability, Infocom Technologies and Optimization (Trends and Future Directions) (ICRITO), pp. 511–514 (2020). https://doi.org/10.1109/ICRITO48877.2020.9197805
28. Lange, S., Ulbrich, F., Goehring, D.: Online vehicle detection using deep neural networks and lidar based preselected image patches. In: 2016 IEEE Intelligent Vehicles Symposium (IV), Aug 2016, pp. 954–959 (2016). https://doi.org/10.1109/IVS.2016.7535503
29. Zhang, X., Zhu, X.: Vehicle detection in the aerial infrared images via an improved yolov3 network. In: 2019 IEEE 4th International Conference on Signal and Image Processing (ICSIP), ICSIP 2019, pp. 372–376 (2019). https://doi.org/10.1109/SIPROCESS.2019.8868430
30. Chen, X., Xiang, S., Liu, C.L., Pan, C.H.: Vehicle detection in satellite images by hybrid deep convolutional neural networks. IEEE Geosci. Remote Sens. Lett. **11**, 1797–1801 (2014). https://doi.org/10.1109/LGRS.2014.2309695
31. Dong, W., Yang, Z., Ling, W., Yonghui, Z., Ting, L., Xiaoliang, Q.: Research on vehicle detection algorithm based on convolutional neural network and combining color and depth images. In: 2019 2nd International Conference on Information Systems and Computer Aided Education (ICISCAE), ICISCAE 2019, pp. 274–277 (2019). https://doi.org/10.1109/ICISCAE48440.2019.221634
32. Geiger, A., Lenz, P., Urtasun, R.: Are we ready for autonomous driving? The KITTI vision benchmark suite. In: 2012 IEEE Conference on Computer Vision and Pattern Recognition, pp. 3354–3361 (2012). https://doi.org/10.1109/CVPR.2012.6248074
33. Sivaraman, S., Trivedi, M.M.: A general active-learning framework for on-road vehicle recognition and tracking. IEEE Trans. Intell. Transp. Syst. **11**, 267–276 (2010). https://doi.org/10.1109/TITS.2010.2040177
34. Tzutalin, D.: LabelImg–LabelImg is a graphical image annotation tool. Github Repository (2018)

FCNet: Flower Classification Using Custom-Made Convolution Neural Network and Transfer Learning

Roma Vardiyani and Satya Prakash Sahu

Abstract Although much work has been done in the field of machine learning, there are some limitations in its real-world applications. The optimal situation for machine learning requires the abundantly labeled testing instances of the same distribution as of the test results. Collecting sufficient data requires lots of efforts and consumes a large amount of time and sometimes its unrealistic to collect the data. However, training such vast volumes of data takes time and often has complex hardware requirements, which sometimes prove expensive. Transfer Learning, which centers around communicating data across domains, is a promising AI approach for tackling the previously mentioned problem. We also compared the methods CNN, pre-trained models RESNET50, and RESNET18 to predict the correct class of the flower. It was observed that CNN classified it with 77.679% accuracy, RESNET50 with 79.0178% accuracy, and RESNET18 with 90.1786%. The link to the implementation can be found at: https://colab.research.google.com/drive/1Oae3e1yHoR8kbjATYI6YokfuLFMCR-4-#scrollTo=IJNCVoCe_UNH

Keywords Image classification · Transfer learning · Convolution neural network · RESNET18 · RESNET50

1 Introduction

Transfer learning is a sort of AI calculation where an example is reused as the beginning stage for a model similar yet another mission. It is a mainstream technique in deep learning where per-prepared models are utilized as a beginning stage for computer vision and natural language handling tasks, provided the gigantic computational and tedious assets needed to build neural network models are the issues, and the huge skill gaps for the same. Hypothesis of transfer, as recommended by the therapist C. H. Judd, is figuring out how to transfer the outcome for the speculation of experience [1]. According to the theory of transfer learning, there must be some

R. Vardiyani · S. P. Sahu (✉)
National Institute of Technology Raipur, G.E. Road, Raipur, Chattisgarh, India
e-mail: spsahu.it@nitrr.ac.in

A. Noor et al. (eds.), *Proceedings of Emerging Trends and Technologies on Intelligent Systems*, Advances in Intelligent Systems and Computing 1371,
https://doi.org/10.1007/978-981-16-3097-2_10

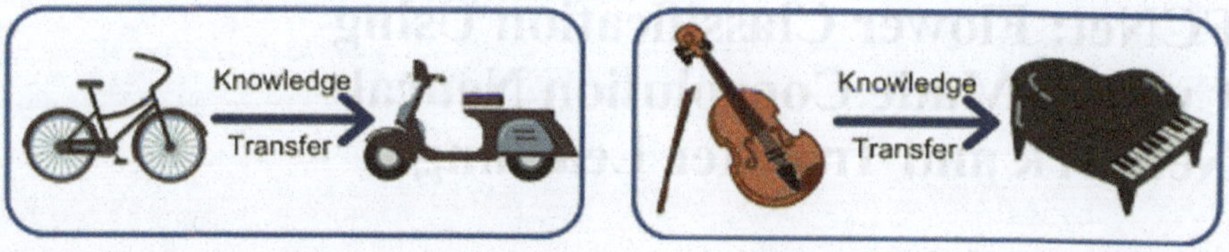

Fig. 1 Instinctive example of transfer learning

connection between two learning activities. Real-life examples could be a person who knows how to ride a bicycle easily learns to ride a motorcycle [2]. Similarly, a person who knows how to violin easily learns to play a piano because of some common knowledge in both the examples (Fig. 1).

Additionally, likenesses between spaces do not generally encourage learning, on the grounds that occasionally similitudes can be misdirecting. For instance, albeit Spanish and French have close connections to each other and both have a place with a gathering of dialects, individuals who learn Spanish may encounter troubles in learning French, for example, abusing jargon or formation. This is on the grounds that past involvement with Spanish can meddle with the learning of word arrangement, use, elocution, formation, and so forth in French. In the theory of psychology, when the previous knowledge interferes with current learning this is known as negative transfer. Hence, the new problem (data), which the model is going to predict, must be similar to the original problem on which the data were trained.

Deep learning approaches are algorithms used to solve the machine problem. The Picture Recognition Area is one of the fields where deep learning is used. Image recognition is a matter of image processing and is very correct and its comprehensive. It is a critical technology that many people use in any area of human life. Classification is the period of the delivery where the images that do not belong to either class are included in their class [3]. Classification can be either binary or multiple. There are several articles in the literature for the classification of pictures. CNN architecture is the most chosen form. However, feature extraction has been utilized in various image processing-based domains and in recent years, computer vision-based image processing [4, 5] and supervised learning with convolutional networks (CNNs) have seen huge adoption in intelligent applications [6–8].

2 Literature Review

Important research has been devoted to the classification problem. Previous works include the different feature-based methods for flower classification like text features [9] and gray level co-matrix [10]. Some recent works include textual labels to help deep Convolutional Neural Networks for recognition [11]. Reference [12] involves a grouping framework for flower pictures by utilizing Deep CNN and Data Augmentation. It likewise proposed a grouping model to develop the exhibition of the ordering of flower pictures by utilizing Deep CNN for extricating the highlights and different AI calculations for characterizing purposes and showed the utilization of image augmentation for accomplishing better execution results. Reference [13] proposed an algorithm on the classification of flowers dependent on just the whorl of the flower. It utilized Gabor reactions to distinguish the whorl parts of the bloom pictures. The appropriate surface highlights are investigated with the end goal of flower characterization. The proposed whorl-based classification algorithm accomplishes generally a decent order exactness when contrasted with whole flower-based grouping. Reference [14] endeavors to investigate a novel morphology to include extraction, and the pertinence of emblematic portrayal conspires alongside various arrangement systems for viable multi-name grouping of flower categories. Also, it proposed a mixture procedure utilizing MKL–SVM with a multi-name characterization that probed a dataset containing 25,000 blossom pictures of 102 distinct flavors. Fundamental and morphology highlights including shading, size, surface, petal-type, petal check, circle blossom, crown, aestivation of bloom, and blossom class are extricated to expand the order of exactness. Different classifiers are applied to the separated list of capabilities, and their presentation is talked about. All these methods involve classification of flowers using Machine Learning Algorithms or Deep Learning Algorithms.

Flowers are also classified using Transfer Learning Technique. [15] build up a successful flower characterization approach utilizing convolution neural organization and transfer learning. VGG-16, VGG-19, Inception-v3, and ResNet50 models were utilized to contrast the organization statement model and the transfer learning model. The outcomes show that transfer learning can adequately maintain a strategic distance from profound convolution networks that are inclined to neighborhood optimization issues and over-fitting issues. Contrasted and the customary techniques, the exactness of flower categorization on Oxford flower dataset is clearly improved and has better strength and speculation capacity. Reference [16] proposed the categorization of flowers using transfer learning and Adam's profound learning improvement calculation for the imperfections of the current standard convolutional neural organization with profound profundity and long boundaries, long preparing time, and moderate union. The VGG16 model is altered and enhanced. Simultaneously, the transfer learning strategy and the Adam optimization are utilized to quicken network convergence. Some comparisons using ALexNet [17] and GoogleNet [18] have been done [19].

In this paper, we have compared the performance of different models namely CNN built from scratch, pre-trained RESNET50, and RESNET18 for classification of flowers. RESNET50 and RESNET18 have set up a benchmark for image classification problems.

3 Proposed Methodology

For a given set of data of RGB images, we have used 3 architectures to classify the categorical data and have compared the results using the accuracy of the models.

3.1 Dataset Details

The dataset used for the purpose is a flower dataset available in kaggle [20]. This dataset contains five classes of flowers namely Daisy, Dandelion, Rose, Sunflower, and Tulips. There are a total 3670 images in it belonging to different classes. The dataset is divided into two groups: training set and validation set. The validation set contains 500 images, while the test set contains 3170 images.

Each of the models trained is for 50 epochs, and Adam Optimizer is used as Optimizer to reduce the losses.

3.2 Classification

Custom-Made Convolution Neural Network

Designed a CNN on my own and trained it from scratch to classify the flowers. The ReLu layer was used after each layer to convert negative data to non-negative as we might encounter negative pixel values practically. The design of the network is shown below (Fig. 2).

In the architecture, we can observe that for every series of Convolution Layer and MaxPool Layer the image is reduced by two and finally converges to a Fully Connected Layer of 256 * 28 * 28(=200,704) whose features are the further reduced to 1024 and then to 512. The input of the last fully connected layer flat image of 512 features gives five-dimensional output. The index which has the maximum value gives the corresponding flower. For example, id index 2 has the maximum value and the input image is classified as "ROSE". The indexes are mapped as follows: 0: 'daisy', 1: 'dandelion', 2: 'rose', 3: 'sunflower', 4: 'tulip'.

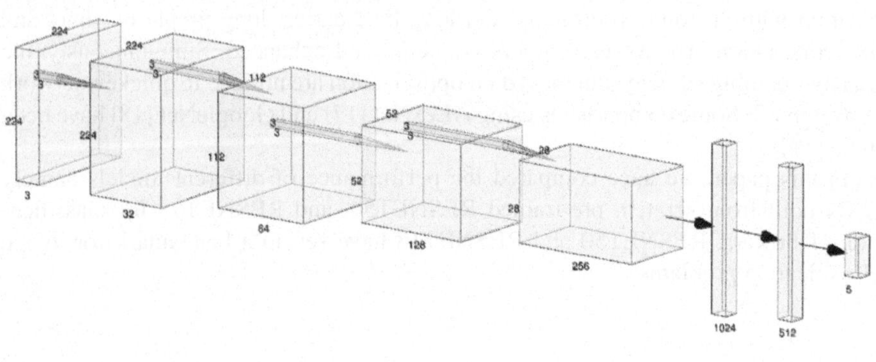

Fig. 2 Architecture of CNN

layer name	output size	18-layer	34-layer	50-layer	101-layer	152-layer
conv1	112×112	7×7, 64, stride 2				
conv2_x	56×56	3×3 max pool, stride 2				
		$\begin{bmatrix}3\times3, 64\\3\times3, 64\end{bmatrix}\times2$	$\begin{bmatrix}3\times3, 64\\3\times3, 64\end{bmatrix}\times3$	$\begin{bmatrix}1\times1, 64\\3\times3, 64\\1\times1, 256\end{bmatrix}\times3$	$\begin{bmatrix}1\times1, 64\\3\times3, 64\\1\times1, 256\end{bmatrix}\times3$	$\begin{bmatrix}1\times1, 64\\3\times3, 64\\1\times1, 256\end{bmatrix}\times3$
conv3_x	28×28	$\begin{bmatrix}3\times3, 128\\3\times3, 128\end{bmatrix}\times2$	$\begin{bmatrix}3\times3, 128\\3\times3, 128\end{bmatrix}\times4$	$\begin{bmatrix}1\times1, 128\\3\times3, 128\\1\times1, 512\end{bmatrix}\times4$	$\begin{bmatrix}1\times1, 128\\3\times3, 128\\1\times1, 512\end{bmatrix}\times4$	$\begin{bmatrix}1\times1, 128\\3\times3, 128\\1\times1, 512\end{bmatrix}\times8$
conv4_x	14×14	$\begin{bmatrix}3\times3, 256\\3\times3, 256\end{bmatrix}\times2$	$\begin{bmatrix}3\times3, 256\\3\times3, 256\end{bmatrix}\times6$	$\begin{bmatrix}1\times1, 256\\3\times3, 256\\1\times1, 1024\end{bmatrix}\times6$	$\begin{bmatrix}1\times1, 256\\3\times3, 256\\1\times1, 1024\end{bmatrix}\times23$	$\begin{bmatrix}1\times1, 256\\3\times3, 256\\1\times1, 1024\end{bmatrix}\times36$
conv5_x	7×7	$\begin{bmatrix}3\times3, 512\\3\times3, 512\end{bmatrix}\times2$	$\begin{bmatrix}3\times3, 512\\3\times3, 512\end{bmatrix}\times3$	$\begin{bmatrix}1\times1, 512\\3\times3, 512\\1\times1, 2048\end{bmatrix}\times3$	$\begin{bmatrix}1\times1, 512\\3\times3, 512\\1\times1, 2048\end{bmatrix}\times3$	$\begin{bmatrix}1\times1, 512\\3\times3, 512\\1\times1, 2048\end{bmatrix}\times3$
	1×1	average pool, 1000-d fc, softmax				
FLOPs		1.8×10^9	3.6×10^9	3.8×10^9	7.6×10^9	11.3×10^9

Fig. 3 A general architecture table for different RESNET(s)

RESNET

CNN architecture has demonstrated considerable progress in the identification of objects [21, 22] and classifying images using it since more discriminatory features can be derived from CNN architectures. There are several pre-trained CNNs such as AlexNet [23], VGGNet [24], CaffeNet [25], GoogleNet [18], and Resnet [26]. It has been observed that RESNET performs better than other networks [27] in the classification of images as it increases the depth of the network and brings out the features with more semantic information (Fig. 3).

RESNET50

The architecture comprises 50 layers as defined in Fig. 2. Due to the concentration layer, the function map in Resnet 50 is reduced in size as presented in the figure. The ratio of the size of the initial image to the final feature map after four phases is 16-1. A coding with a kernel size of 7 * 7 and 64 kernels, both of which have a phase 2 and give us the first layer. Next, we see total pooling with a capacity of 2 as well. In the next stage, a kernel of 1 * 1 * 64 follows a kernel of 3 * 3 * 64 and at long last a kernel of 1 * 1 * 256, these three layers are replicated 3 times over and give us 9 layers in this stage. Next, we see 1 * 1 * 128 kernels, after that a 3 * 3 * 128 kernel and at last a 1 * 1 * 512 kernel, this process was replicated four times and used 12 layers in this step. After that, there are 1 * 1 * 256 kernels with 2 additional 3 * 3 * 256 and 1 * 1 * 1024 kernel, which is replicated 6 times which brings us 18 layers. And again, a kernel of 1 * 1 * 512, with a further two of 3 * 3 * 512 and 1 * 1 * 2048, which was replicated three times making us a total of 9 layers. After that, we build an average pool and finish it with a completely connected layer with 1000 nodes and at the end a softmax feature so we have 1 layer. This gives us a Large Convolutional network of $1 + 9 + 12 + 18 + 9 + 1 = 50$ layers. The findings on the ImageNet validation set were pretty strong, ResNet 50 models achieved a top-1 error rate of 20.47% and a high-5 error rate of 5.25%, which was recorded for a single model consisting of 50 layers and not a set of it (Fig. 4).

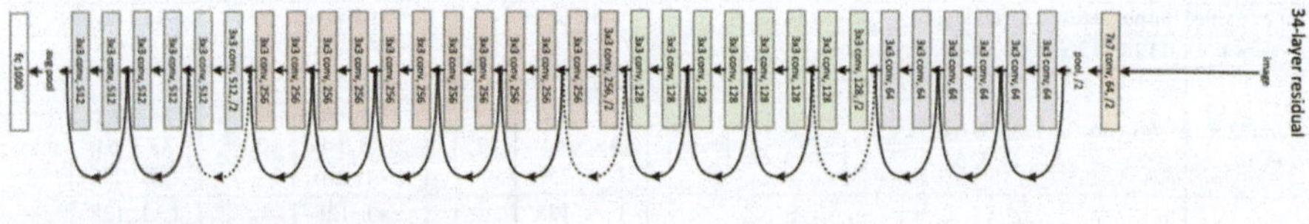

Fig. 4 RESNET50 Architecture

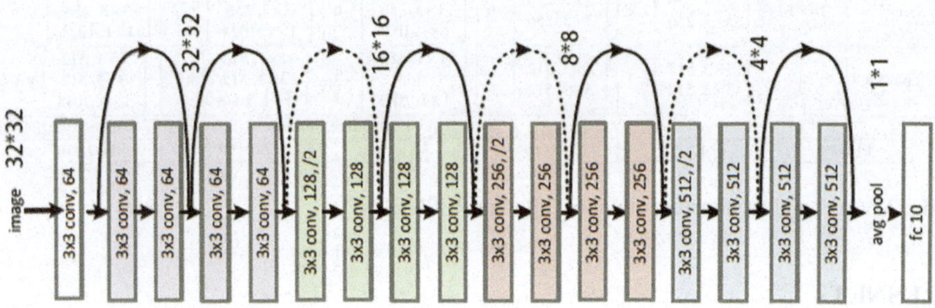

Fig. 5 Architecture of RESNET18

RESNET18

ResNet-18 is a neural network with over one million images trained in the ImageNet database. There are 18 layers in its architecture Fig. 3. The picture classification is highly precise and can classify images into 1000 artifact forms. The network's image input size is 224 × 224. The residual block output is then added to the output of two convolution layers with 3 × 3 and 128 kernel filter sizes. This was the second stone left. The third residual block consists of the output of a skip reference for the second block and the output of two 3 × 3 and 256 convertible filter convolution layers. The fourth and last residual block include the performance of the third block, via the overhead connections and output of two 3 × 3 and 512 filter calculation convolution layers. Finally, the output of the final residual block is summed and the obtained feature map is applied to the entirely connected layers followed by the final output Softmax function. The diagram demonstrates the output of each layer and modifies the feedback to the overlay connections (Fig. 5).

4 Experimental Results

All the models were trained for 3170 train set images and for 50 epochs. The accuracy of RESNET18 (90.1786%) beats both RESNET50 (79.0178%) and CNN model built from scratch (77.679%).

4.1 Accuracy Versus Number of Epochs

From Fig. 6a–c, we observe that the fluctuations in accuracy with respect to epoch are least for CNN. Though the fluctuations are least, but for the same number of epochs the maximum accuracy custom CNN can achieve is less than both RESNET50 and RESNET18. On comparing plots for RESNET50 and RESNET80, we observe that RESNET18 outperforms RESNET50 for the same number of epochs.

4.2 Loss Versus Number of Epochs

From Fig. 7a–c, we can observe the gap between the training curve and validation curve in each plot. From the observation, we infer that the gap is maximum for CNN and least for RESNET18. If we increase the number of epochs, the gap between the two curves for CNN increases. This shows that over-fitting takes place in case on CNN, lesser on RESNET50 and least in RESNET18. Therefore, RESNET18 outperforms the other two.

4.3 Comparison Table

Since the number of epochs and other hyperparameters are similar, this models can be only compared with two factors (i.e., time and accuracy) (Table 1).

5 Conclusion and Future Work

Transfer learning was introduced using PyTorch as a backend. Our implementation reveals that it will take time and commitment to build models from scratch too. These may therefore be minimized with the aid of transfer learning. The dataset was split into two halves: the train set and validation set for each model. We also classified picture training into three most justified and bench-marked CNN strategies with accurate results of 77.679%, pre-trained RESNET50 with accurate results of 79.0178% and pre-trained RESNET18 with accurate results of 90.1786%. RESNET18 and RESNET50 can be considered as the strongest example of transfer learning, and they all surpass the efficiency of scratch-built CNN.

Transfer learning is beneficial if we have insufficient data for a new domain you want to navigate with a neural network and a large pre-existing data pool can be transferred to our demand. However, the data of the new problem (target domain) must be very similar to the previous problem (source domain). If this is not the case, the results of the new problem will not be satisfactory. For example: Biomedical engineers then

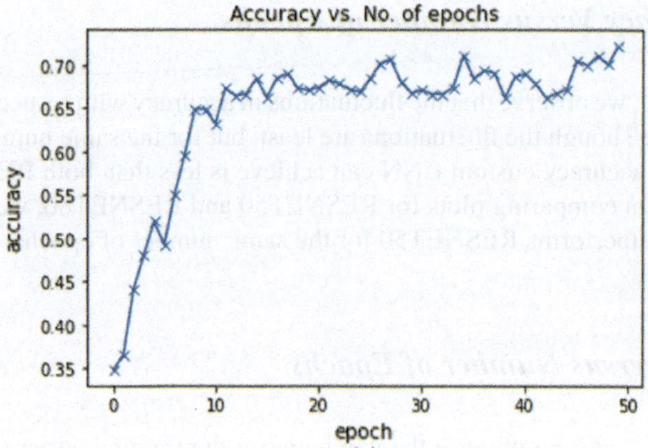

(a)

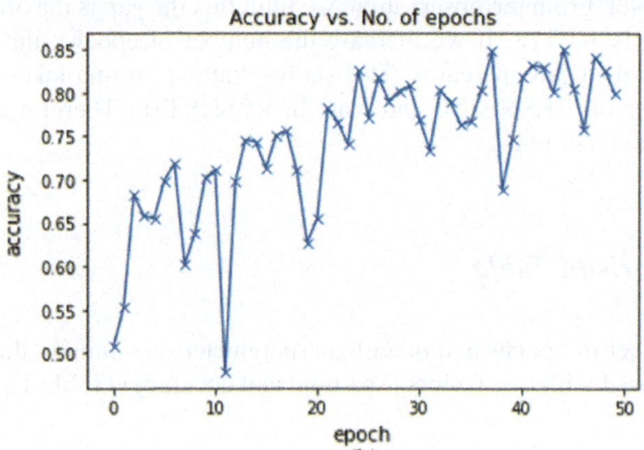

(b)

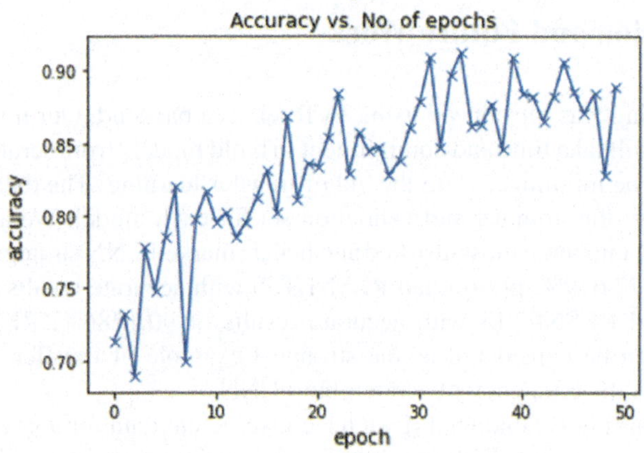

(c)

Fig. 6 Accuracy versus number of epochs **a** for CNN, **b** for RESNET50, **c** for RESNET18

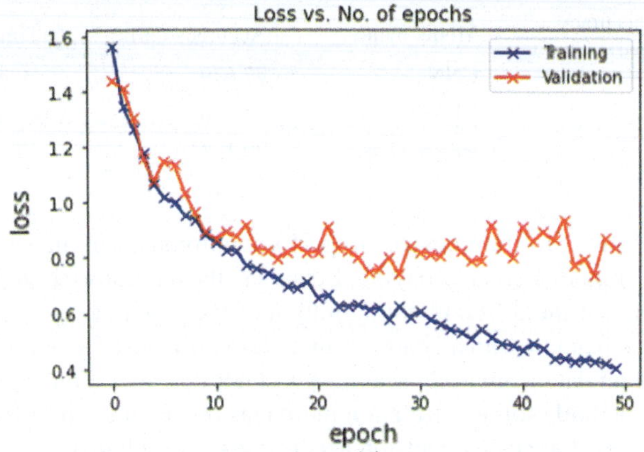

(a)

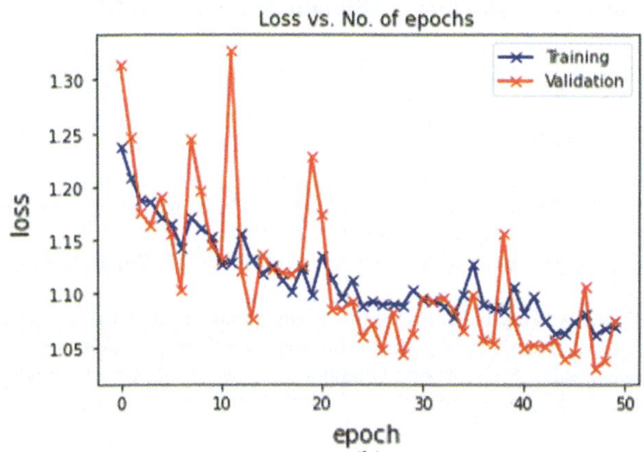

(b)

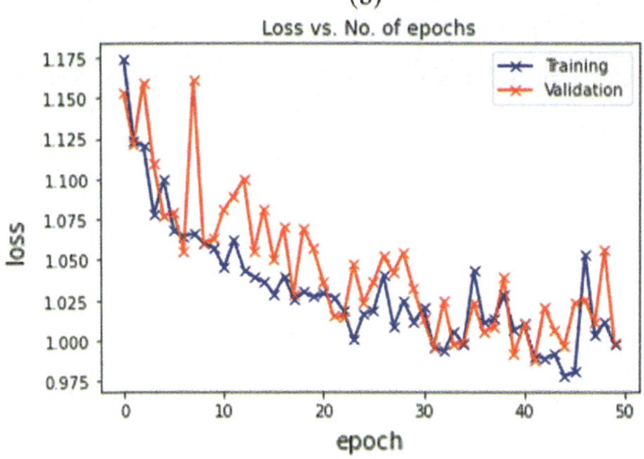

(c)

Fig. 7 Loss versus number of epochs, **a** For CNN, **b** for RESNET50 (c) For RESNET18

Table 1 Model accuracy versus time comparison table

Model name	Accuracy (%)	Time taken
CNN	77.679	60 min 12 s
RESNET50	79.018	43 min 11 s
RESNET18	90.1786	17 min 38 s

use these annotations to train devices to perform automated tissue segmentation or pathology detection in medical imaging. Let us say the laboratory bought a new MRI scanner. Because of the different configuration of the new machine, it will produce different images from the other scanners that were previously installed. As a result, devices trained on data from other scanners will not do well on the new scanner. An adaptive machine finds some correlation in images obtained from different scanners (new and old) and changes its decision accordingly. In such instances, another area of machine learning falls into play: "Domain Adaptation [28]".

References

1. Zhuang, F., Qi, Z., Duan, K., Xi, D., Zhu, Y., Zhu, H., Xiong, H., He, Q.: A comprehensive survey on transfer learning. arXiv, arXiv:1911.02685v3, 31 (2020)
2. Pan, S.j., Yang, Q.: A survey on transfer learning. IEEE **15** (2010). https://doi.org/10.1109/TKDE.2009.191
3. Sonka, M., Hlavac, V., Boyle, R.: Image processing, analysis, and machine vision. In: Cengage Learning. Springer, Berlin (2014). https://doi.org/10.1007/978-1-4899-3216-7
4. Dewangan, D.K., Sahu, S.P.: Driving behavior analysis of intelligent vehicle system for lane detection using vision-sensor. IEEE Sens. J. **21**(5), 6367–6375 (2021). https://doi.org/10.1109/JSEN.2020.3037340
5. Dewangan, D.K., Sahu, S.P.: Real time object tracking for intelligent vehicle. In: 2020 First International Conference on Power, Control and Computing Technologies (ICPC2T), pp. 134–138. IEEE (2020)
6. Dewangan, D.K., Sahu, S.P.: Deep learning-based speed bump detection model for intelligent vehicle system using Raspberry Pi. IEEE Sens. J. **21**(3), 3570–3578 (2021). https://doi.org/10.1109/JSEN.2020.3027097
7. Dewangan, D.K., Sahu, S.P.: PotNet: pothole detection for autonomous vehicle system using convolutional neural network. Electron. Lett. (2021). https://doi.org/10.1049/ell2.12062
8. Dewangan, D.K., Sahu, S.P.: RCNet: road classification convolutional neural networks for intelligent vehicle system. Intel. Serv. Robot. (2021). https://doi.org/10.1007/s11370-020-00343-6
9. Guru, D.S., Sharath, Y.H., Manjunath, S: Texture features and KNN in classification of flower images. RTIPPR, IJCA Special Issue on RTIPPR, pp. 21–29 (2010)
10. Guru, D.S., Sharath Kumar, Y.H., Manjunath, S.: Textural features in flower classification. Math. Computer Model. **7** (2010). https://doi.org/10.1016/j.mcm.2010.11.032
11. Ba, J., Swersky, K., Fidler, S., Salakhutdinov, R.: "Predicting deep zero-shot convolutional neural networks using textual descriptions. arXiv, arXiv:1506.00511v2, 15, (2015)
12. Mete, B.R., Ensari, T.: Flower classification with deep cnn and machine learning algorithms. In: 2019 3rd International Symposium on Multidisciplinary Studies and Innovative Technologies (ISMSIT), Ankara, Turkey, pp. 1–5 (2019). https://doi.org/10.1109/ISMSIT.2019.8932908
13. Guru, D., Kumar, Y.H., Shantharamu, M.: Classification of flowers based on Whorl region (2011). https://doi.org/10.13140/2.1.3528.3204

14. Patel, S., Patel, S.: Flower identification and classification using computer vision and machine learning techniques. Int. J. Eng. Adv. Technol. (IJEAT) **8**(6), 1–9 (2019). ISSN: 2249-8958
15. Wu, Y., Qin, X., Pan, Y., Yuan, C.: Convolution neural network based transfer learning for classification of flowers. In: 2018 IEEE 3rd International Conference on Signal and Image Processing (ICSIP), Shenzhen, pp. 562–566 (2018). https://doi.org/10.1109/SIPROCESS.2018.8600536
16. Feng, J., Wang, Z., Zha, M., Cao, X.: Flower recognition based on transfer learning and adam deep learning optimization algorithm. In: RICAI 2019: Proceedings of the 2019 International Conference on Robotics, Intelligent Control and Artificial Intelligence, Sept 2019, pp 598–604. https://doi.org/10.1145/3366194.3366301
17. Krizhevsky, A., Sutskever, I., Hinton, G.E.: ImageNet classification with deep convolutional neural networks. Commun. ACM **7** (2017). https://doi.org/10.1145/3065386
18. Szegedy, C., Liu, W., Jia, Y., Sermanet, P., Reed, S., Anguelov, D., Erhan, D., Vanhoucke, V., Rabinovich, A.: Going deeper with convolutions, arXiv, arXiv:1409.4842v1, 12, (2014)
19. Gurnani, A., Mavani, V., Gajjar, V., Khandhediya, Y.: Flower categorization using deep convolutional neural networks. arXiv, arXiv:1708.03763v2, 4, (2017)
20. Flower Dataset: Flower Classification in 2019 spring semester at Peking University. Kaggle (2019). https://www.kaggle.com/c/flower-classification-2019/overview
21. Ha, J.G., Moon, H.J., Kwak, J.T., Hassan, S.I., Dang, M., Lee, O.N., Park, H.Y.: Deep convolutional neural network for classifying Fusarium wilt of radish from unmanned aerial vehicles. J. Appl. Remote Sens. **14** (2017). https://doi.org/10.1117/1.JRS.11.042621
22. Dang, L.M., Hassan, S.I., Soo yeon, I., Sangaiah, A.K., Mehmood, I., Rho, S., Seo, S., Moon, H.: UAV based wilt detection system via convolutional neural networks. Sci. Direct **65** (2018). https://doi.org/10.1016/j.suscom.2018.05.010
23. Krizhevsky, A., Sutskever, I., Hinton, G.E.: Imagenet classification with deep convolutional neural networks, Commun, ACM **7** (2017). https://doi.org/10.1145/3065386
24. Simonyan, K., Zisserman, A.: Very deep convolutional networks for large-scale image recognition, arXiv, arXiv:1409.1556v6, 14 (2014)
25. Jia, Y., Shelhamer, E., Donahue, J., Karayev, S., Long, J., Girshick, R., Guadarrama, S., Darrell, T.: Caffe: convolutional architecture for fast feature embedding. arXiv, arXiv:1408.5093v1, 4 (2014)
26. He, K., Zhang, X., Ren, S., Sun, J.: Deep residual learning for image recognition. arXiv, arXiv: 1512.03385v1, 12 (2015)
27. Mo, N., Yan, L., Xie, H.: Class-specific anchor based and context-guided multi-class object detection in high resolution remote sensing imagery with a convolutional neural network. Remote Sens. **23** (2019). https://doi.org/10.3390/rs11030272
28. Kouw, W.M., Loog, M.: An introduction to domain adaptation and transfer learning. arXiv, arXiv:1812.11806v2, 42 (2019)

Application of Random Vector Functional Link Network for Software Defect Prediction

Ruchika Malhotra, Deepti Aggarwal, and Priya Garg

Abstract Software defect prediction (SDP) aims to develop predictive techniques that identify software modules' default proneness using structural and quality attributes. The use of randomized neural networks, especially random vector functional link (RVFL) network, has been limited in SDP. This study proposes a novel SDP model based on principal component analysis (PCA) and RVFL. The model uses PCA for dimensionality reduction of the data and employs RVFL to classify the defective modules. Extensive experiments are conducted on 17 PROMISE repository datasets, and the proposed model is compared with ten classic machine learning (ML) techniques for within-project and inter-release scenarios. The experimental results indicate that PCA–RVFL outperforms the classic ML techniques for most of the datasets, proving the predictive capability of the proposed model in the field of SDP.

Keywords Software defect prediction · Principal component analysis · Neural network · Machine learning · Random vector functional link network

1 Introduction

Software defect prediction (SDP) is a field of study that aims at developing predictive models to identify defect-prone software modules using code quality attributes. SDP helps in the timely identification of defective modules, which benefits the industries in effectively allocating their limited software testing resources. The researchers have developed various ML techniques to predict defect proneness of the software modules [1]. Various SDP studies have also been developed to address challenges like feature selection and class imbalance in the datasets. However, these models face several challenges like low prediction performance and high training time [2].

R. Malhotra · D. Aggarwal · P. Garg (✉)
Department of Software Engineering, Delhi Technological University, Bawana Road, New Delhi, India

R. Malhotra
e-mail: ruchikamalhotra@dtu.ac.in

A. Noor et al. (eds.), *Proceedings of Emerging Trends and Technologies on Intelligent Systems*, Advances in Intelligent Systems and Computing 1371,
https://doi.org/10.1007/978-981-16-3097-2_11

The issue of obtaining an effective feature representation [3] is also common among these SDP models [4, 5]. This study tries to address these issues by proposing a novel defect prediction model PCA–RVFL.

The proposed model, PCA–RVFL, uses PCA for dimensionality reduction and RVFL to classify the defective and non-defective classes from the features obtained by PCA. RVFL is a randomized neural network architecture (RNNA) with direct links between the input and output layers. Unlike the iterative neural networks, in RVFL, the weights between the input and the hidden layer are randomly assigned while the weights of the input and hidden layers to the output layer are calculated analytically. These structural variations in RVFL make it computationally faster than traditional neural networks like multi-layer perceptron (MLP), without compromising with the prediction performance [6]. RVFL has been widely used in many different research areas like forecasting [7], online learning [8], classification [9], regression [10], etc. However, its use in SDP has not been explored. Hence, the motivation of this paper is to study the predictive capability of RVFL in SDP.

The research questions addressed in this study are listed below.

RQ1. What is the prediction performance of PCA–RVFL in comparison with other ML techniques for within-project scenarios in SDP?
RQ2. What is the prediction performance of PCA–RVFL in comparison with other ML techniques for inter-release scenarios in SDP?
RQ3. What is the computational efficiency of PCA–RVFL in comparison with other traditional neural network architectures?

The rest of the paper is organized in this format. Section 2 gives the previous literature review on SDP and RVFL, Sect. 3 presents PCA–RVFL, the proposed model in detail, and the pseudo-code, Sect. 4 presents the experimental framework, Sect. 5 presents the results and observations of the experiment, and Sect. 6 concludes the study.

2 Related Work

In various previous studies, it has been evident that neural network architectures tend to perform better in terms of predictive performance but lag in computation time efficiency [1, 11]. RNNA was introduced to overcome this barrier by using analytical functions rather than backward passes to update the weights and biases [12]. One such category of RNNA is RVFL. RVFL has performed better than other techniques in other fields [7–10]. However, the prediction capability of RVFL has not been assessed in SDP. Hence in this study, the performance of RVFL has been observed.

In previous studies, many different ML techniques have been employed for SDP. Dhamayanthi and Lavanya [13] explored PCA's performance with Naive Bayes (NB) for seven NASA datasets and achieved an improvement of 10.3% in terms of accuracy. Malhotra also conducted various studies [14, 15] to observe ML techniques for SDP on Android projects. The study concluded that ML techniques like NB, LogitBoost, and MLP outperform other ML techniques. Ghotra et al. [11] applied various machine learning techniques on cleaned versions of NASA and PROMISE datasets. ML techniques like NB, SVM, K-nearest neighbors (KNN), ANN, DTree, and ensemble technique were evaluated in the study. We selected different ML techniques from these studies that have previously outperformed others for comparison with the proposed model PCA–RVFL.

3 PCA–RVFL: The Proposed Model for Software Defect Prediction

3.1 Dimensionality Reduction Using Principal Component Analysis

PCA [16] is a popular statistical technique that extracts essential observations from a given dataset and represents it into a set of mutually perpendicular features called principal components formed by the linear combinations of the original features of the given dataset. PCA compresses the dataset by representing it using the principal components with the highest variance.

3.2 Classification Using Random Vector Functional Link Network

RVFL [17] is a feed-forward RNNA with randomized weights and biases. Figure 1 demonstrates the structure of RVFL. The weights between the input layer and the hidden layer are randomly initialized from a particular range. These weights and biases are not manipulated or updated throughout the training. The weights between

Fig. 1 Structure of RVFL

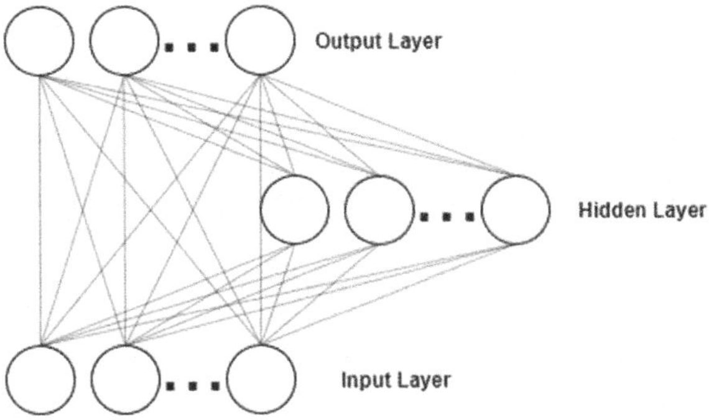

the hidden and output layers and the weights between the input and output layers are analytically calculated.

Given N arbitrary samples (x_i, y_i), $i = 1, 2, 3, \ldots N$, where x_i is a d-dimensional feature vector and y_i is the encoded class label. All the samples belong to one of the output class labels, i.e., {0,1}.

The equation of RVFL for calculating the output labels is defined as:

$$\sum_{t=1}^{k} \beta_t g(w_t x_i + b_t) + \sum_{t=k+1}^{k+d} \beta_t x_{i(t-k)} = o_i \tag{1}$$

where $i = 1, 2, 3, 4 \ldots N$, β_t is the output weights vector that connects the tth hidden layer node to the output layer nodes. $\mathrm{w_t}$ is the vector of input weights connecting the input layer nodes to the tth hidden layer node. b_t is the bias for the tth hidden node, and o_i is the output label for ith sample.

Equation (1) can also be rewritten as:

$$H \cdot \beta = O \tag{2}$$

where β is the output weight matrix, O is the predicted output class label matrix and $H = \{h_i\}, I = 1, 2, \ldots N$ is written as:

$$H = \begin{bmatrix} g(w_1 x_1 + b_1) & \cdots & g(w_k x_1 + b_k) & x_{11} & \cdots & x_{1d} \\ \vdots & \vdots & \vdots & \vdots & \vdots & \vdots \\ g(w_1 x_N + b_1) & \cdots & g(w_k x_N + b_k) & x_{N1} & \cdots & x_{Nd} \end{bmatrix}_{NX(k+d)} \tag{3}$$

where H is formed by combining the values of both the input layer nodes and the hidden layer nodes to form a matrix of size $N * (k + d)$.

The goal is to minimize the classification error, i.e.,

$$|O - Y| = 0 \tag{4}$$

where Y is actual output class labels.

Using Eq. (2) in Eq. (4), we get

$$|H \cdot \beta - Y| = 0. \tag{5}$$

In this study, we are using ridge regression [18] or l2 least square method with a regularization parameter λ for calculation of optimum value of β:

$$\sum_{i=1}^{N} (y_i - h_i^T \beta)^2 + \lambda \beta^2 = 0. \tag{6}$$

The optimal β derived from Eq. (6) would be:

$$\beta = H \cdot (H^T \cdot H + \lambda I)^{-1} \cdot Y. \tag{7}$$

This optimized β value is now the trained weights between the hidden and the output layer nodes and between the input and the output layer nodes. To predict the trained model's output, Eq. (2) is used where β is now known, and the equivalent result would be the predicted class labels.

3.3 *Pseudo-code for PCA–RVFL*

The pseudo-code of PCA–RVFL is shown in Fig. 2.

4 Experimental Setup and Framework

Figure 3 shows the experimental framework used in this study. The following subsections describe the various components of this framework.

```
Input: X = {xi}, Y = {yi}, i = 1, 2, ..., N
Output: Class labels O
Initialization:
  k = Number of hidden layer nodes
  Randomly initializing weights and biases of RVFL using uni-
  form distribution:
      W = {wt}, t = 1, 2, ... k wt ∈ [-1,1]
      B = {bt}, i = 1, 2, ...k bt ∈ [0,1]
  Number of hidden layers = 1
  λ = Regularized parameter for ridge regression
  n_components = number of components in PCA
Algorithm:
    1. Applying PCA for feature reduction on input dataset X to
       get Xred (dataset with n_components features)
    2. Divide {Xred, Y} to training data {Xtrain, Ytrain} and test-
       ing data {Xtest, Ytest}
    3. Calculate Htrain for {Xtrain, Ytrain} using equation (3)
    4. Calculate the parameter vector β using equation (7)
    5. Calculate Htest for {Xtest, Ytest} using equation (3)
    6. Calculate the class labels O for the {Xtest, Ytest} from
       equation (2)
```

Fig. 2 Pseudo-code for PCA–RVFL

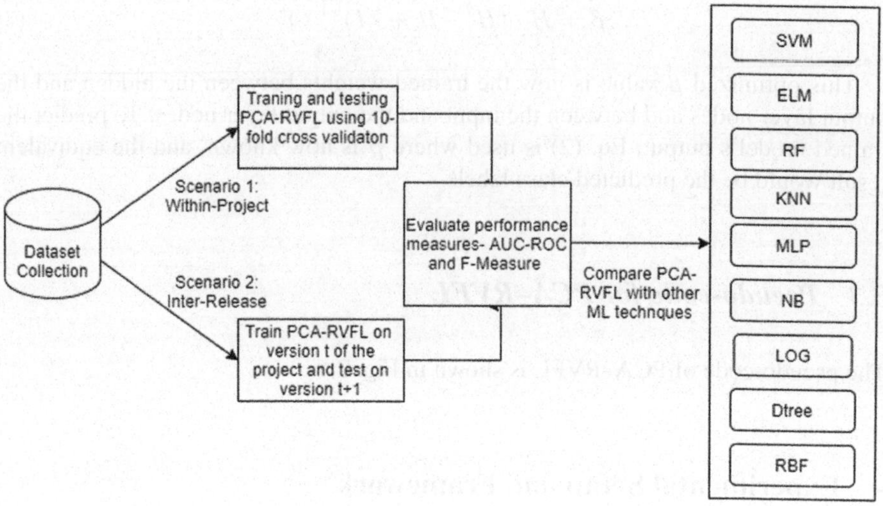

Fig. 3 Experimental framework of this study

4.1 Dataset

For the experiment, 17 datasets are taken from the open-source PROMISE repository [19]. Different releases of the projects are collected to analyze the proposed model PCA–RVFL for both within-project and inter-release scenarios. Table 1 shows the statistics and properties of the datasets. For every dataset, the number of samples, the number of defective samples, and the percentage of defective samples are mentioned.

4.2 Variables

The datasets described in Sect. 4.1 consist of 20 independent variables taken from the object-oriented (OO) metrics suite and one binary dependent variable which shows if the class in the project is defective (1) or non-defective (0). The 20 OO metrics defined by various researchers [20–24] are shown in Table 2. These independent variables have been used widely in previous studies [14] for training and defect prediction.

4.3 Performance Measures

The performance of PCA–RVFL is evaluated using AUC–ROC and F-measure. The performance measures used in the study are defined below.

Table 1 Statistics of dataset used in the study

Dataset	Version	Total samples	Number of defective samples	Defective samples (%)
ant_1.7	1.7	745	166	22.28
ant_1.6	1.6	351	92	26.21
ant_1.5	1.5	293	32	10.92
ant_1.4	1.4	178	40	22.47
arc	1.0	225	29	12.88
camel_1.6	1.6	965	188	19.48
camel_1.4	1.4	872	145	16.63
camel_1.2	1.2	608	216	35.53
camel_1.0	1.0	339	13	3.83
ivy_2.0	2.0	352	40	11.36
lucene_2.4	2.4	339	202	59.59
lucene_2.2	2.2	247	144	58.30
lucene_2.0	2.0	195	91	46.67
poi_3.0	3.0	439	279	63.55
poi_2.5	2.5	385	248	64.42
poi_2.0	2.0	314	37	11.78
prop_6	6	644	61	9.47

F-measure is the harmonic mean of precision and recall. Recall is the correctly predicted defective samples to the total samples which are defective. Precision is the ratio of the correctly predicted defective samples to the total predicted defective samples.

AUC–ROC is the area under the receiver operating characteristics curve. This curve shows the relationship between recall and false positive rate (FPR). The AUC–ROC indicates the ability of techniques to distinguish between defective and non-defective samples. Higher AUC–ROC values indicate better performance of the techniques.

4.4 Parameter Settings

For PCA–RVFL, the average AUC–ROC variation with the number of PCA components is shown in Fig. 4. The highest average AUC–ROC value is achieved for 12 PCA components, so it is used in the experiments.

The parameter settings for RVFL are decided by conducting experiments and based on previous studies [6]. In this study, RVFL is implemented using ridge regression with the sigmoid activation function. The randomization range of weights and

Table 2 Object-oriented metrics used in the study

Abbreviation	Full form
WMC	Weighted methods per class
RFC	Response for a class
NPM	Number of public methods
NOC	Number of children
MOA	Measure of aggregation
MFA	Measure of functional abstraction
Max_CC	Maximum McCabe cyclomatic complexity
LOC	Lines of code
LCOM3	Lack of cohesion in method
LCOM	Lack of cohesion in methods
IC	Inheritance coupling
DIT	Depth of inheritance tree
DAM	Data access metric
Ce	Efferent couplings
CBO	Coupling between objects
CBM	Coupling between methods
CAM	Cohesion among methods of class
Ca	Afferent couplings
Avg_CC	Average McCabe cyclomatic complexity
AMC	Average method complexity

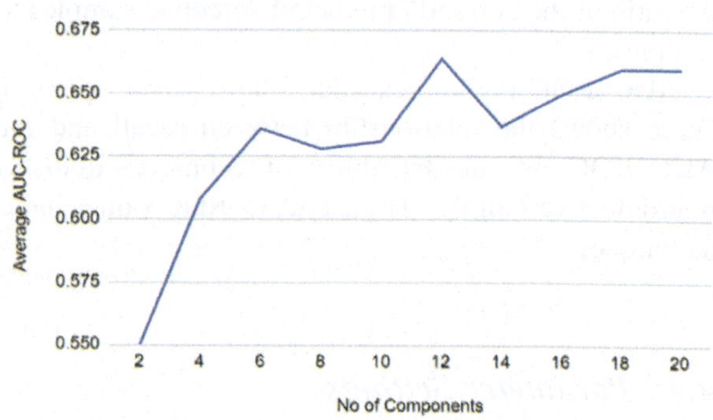

Fig. 4 Variation of average AUC–ROC with number of components in PCA

biases is [−1,1] from a uniform distribution, and the number of hidden layer nodes is 80.

PCA–RVFL has been compared with various ML techniques—KNN, MLP, RBF, NB, DTree, LOG, SVM, RF [25], and ELM [26]. DTree, LOG, SVM, RF, and NB techniques are implemented using the open-source Scikit-learn library of Python with default parameter settings. MLP and RBF networks are implemented using the Keras library of Python. The parameters for MLP, RBF, and KNN are chosen by conducting experiments on the datasets described in Table 1. For MLP and RBF, the learning rate has been taken as 0.1, with RMSProp optimizer, the number of hidden nodes 50, and batch size 50. For MLP, the activation function sigmoid is used. For ELM, the number of hidden layer nodes is 15, the activation function is sigmoid, and the randomization range of weights and bias is [−1,1] and [0,1], respectively.

5 Experimental Results and Analysis

5.1 *RQ1. What is the Prediction Performance of PCA–RVFL in Comparison with Other ML Techniques for Within-Project Scenarios in SDP?*

To answer this RQ, experiments are conducted according to the framework described in Sect. 4. Tables 3 and 4 show the comparison of AUC–ROC and F-measure of the ML techniques in the within-project scenario. The highest value for every dataset is highlighted in the tables.

The proposed model has achieved the best AUC–ROC value for most datasets, i.e., 15 out of 17 datasets. The average AUC–ROC value for PCA–RVFL is 0.753 which is the highest among other ML techniques with average AUC–ROC values as—0.656 (for PCA–MLP), 0.652 (for PCA–NB), 0.652 (for RVFL), 0.693 (for PCA–ELM), 0.611 (for PCA–SVM), 0.645 (for PCA–RF), 0.672 (for PCA–DTree), 0.626 (for PCA–RBF), 0.644 (for PCA–KNN), and 0.631 (for PCA–LOG). PCA–RVFL has given a significantly better performance than RVFL, which shows that dimensionality reduction can play a crucial role in a model's prediction performance. The F-measure values in Table 4 indicate that PCA–RVFL has achieved an average F-measure value of 0.613. It has shown an increase of 12.47% (for PCA–MLP), 15.44% (for PCA–NB), 13.1% (for RVFL), 16.3% (for PCA–ELM), 17.54% (for PCA–SVM), 14.7% (for PCA–RF) 18.2% (for PCA–DTree), 16.3% (for PCA–RBF), 17.6% (for PCA–KNN), and 30.34% (for PCA–LOG) in F-measure. The better performance of PCA–RVFL indicates that it has good prediction capability in the within-project scenarios of SDP.

5.2 *RQ2: What is the Prediction Performance of PCA–RVFL in Comparison with Other ML Techniques for Inter-release Scenarios in SDP?*

To answer this RQ, experiments are conducted according to the framework described in Sect. 4. To perform the defect prediction on an inter-release scenario, one release of a dataset is selected for training, and the next release is selected for testing. Table 5

Table 3 AUC–ROC comparison between ML techniques for within-project scenario

DATASET	PCA–RVFL	RVFL	PCA–ELM	PCA–SVM	PCA–RF	PCA–MLP	PCA–RBF	PCA–KNN	PCA–DTree	PCA–NB	PCA–LOG
ant_1.7	**0.744**	0.728	0.729	0.601	0.629	0.664	0.512	0.672	0.648	0.666	0.514
ant_1.6	**0.784**	**0.784**	0.691	0.554	0.601	0.740	0.712	0.675	0.714	0.773	0.539
ant_1.5	**0.889**	0.686	0.673	0.614	0.650	0.759	0.759	0.632	0.722	0.722	0.712
ant_1.4	**0.721**	0.700	0.622	0.608	0.682	0.671	0.654	0.543	0.505	0.580	0.616
arc	**0.875**	0.692	0.667	0.625	0.679	0.523	0.692	0.565	0.512	0.580	0.601
camel_1.6	**0.741**	0.627	0.638	0.612	0.628	0.623	0.601	0.524	0.503	0.606	0.512
camel_1.4	**0.754**	0.655	0.649	0.599	0.523	0.636	0.678	0.674	0.537	0.565	0.613
camel_1.2	**0.722**	0.648	0.647	0.605	0.576	0.712	0.666	0.708	0.641	0.698	0.600
camel_1.0	0.624	0.660	0.636	0.602	0.660	0.623	0.614	0.643	0.612	**0.699**	0.601
ivy_2.0	**0.761**	0.726	0.700	0.600	0.659	0.661	0.612	0.651	0.654	0.690	0.567
lucene_2.4	**0.777**	0.680	0.727	0.687	0.633	0.602	0.644	0.657	0.657	0.645	0.530
lucene_2.2	0.643	0.657	0.607	0.575	0.600	0.618	0.623	0.619	0.604	**0.677**	0.603
lucene_2.0	**0.722**	0.620	0.632	0.605	0.638	0.611	0.617	0.558	0.628	0.603	0.576
poi_3.0	**0.791**	0.774	**0.791**	0.638	0.757	0.729	0.638	0.690	0.689	0.665	0.605
poi_2.5	**0.772**	0.769	0.632	0.670	0.755	0.757	0.726	0.723	0.730	0.686	0.667
poi_2.0	**0.743**	0.741	0.706	0.600	0.664	0.594	0.627	0.650	0.678	0.637	0.667
prop_6	**0.747**	0.639	0.679	0.600	0.628	0.623	0.567	0.536	0.612	0.595	0.610

The bold represents the highest value of the performance measure for each dataset

Table 4 *F*-measure comparison between ML techniques for within-project scenario

DATASET	PCA–RVFL	RVFL	PCA–ELM	PCA–SVM	PCA–RF	PCA–MLP	PCA–RBF	PCA–KNN	PCA–DTree	PCA–NB	PCA–LOG
ant_1.7	**0.553**	0.543	0.500	0.545	0.552	0.516	0.509	0.530	0.543	0.481	0.380
ant_1.6	**0.638**	0.629	0.600	0.588	0.607	0.490	0.566	0.613	0.620	0.539	0.334
ant_1.5	**0.524**	0.364	0.304	0.352	0.347	0.423	0.390	0.222	0.320	0.345	0.123
ant_1.4	**0.550**	0.401	0.385	0.377	0.367	0.450	0.402	0.440	0.340	0.375	0.109
arc	**0.544**	0.428	0.522	0.362	0.540	0.533	0.409	0.482	0.302	0.486	0.450
camel_1.6	**0.523**	0.412	0.426	0.347	0.435	0.445	0.501	0.385	0.382	0.417	0.223
camel_1.4	**0.542**	0.484	0.464	0.452	0.402	0.500	0.490	0.530	0.472	0.474	0.424
camel_1.2	**0.652**	0.596	0.533	0.563	0.578	0.574	0.520	0.525	0.519	0.542	0.425
camel_1.0	0.490	0.411	0.405	**0.522**	0.420	0.470	0.460	0.380	0.350	0.436	0.307
ivy_2.0	0.492	0.412	**0.512**	0.207	0.207	0.386	0.390	0.250	0.302	0.355	0.176
lucene_2.4	0.741	0.710	0.750	0.764	0.713	0.731	0.763	0.711	0.716	**0.801**	0.744
lucene_2.2	0.724	0.680	0.636	**0.747**	0.677	0.710	0.704	0.660	0.687	0.705	0.698
lucene_2.0	**0.759**	0.752	0.576	0.640	0.605	0.600	0.712	0.549	0.583	0.754	0.560
poi_3.0	**0.852**	0.782	0.724	0.797	0.830	0.790	0.771	0.767	0.781	0.723	0.786
poi_2.5	0.811	0.818	**0.848**	0.825	0.835	0.820	0.802	0.794	0.800	0.805	0.775
poi_2.0	**0.610**	0.455	0.444	0.434	0.600	0.500	0.504	0.411	0.520	0.406	0.423
prop_6	**0.417**	0.342	0.324	0.345	0.363	0.320	0.317	0.329	0.302	0.383	0.323

The bold represents the highest value of the performance measure for each dataset

Table 5 AUC–ROC comparison between ML techniques for inter-release scenario

Train On	Test on	PCA-RVFL	RVFL	PCA-ELM	PCA-SVM	PCA-RF	PCA-MLP	PCA-RBF	PCA-KNN	PCA-DTree	PCA-NB	PCA-LOG
ant												
1.4	1.5	**0.671**	0.670	0.665	0.554	0.669	0.490	0.646	0.495	0.506	0.629	0.500
1.5	1.6	**0.754**	0.699	0.714	0.590	0.659	0.527	0.500	0.588	0.500	0.687	0.510
1.6	1.7	0.730	0.694	0.627	0.683	0.565	**0.788**	0.698	0.658	0.716	0.700	0.614
camel												
1.0	1.2	**0.659**	0.539	0.559	0.600	0.602	0.503	0.645	0.503	0.662	0.534	0.500
1.2	1.4	**0.705**	0.607	0.619	0.590	0.671	0.607	0.619	0.587	0.539	0.590	0.505
1.4	1.6	**0.667**	0.642	0.581	0.603	0.624	0.670	0.570	0.557	0.554	0.559	0.500
lucene												
2.0	2.2	**0.702**	0.588	0.688	0.589	0.638	0.615	0.573	0.627	0.619	0.598	0.579
2.2	2.4	**0.703**	0.673	0.647	0.570	0.625	0.637	0.552	0.574	0.548	0.586	0.547
poi												
2.0	2.5	**0.672**	0.597	0.546	0.670	0.614	0.610	0.621	0.487	0.500	0.568	0.500
2.5	3.0	0.708	0.691	0.682	**0.713**	0.657	0.659	0.656	0.634	0.629	0.656	0.601

The bold represents the highest value of the performance measure for each dataset

represents the comparison of the AUC–ROC values of various ML techniques. PCA–RVFL shows the best results for 8 out of 10 datasets with AUC–ROC value more than 0.7 for 6 out of 10 datasets. The proposed model has shown significant percentage increase in average AUC–ROC with other ML techniques—14.07% (for PCA–MLP), 14.07% (for PCA–NB), 20.79% (for PCA–DTree), 22.1% (for PCA–KNN), 30.03% (for PCA–LOG), 14.63% (for PCA–RBF), 8.9% (for RVFL), 10.11% (for PCA–ELM), 13.14% (for PCA–SVM), and 10.28% (for PCA–RF). This indicates that PCA–RVFL can handle a class imbalance in the datasets. Table 6 represents the comparison of the F-measure values of various ML techniques. The proposed model has shown the second-highest average F-measure (0.512), where PCA–NB (0.587) has shown the highest average *F*-measure.

The model has achieved high AUC–ROC values for most datasets in the experiment, which indicates that PCA–RVFL is suitable for the inter-release scenario of defect prediction. Since the difference between the number of samples and the statistics of two adjacent releases is relatively high, this might be why the low performance of the ML techniques in the inter-release scenario compared to the within-project scenario. In the case of ant_1.5, the highest value of AUC–ROC in the within-project scenario is 0.889, while in the inter-release scenario, it is 0.671 when trained on ant_1.4. These variations indicate the difference in the subsequent releases and affect ML techniques' performance in inter-release scenarios.

5.3 *RQ3: What is the Computational Efficiency of PCA–RVFL in Comparison with Other Traditional Neural Network Architectures?*

Apart from predictive capability, one of the most important aspects that differentiate PCA–RVFL from traditional neural network architectures is the computational efficiency of RVFL due to its randomization nature, unlike MLP and RBF that have iterative nature. This section investigates the training time efficiency of PCA–RVFL. The training time of two neural network techniques—MLP and RBF—is compared to prove the computation efficiency of PCA–RVFL. The results shown in Table 7 indicate that PCA–RVFL is a clear winner in terms of computational efficiency. The average training time for PCA–RVFL is 13.7 ms far lower than the average training time for RBF (52.9) and MLP (195.1).

The non-iterative nature of RVFL makes it a computationally efficient ML technique compared to the traditional neural network architectures and makes it suitable for application in problem domains that involve training on large datasets.

Table 6 F-measure comparison between ML techniques for inter-release scenario

Train On	Test On	PCA–RVFL	RVFL	PCA–ELM	PCA–SVM	PCA–RF	PCA–MLP	PCA–RBF	PCA–KNN	PCA–DTree	PCA–NB	PCA–LOG
ant												
1.4	1.5	**0.277**	0.275	0.217	0.212	0.233	0.010	0.020	0.113	0.053	0.253	0.020
1.5	1.6	**0.579**	0.561	0.571	0.226	0.234	0.010	0.060	0.319	0.080	0.533	0.010
1.6	1.7	0.443	0.433	0.443	0.314	0.308	0.347	0.034	0.345	0.390	**0.717**	0.075
camel												
1.0	1.2	**0.394**	0.235	0.296	0.277	0.018	0.063	0.010	0.018	0.020	**0.394**	0.080
1.2	1.4	0.363	0.341	0.297	0.062	**0.429**	0.319	0.040	0.318	0.209	0.327	0.049
1.4	1.6	**0.361**	0.293	0.315	0.011	0.282	0.020	0.050	0.245	0.255	0.323	0.020
lucene												
2.0	2.2	0.383	0.493	0.582	0.676	0.618	0.060	0.671	0.622	0.593	**0.709**	0.604
2.2	2.4	**0.757**	0.752	0.669	**0.757**	0.347	0.742	0.750	0.691	0.680	0.739	0.745
poi												
2.0	2.5	**0.689**	0.281	0.345	0.176	0.177	0.039	0.030	0.135	0.020	0.682	0.030
2.5	3.0	0.720	0.752	0.763	0.633	0.738	0.773	0.787	0.734	0.750	**0.804**	0.786

The bold represents the highest value of the performance measure for each dataset

Table 7 Training time comparison between neural network techniques

Dataset	RBF	MLP	PCA–RVFL
ant_1.7	100.1	195.7	18.3
ant_1.6	29.5	219.5	9.8
ant_1.5	14.6	157.6	8.8
ant_1.4	10.7	207.5	8.4
arc	14.3	97.2	8.0
camel_1.6	207.1	335.7	14.4
camel_1.4	162.4	397.3	13.5
camel_1.2	106.1	100.6	11.4
camel_1.0	13.0	136.2	9.2
ivy_2.0	23.6	174.2	14.2
lucene_2.4	33.5	175.8	9.7
lucene_2.2	20.4	82.9	8.7
lucene_2.0	12.6	21.0	7.8
poi_3.0	42.6	284.7	12.1
poi_2.5	33.1	174.9	9.7
poi_2.0	24.2	263.6	56.5
prop_6	51.4	292.2	12.8

6 Conclusion

The study proposed a novel SDP model PCA–RVFL based on randomized neural network RVFL and dimensionality reduction technique PCA. The experiments were conducted to assess the performance of PCA–RVFL for within-project and inter-release scenarios in SDP. The results indicate that PCA–RVFL has good prediction capability compared to ten classic ML techniques previously used in SDP in both scenarios. The AUC–ROC values for PCA–RVFL range from 0.624 to 0.889 for within-project scenario and from 0.671 to 0.754 for the inter-release scenario. The training time results are also in favor of the proposed model, making it computationally efficient compared to neural network architectures like MLP and RBF. The study concludes that randomized neural networks like RVFL can be employed in the field of SDP to achieve good prediction performance. Another important conclusion is that appropriate feature representation is necessary for building effective prediction models in SDP, which is evident from the impact of PCA on the model's performance.

References

1. Omri, S., Sinz, C.: Deep learning for software defect prediction: a survey. In: IEEE/ACM 42nd International Conference on Software Engineering Workshops (ICSEW), pp. 209–214 (2020). https://doi.org/10.1145/3387940.3391463
2. Li, L., Lessmann, S., Baesens, B.: Evaluating software defect prediction performance: an updated benchmarking study (2019). arXiv:1901.01726 [cs.SE]
3. Cui, M., Sun, Y., Lu,Y., Jiang, Y.: Study on the influence of the number of features on the performance of software defect prediction model. In: Proceedings of the 2019 3rd International Conference on Deep Learning Technologies, pp. 32–37 (2019). https://doi.org/10.1145/3342999.3343010
4. Arora, I., Tetarwal, V., Saha, A.: Open issues in software defect prediction. Procedia Computer Sci. **46**, 906–912 (2015). https://doi.org/10.1016/j.procs.2015.02.161
5. Malhotra, R.: A systematic review of machine learning techniques for software fault prediction. Appl. Soft Comput. **27**, 504–518 (2015). https://doi.org/10.1016/j.asoc.2014.11.023
6. Zhang, L., Suganthan, P.N.: A comprehensive evaluation of random vector functional link networks. Inf. Sci. **367–368**, 1094–1105 (2016). https://doi.org/10.1016/j.ins.2015.09.025
7. Bisoi, R., Dash, P., Mishra, S.P.: Modes decomposition method in fusion with robust random vector functional link network for crude oil price forecasting. Appl. Soft Comput. **80**, 475–493 (2019). https://doi.org/10.1016/j.asoc.2019.04.026
8. Zhang, L., Suganthan, P.N.: Visual tracking with convolutional random vector functional link network. IEEE Trans. Cybern. **47**(10), 3243–3253 (2017). https://doi.org/10.1109/TCYB.2016.2588526
9. Zhang, Y., Wu, J., Cai, Z., Duc, B., Philip, S.Y.: An unsupervised parameter learning model for RVFL neural network Author links open overlay panel. Neural Netw. **112**, 85–97 (2019). https://doi.org/10.1016/j.neunet.2019.01.007
10. Vukovíc, N., Petrovíc, M., Miljkovíc, Z.: A comprehensive experimental evaluation of orthogonal polynomial expanded random vector functional link neural networks for regression. Appl. Soft Comput. **70**, 1083–1096 (2018). https://doi.org/10.1016/j.asoc.2017.10.010
11. Ghotra, B., McIntosh, S., Hassan, A.E.: Revisiting the impact of classification techniques on the performance of defect prediction models. In: IEEE/ACM 37th IEEE International Conference on Software Engineering (2015). https://doi.org/10.1109/ICSE.2015.91
12. Cao, W., Wang, X., Ming, Z., Gao, J.: A review on neural networks with random weights. Neurocomputing **275**(31), 278–287 (2018). https://doi.org/10.1016/j.neucom.2017.08.040
13. Dhamayanthi, N., Lavanya, B.: Software defect prediction using principal component analysis and naïve bayes algorithm. In: Chaki N., Devarakonda N., Sarkar A., Debnath N. (eds.) Proceedings of International Conference on Computational Intelligence and Data Engineering. Lecture Notes on Data Engineering and Communications Technologies, 28. Springer, Singapore (2019). https://doi.org/10.1007/978-981-13-6459-4_24
14. Malhotra, R.: An empirical framework for defect prediction using machine learning techniques with Android software. Appl. Soft Comput. **49**, 1034–1050 (2016). https://doi.org/10.1016/j.asoc.2016.04.0324
15. Malhotra, R.: A systematic review of machine learning techniques for software fault prediction. Appl. Soft Comput. J. **27**, 504–518 (2015). https://doi.org/10.1016/j.asoc.2014.11.023
16. Abdi, H., Williams, L.J.: Principal component analysis. WIREs. Comput. Statistics **2**(4), 433–459 (2010). https://doi.org/10.1002/wics.101
17. Husmeier, D.: Random vector functional link (RVFL) networks. Neural networks for conditional probability estimation. In: Perspectives in Neural Computing, pp. 87–97. Springer, London (1999). https://doi.org/10.1007/978-1-4471-0847-4_6
18. Hoerl, A.E., Kennard, R.W.: Ridge regression: biased estimation for nonorthogonal problems. Technometrics **12**(1), 55–67 (1970). https://doi.org/10.1080/00401706.1970.10488634
19. Menzies, T., Krishna, R., Pryor, D.: The promise repository of empirical software engineering data. Department of Computer Science, North Carolina State University (2016) [Online] Available: http://promisedata.org/repository

20. Chidamber, S.R., Kemerer, C.F.: A metrics suite for object oriented design. IEEE Trans. Software Eng. **20**(6), 476–493 (1994). https://doi.org/10.1109/32.295895
21. Henderson-Sellers, B.: Object Oriented Metrics: Measures of Complexity in New Jersey, pp. 142–147. Prentice-Hall (1996)
22. Martin, R.: OO design quality metrics—an analysis of dependencies. In: Workshop Pragmatic and Theoretical Directions in Object-Oriented Software Metrics (1994)
23. Tang, M.H., Kao, M.H., Chen, M.H.: An empirical study on object-oriented metrics. In: Proceedings of Metrics, pp. 242–249 (1999). https://doi.org/10.1109/METRIC.1999.809745
24. Thomas, J., McCabe, J.: A complexity measure. IEEE Trans. Software Eng. **SE-2**(4), 308–320 (1976). https://doi.org/10.1109/TSE.1976.233837
25. Alpaydin, E.: Introduction to Machine Learning. MIT press, Cambridge (2014)
26. Huang, G.B., Zhu, Q.Y., Siew, C.K.: Extreme learning machine: theory and applications. Neurocomputing **70**(1–3), 489–501 (2006). https://doi.org/10.1016/j.neucom.2005.12.126

Nikbot: Chatbot for Nikshay Aushadhi

Savita Kumari Pandit, M. Balasubramaniam, Ashutosh Pandey, and Jitendra Singh

Abstract Nikbot is a software application that uses natural language to interact with clients. It is designed to provide information to the users about Nikshay Aushadhi tuberculosis-related frequently asked question without handling telephonic calls which are quite hectic and not scalable for a large number of users. Building Nikbot having challenges like conversational interfaces, technical design principal issues, and data preparation. The Nikbot stores the queries and response in the database, and using the chatterbots machine learning library, the algorithm identifies the input sentence keywords asked by the user. Using N-gram, TF-IDF and cosine comparison, sentence and rank similarities are determined. Each sentence is scored by the closest matching response by finding the closest matching known statement that matches the same input sentences which is selected as a response for the query. The scheduler program at the back-end handles the questions and responses presented to the bot that are not present in the database, and the support team collects that information for further analyzing user's sentiments and preparing those relevant responses if those questions are relevant.

Keywords Chatbot · Machine learning [1] · Natural language processing (NLP) [3] · Term frequency-inverse document frequency (TF-IDF) [2] · Chatterbots [4]

S. K. Pandit (✉) · M. Balasubramaniam · A. Pandey · J. Singh
BDPM Group, Center for Development of Advanced Computing (C-DAC), Noida, India
e-mail: savitakumari@cdac.in

M. Balasubramaniam
e-mail: mbalasubramaniam@cdac.in

A. Pandey
e-mail: ashutoshpandey@cdac.in

J. Singh
e-mail: jitendrasingh@cdac.in

A. Noor et al. (eds.), *Proceedings of Emerging Trends and Technologies on Intelligent Systems*, Advances in Intelligent Systems and Computing 1371,
https://doi.org/10.1007/978-981-16-3097-2_12

1 Introduction

The Chatbot is a kind of software application that interacts with users using natural language, works basically on artificial Intelligence and machine learning. We have implemented this application in to the Patient Management initiative under National Tuberculosis Elimination Programmed (NTEP) website application Nikshay Aushadhi (Ni = End, Kshay = Tuberculosis, Aushadhi = medicine). Nikshay Aushadhi is developed and maintained by the Central Tuberculosis Division (CTD), Ministry of Health and Family Welfare, Government of India in collaboration with National Informatics Center (NIC) and World Health Organization of India.

Across the world, both in the public and private sectors, healthcare practitioners file cases under their supervision, order various types of testing in laboratories across the country, document descriptions of treatment, track compliance with treatment and use the vaccine at different levels across the country, and get the case moved between providers. It also functions as the National Tuberculosis Surveillance System and enables different surveillance data to be published by the Government of India. The proposed project creates a text-to-text conversational agent that solves common questions about Nikshay Aushadhi and explains its solutions in natural language. It uses the Chatbot question and answer protocol to answer these frequently asked questions. It is designed to reduce the users' waiting time, which alleviates the need to attend user calls when needed immediately. The purpose of this paper is to address the need and use of Nikshay Aushadhi Chatbot in the field of healthcare.

2 Proposed System

In this work, we propose our system using the Chatterbot [4] conversational framework. The purpose of the proposed Chatbot is to help map users to resolve their issue, problem or question. The system uses expert systems to respond to inquiries if the answer is not in the database. In the database, Chatbot data is stored in the form of a prototype. Nikbot automate the suggestion of drug inventory (near expiration, expired drugs), drug shortage and drug availability by analyzing drug storage in Nikshay Aushadhi Inventory Database and predicting outcomes based on quarterly/monthly drug statistics [3]. The Arima model is used for predicting outcomes, and before that, it follows the FEFO (First Expiry, First Out) principles which were previously administered by the esteemed employees of the department.

2.1 System Architecture

Our proposed Chatbot architecture is shown in Fig. 1. It is divided into several modules as follows: The client inserts the query in the form of text in the Nikbot

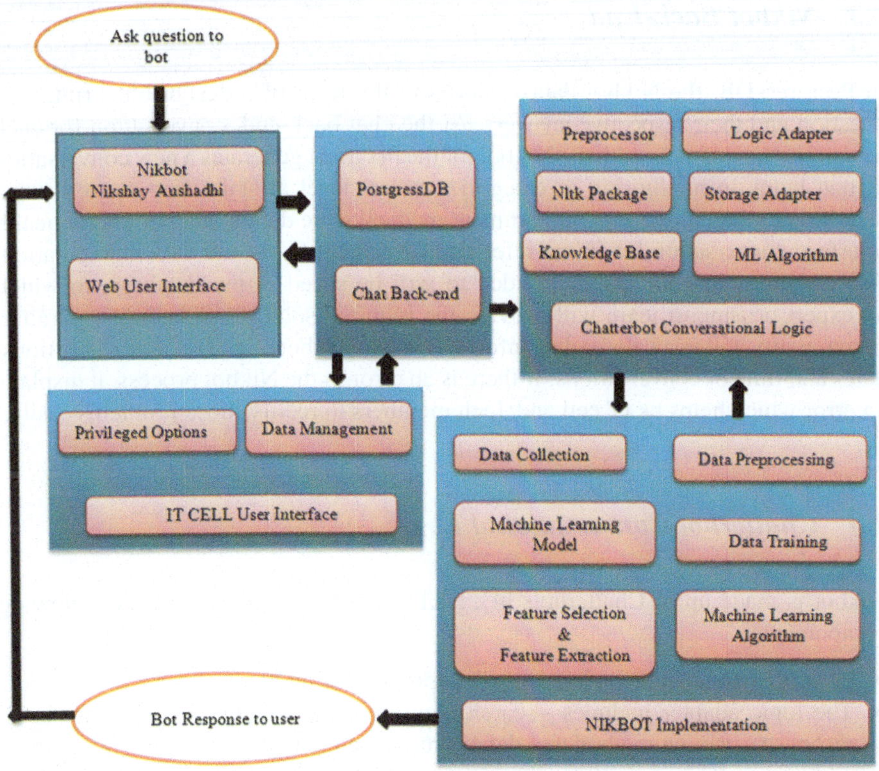

Fig. 1 Nikbot system architecture

user interface. The Nikbot interface receives the client's query and then responds accordingly.

2.2 *Nikbot User Interface*

We have created the Nikbot using a python web framework flask with HTML, CSS and JavaScript [1]. It has an integrated list type for dictionaries and data structures that can be used to create an effective runtime structure. By typing their question, users can get the necessary information. For example, how to add the peripheral health interface, drug stock, etc. The Nikbot verifies whether or not the user's response is available. Chatbots provide users with the same data and solve their issues.

2.3 Nikbot Back-End

In PostgressDB, the Nikbot data is stored in the form of a design blueprint, e.g., question and their respective answers. At the chat back-end, conversation for each new user who interacts with the Nikbot for the first time generates a new conversation id that helps in monitoring to categorize users' level at state drug store, district drug store, tuberculosis unit, government medical store depot, and peripheral health interface conversation for future reference for data analysis and data manipulation. The questions that are not responded to will be saved in the database, to which an expert or support team will provide the required solution. Scheduler operation functionality will track all the data information like Nikbot response, user's questions, dates and time of conversations. If there is an error in the Nikbot process, it displays an error which helps us IT cell and Tech members to rectify those problems easily.

2.4 Chatterbot Conversational Logic

Nikbot is made up of Chatterbots [4]; NLTK and ML engines have the following components:

- Preprocessor adapter for handling data preprocessing works.
- The logic adapter includes—search algorithm, maximum similarity threshold, response selection method and default response method.
- The best matching adapters allow you to set a list of words to match the exclude
- words parameter, which prevents the logical adapter from returning statements that contain text that contains any of those words.
- Storage adapter for MongoDB, SQLite, PostgressDB, MySQL.
- Machine learning comparison algorithm (Jaccard Similarity, Levenshtein Distance, Sentiment Comparison, Synset Distance).
- Machine learning algorithm (Naïve Bayes, supervised learning, reinforcement learning).

2.5 Nikbot Implementation Data Collection

First, we collected data from the IT cell member. Data generally includes all those FAQ queries asked by SDS, DDS, TU, PHI and GMSD.

Data preprocessing

The major work is to arrange all level categories questions with their respective answer using Nikshay Aushadhi documentation where they explain each module how it works and the flow of a process. Preparation of the dataset is manual because it contains images, special Syntax, character, text and most of the things are irrelevant. Steps to follow are the following:

1. Set of questions
2. Set of respective answers
3. Set of respective level (SDS, DDS, TU, PHI, GMSD).

The data is in text format (strings), for example:

Question: How to add a store in Nikshay Aushadhi?
Response: Go to Home menu; click on Admin; Select Approving authority master; Fill mandatory details; 'Record Status' as active;
Click the 'Add' button. Another screen will appear. Fill in the 'Approving Authority Details.'
Select the effective date and click on the 'Save' button.

Model Training

The training process involves loading the dataset into the Chatbot's database. We need to create or build graph data structures that represent answers to known questions. When data is trained with the data set, it creates entries in the knowledge graph of the Nikbot so that the query inputs and answers are correctly represented. These functions update Nikbot's database knowledge graph based on a list of user queries and their answers, the data stored in vector format and for the training process, we use the list of statements based on their assignment of each statement in the document provided in the conversation [4] (Fig. 2).

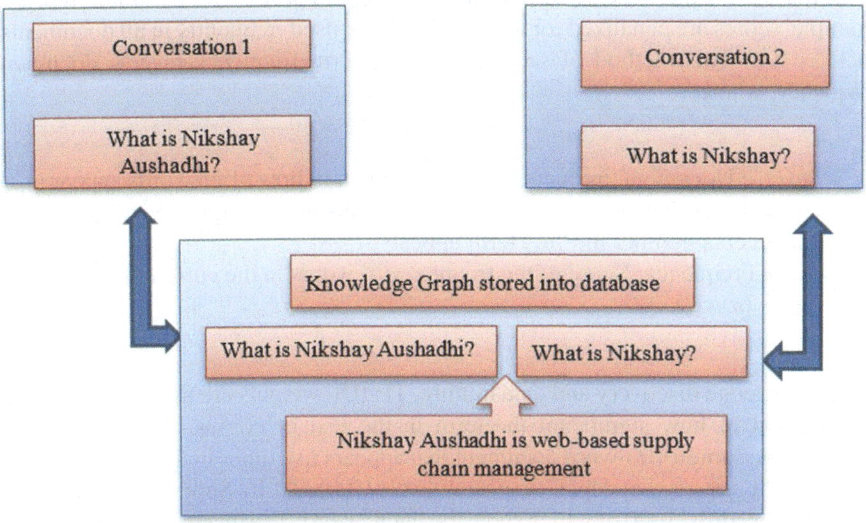

Fig. 2 Nikbot training knowledge graph

Machine learning model

In search algorithms, the attributes that help the Nikbot's to select a response include:

1. The similarity of the entry statements with the well-known statements
2. The frequency with which the known answers are found
3. The possibility of an entry instruction in the category in which the well-known statements are a part which determines whether the input instrument meets a specific set of standards that require a response from this logical adapter.

Feature Selection and Extraction

Feature extraction and selection require some numerical feature vectors to perform the task. So, first data preprocessing [1–3] is performed using a preprocessor (clean white space, remove noise and stop words), then performing Sentence tokenizer uses nltk library Punkt tokenizer Lemmatization [1–3] to find the sentences list from the data stored into the database and from the Sentence tokenizer further Word tokenizer performed using nltk library. Stemming and Lemmatization [1–3] are used in strings to find a list of terms. The Bag of Words Lemmatization [1–3] converts the text into a number vector or array. The occurrence of words in the document is defined. It comprises of two factors: the vocabulary of known words and the existence of known words, and all the details in the text are discarded about the order or word structure, and the model just thinks about whether the keywords that are known appear in the document, not where they appear in the document. If our dictionary, for instance, includes the terms {How, to, adding, store, in, Nikshay, Aushadhi?}, after vectorizing the text is vector-like: (1, 0, 1, 1, 0, 1, 1).

Keyword matching is done by TF-IDF Approach (Lemmatization [1–3]). The bag of words will further rescale so that the frequency of words can be monitored. For example, scores are penalized for the words that are used frequently in all documents, such as 'to.' This method of scoring is called term frequency-reverse document frequency, where:

Inverse Document Frequency: An assessment of how often words appear in a document. IDF = $1 + \log(N/n)$. N is the documents number and n is the documents number the "to" term appears.
Term Frequency: Score of the frequency of words in the current document. TF = $(n)/(N)$.

In knowledge discovery and data mining, TF-IDF weights are used. It is a numerical method of how significant the term in the text or corpus of the sample is. Example: Assume a 100-word document that appears five times in the word 'Nikshay Aushadhi.' Frequency of deposit expectation (16/100) = 0.16. Suppose you now have 10 million records, 1000 of which show the phone word. The inverse frequency of the text (IDF) is then determined as log (10,000,000/1000) = 4. The weight of TF-IDF is therefore the product of the following quantities:

0.16 * 4 = 0.64.TF-IDF [1] is added to the text as follows: in vector space, two real-value vectors. Then, we take a dot product and divide it by a product after obtaining the cosine similarity. This gives the cosine angle between vectors.

A similarity measure between two nonzero vectors is cosine similarity. $d1$ and $d2$ are to calculate the similarity between any of the two documents. Similarity of cosine $(d1, d2)$ = Dot product $(d1, d2)$ /|||$d1$||| * |||$d2$||| where $d1, d2$ are two nonzero vectors.

Machine learning comparison algorithm for the selection for response

Jaccard calculates the similarity of the two statements based on the index. The Jaccard index is made up of a numerator and a denominator [4]. In numerators, we count the number of objects that will be shared between sets. In the denominator, we count the total of elements in both sets. For example, we define the sentences to be equal to 50% or its tokens. Here are two sample sentences:

How to add store in Nikshay Aushadhi?
How to delete store in Nikshay Aushadhi?
{Adding, store, Nikshay, Aushadhi}.
{Deleting, store, Nikshay, Aushadhi}.

In the example above, our intersection {store, Nikshay, Aushadhi}, which is accounted for by 3. The union of the sets is {adding, deleting, store, Nikshay, Aushadhi} which is calculated 5. Therefore, our Jacquard similarity index is three divided by 5, or 50%. Given our equality threshold above, we will consider it a match.

Similarly, the Levenshtein algorithm compares two statements based on the Levenshtein distance of the text of each statement. For example, "How to adding store in Nikshay Aushadhi?" There are 65% similarities in the statements. Calculate the similarity of the two statements based on the proximity of the calculated value to each statement [4].

Model Prediction Result

The Nikbot application uses Q&A protocol. Nikbot will show the answers to the question if they are available in the system or show the same query responses. To check how well Nikbot checked for errors, we saved the text files of Nikbot's conversation of users with different answers. Due to the natural language processing approach in Nikbot, it detects errors such as misspellings and abbreviations, which is an important part of it when we want the right answer (Figs. 3 and 4).

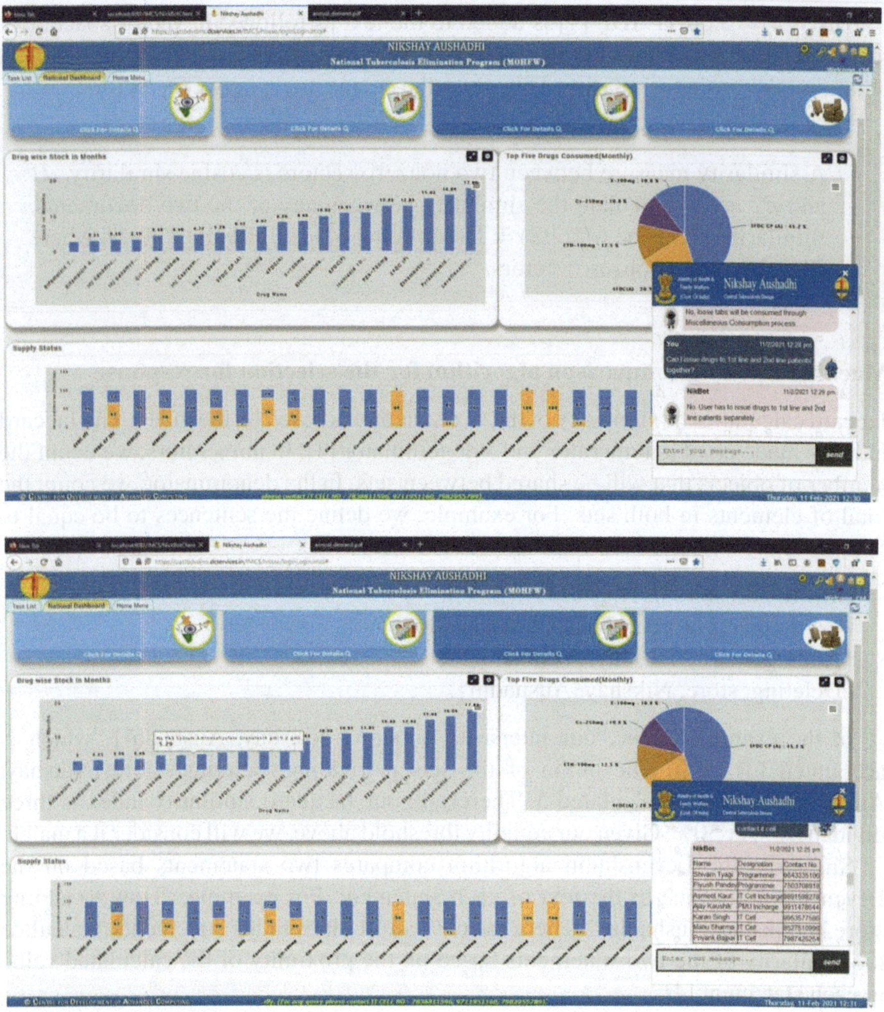

Fig. 3 Nikbot user interface

2.6 *IT Cell User Interface*

IT cell members work on the data management task. IT cell members will update the data as per user's queries. IT cell member will have the privilege to decide whether to display interface of Nikbot or display hidden interface of Nikbot. Only IT cell members have the privileged to authorize users (SDS, DDS, PHI, and GMSD, TU level) to interact with Nikbot. IT cell members handles the unanswered question and response presented to the bot that is not present in the database and collects this information for further analyzing user's sentiments and preparing those answers if

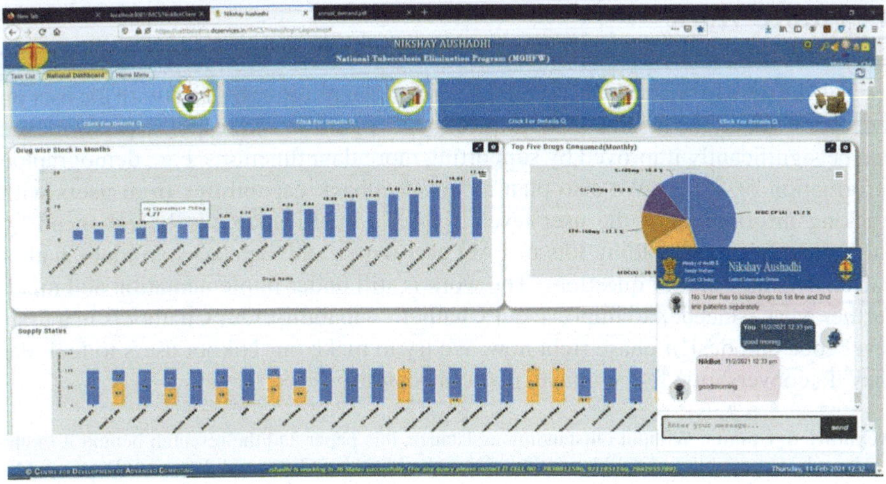

Fig. 4 Nikbot greeting interface

those questions are relevant and at the Nikbot back-end the Job Scheduling operation perform task like training Nikbot with the updated FAQ queries at particular interval of time.

3 Conclusion

The Nikbot application has been enhanced with security and efficiency updates, ensuring user protection and retrieving answers to questions. Due to the small dataset, accuracy of the Nikbot model will not as much as expected. We are working on the model and training part to enhance the model accuracy. Currently, the work is on collecting more data from the Nikbot users via the Nikbot and user conversations, and this information is stored in the form of text files. This will be trained multiple times to enhanced accuracy. The wide range of algorithms for the implementation of this Chatbot mostly is related to nlp and ml. The selection of an appropriate technique depends on the functions given by Chatbot, and also the scope under which the functionality will be offered. In choosing the algorithm, the structure of the data plays an important role. We use NLP and chatterbots because we want the computer to communicate with users on their terms [4].

4 Future Work

It would be used in the future as facial and voice recognition to interact with the user on a deeper level. In the future, the bots question recognition and response performance can be significantly improved by supporting more data functions, e.g., demographic information of a user. We also plan to link feedback capabilities from users with tracking information at the user level. Nikbot's intelligence can be enhanced by further studying and adding lots of FAQ user data to the database, so that Chatbot can answer all types of questions. The work is still under implementation and much further work is needed to improve our Chatbot's capability. Our Chatbot is in a first-level code and design phase right now. We try to make our Nikbot users to feel like they do conversation to the human not to machine.

Acknowledgements Without outstanding assistance, this paper and the research behind it might not have been possible and the idea of Joint Director, Mr. Ajay Gupta and Associate Director, Mr. Jitendra Singh. I am also grateful for the informative insights provided by Ms. Himani Goel and Dr. Sumit Soman. This research has been enriched in countless ways by the kindness and knowledge of one and all. I am extremely grateful to CDAC, Noida, for giving me this opportunity to work on this project.

References

1. Athota, L., Pandey, N., Rana, A., Shukla, V.K.: Chatbot for healthcare system using artificial intelligence. In: 2020 8th International Conference on Reliability, Infocom Technologies and Optimization (Trends and Future Directions) (ICRITO) Amity University, Noida, India, June 4–5, 2020
2. Dharwadkar, R., Deshpande, N.A.: A medical ChatBot. Int. J. Comput. Trends Technol. (IJCTT) **60**(1) (2018)
3. Shangrapawar, A., Ravekar, A., Kale, S., Kumari, N., Shende, A., Taklikar, P.: Artificial intelligence based healthcare chatbot system. Int. Res. J. Eng. Technol. (IRJET) **07**(02) (2020). e-ISSN: 2395–0056
4. Chatterbot link: https://chatterbot.readthedocs.io/en/stable/tutorial.html.

Predicting Hospital Bed Occupancy: A Pilot Evaluation for Tertiary Hospitals in India

Amit Kumar Ateria, Priyesh Ranjan, Sumit Soman, and Amarjeet Singh Cheema

Abstract Managing hospital resources has been one of the key challenges in the domain of health informatics and Hospital Management Information Systems (HMISs). With increasing patient loads and limited health infrastructure, optimizing hospital resources for efficient service delivery is of paramount importance. One of the problems of interest is the bed occupancy rate at hospitals, which in turn determines the availability of beds for new patients. With recent advances in data analytics and machine learning, it has become possible to develop models to determine bed occupancy and assess the factors that contribute to determining the same. This would be beneficial for hospitals in estimating and planning future resources as well. This paper presents a pilot study on evaluating models for predicting bed occupancy as well as ranking possible contributing factors. We present results on data from public tertiary hospitals to demonstrate the utility and validity of employing learning models for determining bed occupancy rates.

Keywords Health informatics · Bed occupancy · Bed availability · Hospital resource management · Regression · Feature ranking

---

A. K. Ateria (✉) · P. Ranjan · S. Soman · A. S. Cheema
Centre for Development of Advanced Computing, Noida, Uttar Pradesh, India
e-mail: amitkumarateria@cdac.in

P. Ranjan
e-mail: priyeshr@cdac.in

S. Soman
e-mail: sumit.soman@gmail.com

A. S. Cheema
e-mail: ascheema@cdac.in
URL: https://cdac.in/

A. Noor et al. (eds.), *Proceedings of Emerging Trends and Technologies on Intelligent Systems*, Advances in Intelligent Systems and Computing 1371,
https://doi.org/10.1007/978-981-16-3097-2_13

1 Introduction

Hospital Management Information System (HMISs) are software systems comprising of clinical and non-clinical modules catering to various workflows in a hospital. The admission, discharge and transfer (ADT) module of HMIS manages bed allocation for indoor patients who are admitted to the hospital for various treatment plans or procedures performed by clinicians. This includes bed allocation during admission, transfer (between wards) and discharge processes, during which patient beds are assigned, freed or transferred. This module closely operates in conjunction with the billing and hospital reporting modules where the data for bed allocation is used for patient billing and hospital resource planning.

Our HMIS [16] has been deployed and is operational at various public sector hospitals, including tertiary hospitals. The HMIS is a comprehensive solution that caters to multiple hospital workflows [13, 14] and is also compliant with electronic health record (EHR) standards [6, 15, 17–19]. Bed allocation and management is a challenging problem that is presently largely managed manually based on a set of business rules catering to patient categories and bed availability in the respective hospital department or unit. This paper seeks to evaluate the feasibility of using machine learning models to determining bed occupancy rates and also rank potential factors that contribute to it. Though the approach of using approaches to predict bed occupancy is not novel, the key contribution of our work lies in comprehensive evaluation of basic approaches on real data obtained from large tertiary care hospitals in India.

2 Related Work

Various works have appeared in the literature presenting approaches for determining hospital bed occupancy rates. Forecasting bed requirements for Burns Ward ICU and step-down areas using basic statistical analysis has been presented in [7]. Optimization-based approaches that take into account various parameters such as arrival process, length of stay, bed blocking and others have been presented in [3]. These works formulate the allocation problem as a constrained optimization problem and use numerical routines to arrive at a solution. Though these approaches have been widely used, these face issues with respect to scalability and solution quality, such as due to the choice of optimization routine used to solve these problems. With the recent advances in machine learning and artificial intelligence, it has been possible to leverage these approaches to predict bed occupancy and other parameters that can be used by hospitals for resource planning and allocation by hospital administrators.

We discuss a few prominent works that have employed machine learning approaches to predict bed occupancy or related parameters. Bayesian models have been adopted in early approaches such as the use of Markov chains [11, 12], queuing theory [4, 5] among others. Alternatives to queuing theory approaches have been presented in [2], while stochastic simulations have been used in [1, 20]. Length of stay (LoS) has been predicted using regression trees and high-dimensional kernel space models in [21]. Review papers discuss various approaches adopted such as [9, 10]. The key benefit in employing machine learning-based approaches is the fact that these are built from past hospital data and can evolve over time, for example, when online learning approaches can be adopted. The use of these approaches thus allows for learning data distributions from past records, making such prediction systems robust and useful.

Most of the work that has appeared in literature has largely been focused on specific hospitals departments where data has been available for analysis. Further, there are few works in the context of development and evaluation of bed allocation models for the Indian healthcare sector. This work focuses on building and evaluation of models that use past data from transactional databases of HMIS operational at tertiary hospitals in India. The novelty lies in using past data to predict bed occupancy and rank factors influencing them, unlike approaches that assume prior distributions for developing forecasting models. As this approach for modeling is in its nascent stages in the Indian public healthcare sector, this work presents a pilot initiative in this direction.

3 Approach and Models Used

This section discusses the existing approach being used at the hospitals for computing bed occupancy. Further, we discuss the dataset and features available for use in computing bed occupancy as well as the approaches followed for feature ranking and building a regressor for the prediction task.

3.1 Baseline Computation of Bed Occupancy

The present model for computing bed occupancy is based on a simple formula that uses the total number of patients admitted in a ward and the number of beds available in the ward. Specifically, for a time period t comprising of observations at intervals $t_1, t_2, \ldots, t_N$, the bed occupancy B_{occ} is computed as

$$B_{\text{occ}} = \frac{\sum_{i=1}^{N} n_p^{t_i}}{B \times N} \times 100, \tag{1}$$

where $n_p^{t_i}$ denotes the number of patients p admitted in the ward at time instant t_i, and B denotes the number of beds in the ward. It can be seen that this is a simple numerical estimate based on past data. It is important for hospitals to understand factors determining bed occupancy and predict the same for the future so that hospital resources can be planned adequately. We now evaluate approaches to rank factors that contribute to determining bed occupancy and develop a regression model for these tasks, respectively.

3.2 *Features for Modeling Bed Occupancy*

We use the following factors that are available in our transactional database associated with the HMIS:

1. Length of stay—Length of stay (LoS) is computed as the time difference between patient admission and discharge. This is added up across the patients for the specific time interval.
2. Discharge is computed as the number of patients discharged during the time period from the specific ward/hospital.
3. Average length of stay is computed as the average of the number of days between patient admission and discharge in the respective hospital/ward for the specified time duration. Numerically, it is determined as LoS/discharge.
4. Number of admissions in the given time period (daily census).
5. Number of beds available in the specific ward(s) of the hospital.
6. Number of days in the specific month.

Using these features, we attempt to model bed occupancy rate as a regression problem. The following sections present results on feature ranking and regression modeling on the dataset.

3.3 *Modeling for Estimating Bed Occupancy*

We discuss the approaches followed for modeling bed occupancy in this section. The problem at hand can be modeled as a supervised learning problem in the regression setting, where a set of numeric features are used to predict a real-valued entity (bed occupancy) which is chosen as the regression label. The approaches used include mutual information-based feature ranking [22] and linear regression [8].

Regression models are used to estimate real-valued outputs $Y = \{y_i \in \mathbb{R} | i = 1, 2, \ldots M\}$ from M training samples $X = \{x_i \in \mathbb{R}^{\kappa} | i = 1, 2, \ldots, M\}$ with n features using the model parameterized by $\{\beta_0, \beta_1, \beta_2, \ldots, \beta_n\}$, where β_0 is the intercept. The model parameters and training data are related to the label as given in Eqn. 2.

$$y_i = \beta_0 + \beta_1 x_i^{(1)} + \beta_2 x_i^{(2)} + \cdots + \beta_n x_i^{(n)} \quad (2)$$

In this paper, we have used the linear regression models[1] provided in Python's $scikit-learn$ package.

Feature Ranking is a common problem in machine learning which aims at determining ranks for the features that relates to their relative importance in predicting the label of interest. Given a set of M training samples $X = \{x_i \in \mathbb{R}^{\mathbb{K}} | i = 1, 2, \ldots, M\}$ with n features and corresponding regression labels $Y = \{y_i \in \mathbb{R} | i = 1, 2, \ldots M\}$, the feature ranking algorithm outputs a set of weights $w = \{w_1, w_2, \ldots, w_n\}$. A higher value of w_i indicates the higher importance of feature i in X for predicting the label Y. In this paper, we have used mutual information based feature selection[2] available in Python's $scikit-learn$.

4 Results

We now present details of our dataset, along with results obtained using the feature ranking and regressor approaches.

4.1 Dataset Description

We collected datasets from two tertiary care hospitals in India (referred to as "*Hospital A*" and "*Hospital B*" in the following sections of this paper) with multiple features and the computed bed occupancy rates. The dataset comprises of the features mentioned with a monthly sampling frequency. A plot of the dataset for the two hospitals are shown in Fig. 1.

The HMIS has began operationalization at these tertiary hospitals recently, as such limited data is available at present to compute bed occupancy. It can be seen from Fig. 1a that data is available from July 2019 to December 2020 for Hospital A, while it is available from July 2018 till December 2020 for Hospital B (as shown in Fig. 1b). The trends presented in the data do not directly reveal any discernible pattern, which motivates the use of machine learning methods to model the effect of these parameters on the target variable of interest.

The intial feature selection and model evaluation presented in this paper have been done on the available data on a pilot basis. However, as more data continues to be acquired over the operational period of the HMIS at these hospitals, the models would be trained incrementally to improve the bed occupancy prediction.

---

[1] https://scikit-learn.org/stable/modules/generated/sklearn.linear_model.LinearRegression.html.

[2] https://scikit-learn.org/stable/modules/generated/sklearn.feature_selection.mutual_info_classif.html.

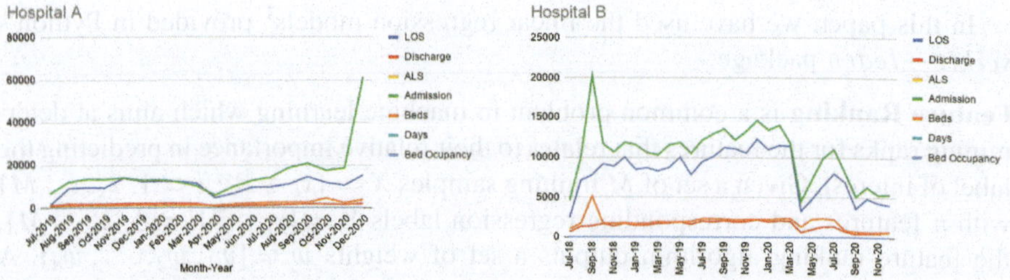

(a) Hospital A (b) Hospital B

Fig. 1 Plot of features and bed occupancy data

4.2 *Feature Selection for Bed Occupancy*

We rank the features using mutual information feature selection for the bed occupancy dataset. Results are shown as bar charts in Fig. 2 for the hospitals.

The ranks assigned to the features have been indicated in Table 1. It can be seen that for *"Hospital A"* the length of stay, discharges and number of beds have higher weights compared to the others, while for *"Hospital B"*, all factors except the sampling time duration are relatively important in determining the bed occupancy rate.

The results using feature selection lead us to conclude that the bed occupancy rate for the individual hospitals depends on different factors based on the data available. As such, predicting bed occupancy should be done by building hospital-specific models which are trained on data from the specific hospital, as opposed to a generic model which would not capture the specific trends affecting bed occupancy of that hospital.

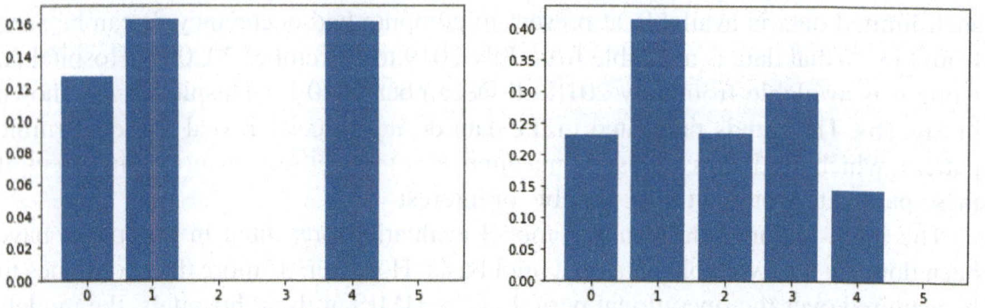

(a) Hospital A (b) Hospital B

Fig. 2 Ranks assigned to features for predicting bed occupancy using mutual information feature selection

Table 1 Feature ranking on the attributes used for bed occupancy prediction

Feature	Description	Weights	
		Hospital A	Hospital B
1	Length of stay	**0.127**	**0.234**
2	Discharge	**0.163**	**0.382**
3	Avg LoS	0.000	**0.275**
4	Admissions	0.000	**0.305**
5	No. of beds	**0.161**	**0.199**
6	Days in month	0.000	0.000

The significance of bold data is to focus on the major contributing factors while others factors remains 0.00

4.3 Quantitative Results on Bed Occupancy Prediction

A linear regression model was trained on the data provided. 80% of the data samples available were taken for training and the remaining was chosen as the test set (20%). The model was trained to predict bed occupancy using the features provided in the dataset. The error was calculated as the mean average error (MAE) and mean squared error (MSE) on the test set.

Let the predictions obtained on the k test samples are denoted by $\hat{Y} = \{\hat{y}_i | i = 1, 2, \ldots, k\}$ and their true labels (bed occupancy rates) are $Y = \{y_i | i = 1, 2, \ldots k\}$, then MAE is calculated as

$$\text{MAE} = \frac{1}{k}\sum_{i=1}^{k} |\hat{y}_i - y_i| \tag{3}$$

MSE is calculated as

$$\text{MSE} = \frac{1}{k}\sum_{i=1}^{k} (\hat{y}_i - y_i)^2 \tag{4}$$

MAE is significant to determine the average error rate across the test samples, while MSE penalizes larger errors highly as it squares the error difference.

Table 2 shows the MAE and MSE obtained using linear regression models on the test set. It can be seen that lower error rates are obtained for "*Hospital B*" when

Table 2 Regression error rates for hospitals

S. No.	Hospital	Regression error metric	
		MAE	MSE
1	Hospital A	**1.734**	5.155
2	Hospital B	**1.561**	3.728

The significance of bold is that MAE is within the tractable limits compared to "*Hospital A*." This can be attributed to the fact that more training data samples are available for the latter compared to the former. The error rates are within tractable limits, indicating that these models can be suitably used to predict bed occupancy rates.

5 Conclusion and Future Work

This paper presented a pilot effort for determining bed occupancy rates and ranking influencing factors using datasets from tertiary care hospitals in India where the HMIS is operational. This is an initial effort using basic methods, as such future work entails:

1. Predicting bed allocation based on type and/or severity of patient diagnosis.
2. Modeling patient arrival rates and determining bed allocation.
3. Implementing bed availability for patient triaging in emergency departments and subsequent transfer to specific wards.

References

1. Baru, R.A., Cudney, E.A., Guardiola, I.G., Warner, D.L., Phillips, R.E.: Systematic review of operations research and simulation methods for bed management. In: IIE Annual Conference. Proceedings, p. 298. Institute of Industrial and Systems Engineers (IISE) (2015)
2. Beeknoo, N., Jones, R.: A simple method to forecast future bed requirements: a pragmatic alternative to queuing theory. Br. J. Med. Med. Res. **18**(4), 1–20 (2016)
3. Bekker, R., Koole, G., Roubos, D.: Flexible bed allocations for hospital wards. Health Care Manage. Sci. **20**(4), 453–466 (2017)
4. Belciug, S., Gorunescu, F.: Improving hospital bed occupancy and resource utilization through queuing modeling and evolutionary computation. J. Biomed. Inform. **53**, 261–269 (2015)
5. Boulton, J., Akhtar, N., Shuaib, A., Bourke, P.: Waiting for a stroke bed: planning stroke unit capacity using queuing theory. Int. J. Healthcare Manag. **9**(1), 4–10 (2016)
6. Cheema, A.S., Srivastava, S., Srivastava, P., Murthy, B.: A standard compliant blood bank management system with enforcing mechanism. In: International Conference on Computing, Communication and Security (ICCCS), pp. 1–7. IEEE (2015)
7. Dias-Amborcar, Y., Parulekar, A.: Forecasting bed requirement using hospital records. J. Health Manage. **15**(1), 75–81 (2013)
8. Freedman, D.A.: Statistical Models: Theory and Practice. Cambridge University Press, Cambridge (2009)
9. Hall, R.: Bed assignment and bed management. In: Handbook of Healthcare System Scheduling, pp. 177–200. Springer, Berlin (2012)
10. He, L., Madathil, S.C., Oberoi, A., Servis, G., Khasawneh, M.T.: A systematic review of research design and modeling techniques in inpatient bed management. Computers Ind. Eng. **127**, 451–466 (2019)
11. Kilinc, D., Saghafian, S., Traub, S.: Dynamic assignment of patients to primary and secondary inpatient units: is patience a virtue? Harvard Kennedy School (2019)
12. Kolesar, P.: A Markovian model for hospital admission scheduling. Manage. Sci. **16**(6), B–384 (1970)

13. Kumar, A., Soman, S., Ranjan, P.: Implementing delta checks for laboratory investigations module of hospital management systems. In: Evolving Technologies for Computing, Communication and Smart World, pp. 451–462. Springer, Berlin (2019)
14. Ranjan, P., Soman, S., Ateria, A.K., Srivastava, P.K.: Streamlining payment workflows using a patient wallet for hospital information systems. In: 2018 IEEE 31st International Symposium on Computer-Based Medical Systems (CBMS), pp. 339–344. IEEE (2018)
15. Soman, S., Ranjan, P., Cheema, A.S., Srivastava, P.K.: Integrating drug terminologies with hospital management information systems. In: 2019 10th International Conference on Computing, Communication and Networking Technologies (ICCCNT), pp. 1–6. IEEE (2019)
16. Soman, S., Ranjan, P., Srivastava, P.: A distributed architecture for hospital management systems with synchronized ehr. CSI Trans. ICT **8**(3), 355–365 (2020)
17. Srivastava, P.K., Soman, S., Sharma, P.: Perspectives on SNOMED CT implementation in indian HMIS. In: Proceedings of the SNOMED CT Expo 2017. SNOMED International (2017)
18. Srivastava, S., Gupta, R., Rai, A., Cheema, A.: Electronic health records and cloud based generic medical equipment interface. arXiv preprint arXiv:1411.1387 (2014)
19. Srivastava, S., Soman, S., Rai, A., Cheema, A., Srivastava, P.K.: Continuity of care document for hospital management systems: an implementation perspective. In: Proceedings of the 10th International Conference on Theory and Practice of Electronic Governance, pp. 339–345. ACM (2017)
20. Torabipour, A., Zeraati, H., Mohammad, A., Rashidian, A., Sari, A.A., Sarzaiem, M.R.: Bed capacity planning using stochastic simulation approach in cardiac-surgery department of teaching hospitals, Tehran, Iran. Iran. J. Public Health **45**(9), 1208 (2016)
21. Turgeman, L., May, J.H., Sciulli, R.: Insights from a machine learning model for predicting the hospital length of stay (los) at the time of admission. Expert Syst. Appl. **78**, 376–385 (2017)
22. Vergara, J.R., Estévez, P.A.: A review of feature selection methods based on mutual information. Neural Comput. Appl. **24**(1), 175–186 (2014)

An Ensemble Approach for Modeling Process Behavior and Anomaly Detection

Ajay S. Chouhan, C. S. Sajeesh, Vineet Sharma, Gopika Vinod, Ajay Kumar, Vinod K. Boppanna, and Gigi Joseph

Abstract In recent years, malware has become more sophisticated. Security solutions based on signature-based detection and known behavior-based detection are not capable of detecting unknown malware. Anomaly detection is an effective strategy against unknown malware. In this paper we present a method to model the normal behavior of processes that run on a computer and detect whether a new process instance conforms with this normal behavior. In this paper we present unsupervised tree-based anomaly detector (UTAD) which is used in combination with autoencoder to detect anomalous process behavior. We evaluated the performance of model in separating behavior of genuine processes from one another. The ensemble model achieved good accuracy with low false positive rate.

Keywords Malware detection · Process behavior modeling · Application behavior fingerprinting · Anomaly detection · Autoencoder

1 Introduction

Nowadays, computers are involved in every aspect of our lives, hence they are attractive targets for attacks. These attacks are increasingly becoming more complex and diverse. Security solutions which are based on signature-based detection and sandboxing cannot be completely relied upon because malware now use code obfuscation, sandbox detection and other bypass techniques. Behavioral anomaly detection has emerged as a key component of attack detection approaches. Once an attacker has penetrated the system through phishing or other means, there is very little labeled data, and the attacker has a huge variety of options, reducing the applicability of supervised machine learning approaches and suggesting use of unsupervised methods that do not require labeled data, such as anomaly detection. The anomaly detection model is based on known normal behavior and has the ability to detect deviations from this known normal behavior. However, a disadvantage of anomaly detection is that it

A. S. Chouhan (✉) · C. S. Sajeesh · V. Sharma · G. Vinod · A. Kumar · V. K. Boppanna · G. Joseph
Bhabha Atomic Research Centre, Trombay, Mumbai, India
e-mail: aschouhan@barc.gov.in

A. Noor et al. (eds.), *Proceedings of Emerging Trends and Technologies on Intelligent Systems*, Advances in Intelligent Systems and Computing 1371,
https://doi.org/10.1007/978-981-16-3097-2_14

generates high number of false positives. Among all the desktop operating systems (OS) Microsoft Windows dominates with nearly 3/4th share [1]. In this paper, we present an approach that uses a combination of autoencoder and unsupervised tree-based anomaly detector to model the behavior of a process, running on Microsoft Windows Operating System, and detect whether a new process instance conforms with this normal behavior.

2 Background

2.1 *Anomaly Detection*

A lot of research work has been done in the field of malware detection. The techniques to detect malware can be broadly classified into static and dynamic analysis. Static analysis techniques examine any given malware sample by determining the signature of the malware binary, without actually executing the code. Li et al. [2] made use of code analysis to detect malwares. The dynamic analysis techniques monitor the behavior of processes during its runtime and are mainly classified as misuse detection and anomaly detection. Misuse detection techniques find common patterns in behavior of known malware samples and match behavior of a new sample with them. Gupta et al. [3] proposed an Immediate Syscall signature structure-based technique to detect malicious program executions in Cloud. Zhang et al. [4] proposed a low-cost feature extraction approach and an effective deep neural network architecture for malware detection. Misuse detection has the advantage that it generates low false alarms but it is unable to detect unknown malware. To overcome this problem, anomaly detection techniques [5] model the normal behavior and then look for behavior that deviate from this model. Tobiyama et al. [6] proposed a method to model the behavior of process by applying a deep learning neural network to extract the features needed from the process and recursive neural network to classify the selected process as a normal or malignant process. In [7] process events were converted to a discrete-time signal in the polar space and then local patterns and the normal co-occurrence relationships between these patterns were learnt. The process behavior included network operations, file operations, inter-process communication, and process events. Tajoddin et al. [8] proposed a way to model the registry behavior of benign programs and then detect malware using this model.

2.2 *Autoencoder*

Autoencoder is a neural network-based method that trains the input vectors to reconstruct as output vectors with an unsupervised approach. Autoencoder learns a function to reconstruct its input through a combination of an encoder and a decoder

$$X^* = D(E(X)) \quad (1)$$

where X is the input data, E is encoder (encoding function) from the input data to the hidden layer, D is decoder (decoding function) from the hidden layer to the output layer, X^* is reconstructed version of the input data.

The encoder and decoder are trained to minimize the difference between X and X^*. Autoencoder generates a reconstruction error which is the loss incurred in recreating the input provided to it. This reconstruction error is less if the input is similar to the data which is used for training the autoencoder and more if the input is different. This reconstruction error can be utilized to detect anomalies by setting a threshold, and if the reconstruction error for an input goes beyond this threshold it can be considered as an anomaly. Aygun et al. [9] proposed a deep learning-based anomaly detection model using autoencoder.

3 Dataset

The data were collected from four systems that use Microsoft Windows Operating System for 2 months. The data contain the information about instances of processes running on these systems. The features used and the preprocessing performed is as follows.

3.1 Features

Feature set one: This feature set is used to train autoencoder. The features used are shown in Table 1.

Feature set two: This includes Registry [10] key and value create and delete operations, registry value modifications, registry key and value rename operations made by a process.

Feature set three: This includes Dynamic Link Library [11] (DLL) accesses made by a process instance.

Feature set four: This includes files created and overwritten by a process instance.

Data Encoding and Normalization: The text values of features, in feature set one, were encoded into numerical form. Different features in feature set one have values which are in a different scale. For example, the CPU usage is in range 0–1 but memory usage is in range of thousands.

Hence, all feature values are normalized before feeding into the model by using following formula:

Table 1 Feature set one

Feature number	Feature name	Remarks
1–3	Path first, path second, path third	If process path is changed, then there is a chance that process may not be genuine. First three levels of the path were used as features. For example, for path "C:\Program Files-\A\B\abc.exe", we considered "C:", "Program Files", "A" as three features
4,5	OS name, file size	These values are needed so that others parameters can be linked with this value. Different versions of a process on different operating systems may behave slightly differently
6	Parent name	Processes are usually created from a given set of parent processes if they deviate from this, then it may be considered anomalous
7–9	Minimum, maximum and average working set size	The usual working set size, CPU usage and thread count of a process, their correlation were modeled so that process deviating from them can be detected
10–12	Minimum, maximum and average CPU usage	
13–15	Minimum, maximum and average thread count	
16	Admin status	This was taken to detect a process running with administrative privileges
17–20	Duration of process, minimum, maximum and average duration of network connection	Some processes run for a very short time and some processes run till the system is on. This is true for duration of network connections also. These features are important for modeling behavior of a process
21–22	Number of TCP connections, number of UDP connections	This is important for behavior modeling and anomaly detection as some processes make only TCP connections, some processes make only UDP connections, some make both and some make no connections
23–24	Number of unique remote IP, number of unique remote ports	These features are considered to differentiate between processes which make network connections. For example, Web browsers may make connections to multiple remote IP and ports but some processes make connection to a fixed IP or fixed port only

(continued)

Table 1 (continued)

Feature number	Feature name	Remarks
25–32	Number of connections in different connection state	These are useful to detect any change in state of a process's network connections. In a given network environment a process' connection state is expected to be similar across multiple instances

$$\text{Normalized value of } X = (X - \text{min})/(\text{max} - \text{min}) \quad (2)$$

where X represents the particular feature value to be normalized, min is the minimum value of that feature and max is the maximum value of that feature.

4 Proposed Method

4.1 Overview

We propose an ensemble model for application behavior fingerprinting and anomaly detection. This ensemble model is created by combining autoencoder and unsupervised tree-based anomaly detector (UTAD) which is described in Sect. 4.2. This ensemble model is trained for each process using already collected instances of that process, and the final trained model is stored. Such a trained model is generated and stored for each process. Hash of the executable of process is used as the unique identity for these models. These models are then used to check the genuineness of a new process instance. A new process instance is checked by passing it through the corresponding model (same hash). The flowchart for generating the model and then using it for prediction is shown in Figs. 1 and 2, respectively.

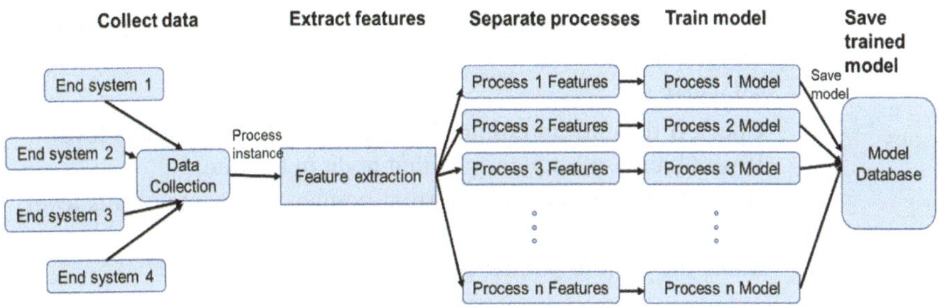

Fig. 1 Store model for each process

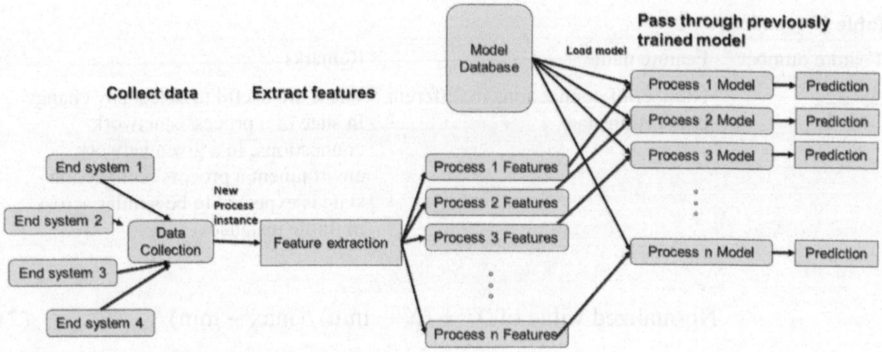

Fig. 2 Prediction for new process instance

4.2 Unsupervised Tree-Based Anomaly Detector (UTAD)

We propose this method to find anomalies in Registry, DLL, File accesses of a process. This method works by constructing a tree for each feature (Registry, DLL, File accesses) of a process and then for a new instance, its feature values will be parsed through the corresponding tree to generate a score which can be used to predict that instance as genuine or anomalous. The construction and parsing of tree for Registry accesses is described in this section. The tree for DLL and File accesses can be created in a similar way.

Construction of Tree: Given a dataset of Registry accesses made by a process the steps to construct the tree are as follows:

1. Create root of the tree with the name of process. For example, "abc.exe".
2. Assign the frequency of root to 0.
3. For each registry access,
    (a) Increase the frequency of root by 1.
    (b) Make the current pointer point to root of the tree.
    (c) Divide the registry into levels. For example, divide "HKLM/SOFTWARE/MICROSOFT" into ("HKLM", "SOFTWARE", "MICROSOFT").
    For each level.
        I. Check if the current pointer has a child with name of this level.
        If yes: Increase frequency of that node in the tree.
        If no: Add a child node to current pointer with name of this level. Assign frequency 1 to it.
        II. Make the current pointer point to this node.

For example, if we want to add "HKLM/SOFTWARE/MICROSOFT" and we have already added "HKLM" and "SOFTWARE", now to add "MICROSOFT" we will parse the tree as root- > " HKLM"- > " SOFTWARE" and now check whether

"SOFTWARE" has a child with the name "MICROSOFT". If yes, we will increase the frequency of "MICROSOFT" in tree. Otherwise add "MICROSOFT" as child of "SOFTWARE" and assign frequency 1 to it.

4. Tree Pruning: This step removes the nodes whose names are temporary, random or user specific. The steps for tree pruning are as follows:
   For each level (depth) "L" of tree greater than 2, repeat the following steps:

   I. Find unique paths from root to each node at level L without considering nodes at the level L-1 (previous level).
   II. Find unique paths from root to each node at level L.
   III. For each path obtained in Step I. find paths obtained in Step II. with matching signature (same nodes at level 0 to L-2 and at level L). If more than one such path exists then for such paths replace node at L-1 level with a common name (For example, Cname) and save this signature for future reference.
   For merging such nodes follow the following steps:
   (a) Add a child "Cname" to the node at level L-2 along the selected path.
   (b) Add the frequency of all the nodes which are to be merged and assign the sum total to Cname.
   (c) Add all the subtrees which have merged nodes as their root to Cname. If there are common nodes within the subtrees then add their frequency and remove the redundant nodes.

5. Increase the frequency of root by 1.
6. Remove the nodes along with their child nodes which have the confidence less than a set threshold.

An example of a subtree of tree constructed using Registry access of a process is shown in Fig. 3.

Parsing the Tree to generate Anomaly Scores: The steps to generate anomaly scores are as follows.

1. Generate the maximum and minimum score.
   For this, we pass the same registry accesses through which the tree was created. The steps are as follows:

   I. For each registry access,
   (a) Make the current pointer point to root node.
   (b) Assign $score = 0$
   (c) Divide the registry into levels.
   Check if this registry matches with the signatures of merged nodes that were saved while tree pruning, if yes replace the corresponding level with common name used while tree pruning.
   Now, for each level:

   - Find the child node of current pointer with the name of this level.

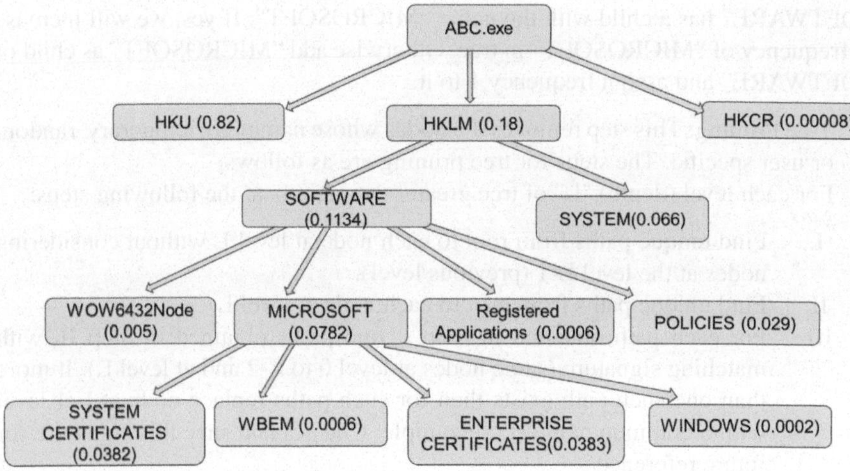

Fig. 3 A subtree of Registry access tree

- Add the confidence of this node to the score.
- Make the current pointer point to this node.

The final score for the given registry access can be written as:

$$\text{score} = \sum_{k=0}^{l} C(P \rightarrow N_k) \tag{3}$$

where
l = number of levels in that registry access.
$C(P \rightarrow N_k)$ = Confidence of Process P to the registry access corresponding to kth node.

II. Find the maximum and minimum of the scores obtained by all the registry accesses and assign those to $S_{\max}$ and $S_{\min}$, respectively.

2. To check if a process instance is genuine or not, the following procedure is used.

A. Assign total score $S_{\text{total}} = 0$
B. For each registry access,

   I. Assign $score = 0$
   II. Make the current pointer point to root node.
   III. Divide the registry into levels.
   For each of this level.

   - Find the child node of current pointer with the name of this level.
   - Add the confidence of this node to the score.
   - Make the current pointer point to this node.

- If this is the last level and current pointer does not have any child then assign value of 1 to the score.
  The final score for ith registry access can be expressed as:

$$\text{Score}_i = \begin{cases} \sum_{k=0}^{l_i} C(P \rightarrow N_k), & \text{If the registry access does not parse a branch of tree completely.} \\ 1, & \text{If the registry access parse a branch of tree completely.} \end{cases} \tag{4}$$

C. Select the lowest "n_p" scores. Then normalize these scores using $S_{\min} \wedge S_{\max}$, then find their average to get the genuineness score S_{genuine}. The final equation is as follows:

$$S_{\text{anomaly}} = \frac{\sum_i \sum_{k=0}^{l_i} (\text{Score}_k - S_{\min})}{n_p \times (S_{\max} - S_{\min})} \tag{5}$$

where
Score_k = Confidence of kth level in the ith registry access.
p = percentile for considering a subset of scores among all the scores (for this work we have chosen $p = 90$).
n_p = number of scores in the pth percentile of all the scores.
i = indexes of scores in the pth percentile of all the scores.
l_i = number of levels in registry access with ith index.

4.3 Prediction Using Ensemble Model

To check whether a new instance is genuine or anomalous, we extract the features for this new process instance and separate it into the four feature sets. Feature set one is passed through autoencoder to generate a reconstruction error, and if this reconstruction error is higher than threshold than the process instance is predicted as anomalous otherwise normal. Feature set two is passed through UTAD trained on feature set two to generate an anomaly score, and if this score is less than threshold, then the process instance is predicted as anomalous. Feature set three and four are passed through corresponding UTAD and prediction is done in a similar manner as for feature set two. Finally, we get four predictions from four models, if at least "n" out these four models predict the process instance as anomalous, then the final prediction is anomalous, otherwise genuine. The value of "n" can be set based on the criticality of the environment. UTAD trained on feature set two, three and four will be referred as UTAD(Reg), UTAD(DLL) and UTAD(File), respectively, in rest of the paper. The flow for prediction using ensemble model is shown in Fig. 4.

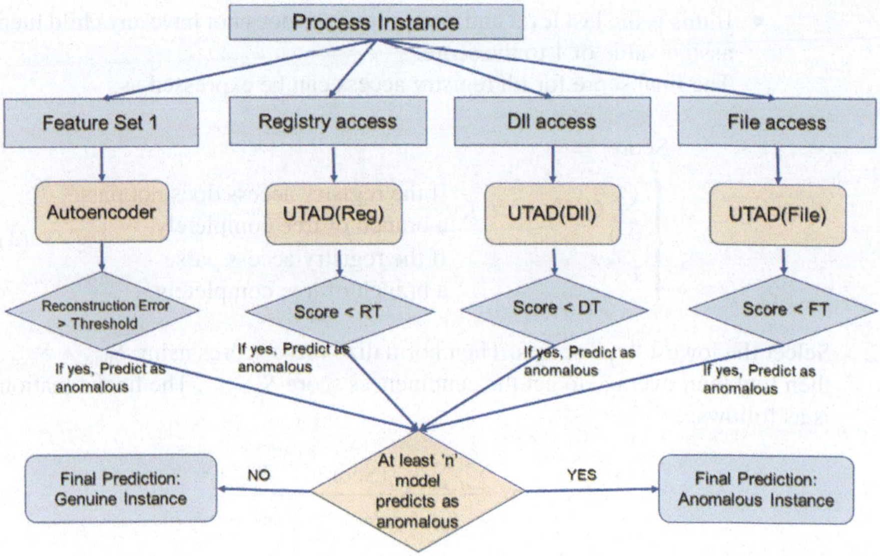

Fig. 4 Prediction using ensemble model

5 Experimental Results

We used around 7000 instances of 30 different processes for the experiment. These processes included system processes, antivirus processes, hypervisor processes, document editors and web browsers.

5.1 *Autoencoder Training and Prediction*

The autoencoder was trained on instances of a web browser process because they show a lot of variation in their usage, and it was tested for detecting instances of other processes as anomalous. A threshold was set at 90th percentile of reconstruction errors obtained on instances for which autoencoder was trained, and any instance for which reconstruction error was above this threshold was considered as anomalous. The autoencoder was trained over various combinations of hyperparameters and structures. The configuration at which autoencoder detected the highest number of anomalous instances was selected. The selected configuration has the structure 32:24:16:8:4:8:16:24:32, where 32 is the number of neurons in the first layer, 24 is number of neurons in the second layer and so on. Tanh was used as activation function in hidden layers and output layer. Learning rate was set at 0.0001. ADAM [12] optimizer was used for training. A dropout [13] layer with probability 0.1 was used after hidden layers.

We trained autoencoder for each of the process, with the above configuration. These 30 trained models stored along with the threshold were set at 90th percentile of reconstruction error obtained on the genuine instances of the concerned process.

5.2 *UTAD Training and Prediction*

UTAD(Reg), UTAD(DLL) and UTAD(File) were trained over feature set two, feature set three and feature set four, respectively, for each of the processes. These models were stored along with a threshold which was set at "mean-3*std" where mean is average and std is the standard deviation for the scores obtained by genuine instances of a process.

5.3 *Ensemble Model*

We checked the performance of ensemble model using one vs other approach where for each process we considered the instances of the given process as genuine and instances of other 29 processes as anomalous. To predict a new instance as anomalous or genuine four cases were considered:

- Case1:An instance is considered anomalous when at least one of the four models predict the instance as anomalous ($n = 1$).
- Case 2: An instance is considered anomalous when at least two of the four models predict the instance as anomalous ($n = 2$).
- Case 3: An instance is considered anomalous when at least three of the four models predict the instance as anomalous ($n = 3$).
- Case 4: An instance is considered anomalous when all the four models predict the instance as anomalous ($n = 4$).

An instance predicted as anomalous was considered as positive prediction, and an instance predicted as genuine was considered a negative prediction.

The performance of the ensemble model for above four cases is given in Table 2.

Table 2 Evaluation of ensemble model

	True positive rate	True negative rate	False positive rate	False negative rate	Accuracy	Precision	Recall	F-1 score
Case 1	100	72.42	27.58	0	99.08	99.05	100	99.06
Case 2	99.75	95.51	4.49	0.25	99.6	99.8	99.7	99.74
Case 3	95.39	100	0	4.61	95.5	100	95.3	97.59
Case 4	52.72	100	0	47.28	54.29	100	52.7	69.02

We obtained the highest accuracy in Case 2. For further evaluation of the model we checked its performance in detecting actual malware instances. We kept the trained models and the threshold same but instead of using instances of other processes as anomalous instances, we used actual malware instances. The accuracy obtained in detecting malware instances as anomalous for Case 2 ($n = 2$) and Case 3 ($n = 3$) is 97.6% and 94%, respectively. Overall accuracy for correct prediction for Case 2 and Case 3 was 96.5% and 94%, respectively.

6 Conclusion

This research work presents a way to model the normal behavior of a process and detect whether a process instance conforms with this normal behavior. For this purpose, we proposed unsupervised tree-based anomaly detector (UTAD) which is used in combination with autoencoder to create an ensemble model. Autoencoder and UTAD were trained on different features of the same process. The ensemble model was trained and tested on 30 different processes, and it was able to separate behavior of genuine processes from one another with 99.6% accuracy. It was able to separate genuine instance of a process from malware instance with 96.5% accuracy. The ensemble model was able to achieve high accuracy with low false positive rate.

7 Further Research

In this work, the threshold of Autoencoder was set at a fixed percentile for all processes. A method can be devised to dynamically set different thresholds for different processes. UTAD has a limitation of giving a good score to an anomalous process if its accesses are a subset of accesses made by the genuine process. Further research can be done to involve some additional checks to mitigate this limitation. Also, research can be done to provide different weights to the decisions made by the four models used to create the ensemble model for making the final decision. Further, some more features can be added to improve the performance.

References

1. Global Market Share Held by Operating Systems for Desktop PCs, from January 2013 to July 2020 [Online]. Available at: https://www.statista.com/statistics/218089/global-market-share-of-windows-7/. Accessed 31 Dec 2020
2. Li, Y., Ma, R., Jiao, R.: A hybrid malicious code detection method based on deep learning. Int. J. Softw. Eng. Appl. **9**, 205–216 (2015). https://doi.org/10.14257/ijseia.2015.9.5.21

3. Gupta, S., Kumar, P.: An immediate system call sequence based approach for detecting malicious program executions in cloud environment. Wirel. Pers. Commun. **81**, 405–425 (2015). https://doi.org/10.1007/s11277-014-2136-x
4. Zhang, Z., Qi, P., Wang, W.: Dynamic Malware Analysis with Feature Engineering and Feature Learning. arXiv:1907.07352 (2019)
5. Chandola, V., Banerjee, A., Kumar, V.: Anomaly detection: a survey. ACM Comput. Surv. **41** (2009). https://doi.org/10.1145/1541880.1541882
6. Tobiyama, S., Yamaguchi, Y., Shimada, H., Ikuse, T., Yagi, T.: Malware Detection with Deep Neural Network Using Process Behavior, pp. 577–582 (2016). https://doi.org/10.1109/COMPSAC.2016.151
7. Fawaz, A., Sanders, W.: Learning Process Behavioral Baselines for Anomaly Detection, pp. 145–154 (2017). https://doi.org/10.1109/PRDC.2017.28
8. Tajoddin, A., Abadi, M.: RAMD: registry-based anomaly malware detection using one-class ensemble classifiers. Appl. Intell. **49**, 2641–2658 (2019). https://doi.org/10.1007/s10489-018-01405-0
9. Aygun, R., Yavuz, A.: Network Anomaly Detection with Stochastically Improved Autoencoder Based Models, pp. 193–198 (2017). https://doi.org/10.1109/CSCloud.2017.39
10. Windows registry information for advanced users [Online]. Available at: https://docs.microsoft.com/en-us/troubleshoot/windows-server/performance/windows-registry-advanced-users. Accessed 31 Dec 2020
11. What is a DLL [Online]. Available at: https://docs.microsoft.com/en-us/troubleshoot/windows-client/deployment/dynamic-link-library. Accessed 31 Dec 2020
12. Kingma, D., Ba, J.: Adam: a method for stochastic optimization. In: International Conference on Learning Representations (2014). arXiv:1412.6980
13. Srivastava, N., Hinton, G., Krizhevsky, A., Sutskever, I., Salakhutdinov, R.: Dropout: a simple way to prevent neural networks from overfitting. J. Mach. Learn. Res. **15**, 1929–1958 (2014)

Predicting Employability of Candidates: Comparative Study of Different Machine Learning Models

K. B. Sai Hitharth and N. M. Dhanya

Abstract For a company interview procedure, the company uses a Classified Dataset, to determine whether or not that person is selected. Each candidate in the database has a set of attributes associated with them, as specified by the interviewer. The number of people who apply for any job is quite large, and the interview panel of a company has a stipulated period of time to complete the process of selecting the right candidate. So, it is essential that if the probability of a person qualifying is very low, he/she need not be interviewed by the interview panel. Instead, the focus of the panel can be on the shortlisted candidates, which leads to saving time. Machine learning models can be created based on the available data that can predict if the candidate will be shortlisted. This paper tries to compare and analyse the performance of different machine learning techniques that are used for the Interview Style Candidate Selection problem. It uses state-of-the-art Deep Learning frameworks such as Google's TensorFlow2.0 and PyCaret. Eight analytical techniques. The chosen techniques include Feature Selection (Logistic Regression), *K* Means Clustering, Quadratic Discriminant Analysis, CatBoost, Extreme Gradient Boosting, Boosting with Bayesian Optimization, Random Forest and Artificial Neural Networks. Models were applied on a Company-given Classified Dataset which contains 1000 records. Results show that though all models have an accuracy greater than 90%, the Bayesian Optimized Model with an accuracy of 98.34% outperforms all the other models with the Deep Learning Model trailing behind with an accuracy of 97%. Both *K* Means and Logistic Regression Models can be recommended with accuracies of 95% and 93%, respectively. CatBoost, though on the lower half, with 92.69% is also a fair choice along with Extreme Gradient Boosting (92.36%), QDA (91.69%) and finally Random Forest (91.03%).

Keywords Machine learning · Business intuition · Predictive analytics · Neural networks · Classification · Fine-tuning · Boosting · Selection

K. B. Sai Hitharth (✉) · N. M. Dhanya
Department of Computer Science and Engineering, Amrita School of Engineering, Amrita Vishwa Vidyapeetham, Coimbatore, India
e-mail: cb.en.u4cse18349@cb.students.amrita.edu

N. M. Dhanya
e-mail: nm_dhanya@cb.amrita.edu

A. Noor et al. (eds.), *Proceedings of Emerging Trends and Technologies on Intelligent Systems*, Advances in Intelligent Systems and Computing 1371,
https://doi.org/10.1007/978-981-16-3097-2_15

1 Introduction

Machine learning is the ability of machines to predict outcomes without being explicitly programmed to do so. The algorithms applied to machine learning programs and software are created to be versatile and enable developers to make changes via fine-tuning of hyperparameters. The machine 'learns' by processing large amounts of data and detecting patterns within this set. It is a form of AI that simulates gaining knowledge and improving the decision-making process through means of a trial-and-error mechanism, similar to how a human learns and improves from mistakes. Machine learning is the foundation for cutting-edge innovations like deep learning, neural networks and self-sustaining vehicle operation. Creating an algorithm that a computer then uses to find a model that fits the data as best as possible and thus, makes very accurate predictions based on this [1]. This is quite sought after, especially when a global product is launched by an MNC, and they have to figure out how their users are responding to it (User Experience). Time-series analysis performed in a business and financial companies play a major role in Sales Forecasting which helps in setting goals for the upcoming quarter. The hotels and resorts in the tourism business with a wide network chain are well known in light of the fact that their systems figure out how to optimize the prices, i.e., when to charge a room for a higher cost. Major production and manufacturing companies also use business intelligence algorithms such as Inventory Management techniques to prevent over-supply and under-supply, thus saving both time and capital. Considering the aforementioned real-life examples, the need for understanding the principles of predictive analytics is a must.

Thus, this paper aims to demonstrate and compare various machine learning models in predicting the employability of candidates while using new and upcoming ML technologies.

1.1 *Dataset Description*

The dataset used in this research is collected from the Kaggle platform [2], namely Classified Company Data (Interview Style Data) published by Mr. Aayush Mishra. You have a.csv with 1000 rows summarizing the data. This Data set has 10 columns that work as features and one target column. Table 1 contains the first 5 records (upto 2 decimal places) of the Classified Dataset to give readers an overview of how the dataset looks.

These are the random features (WTT, PTI, EQW, SBI, LQE, QWG, FDJ, PJF, HQE, NXJ) which the company uses as the different criteria, on the basis of which, a candidate is shortlisted. The targets are a Boolean variable (0 or 1). The data in these columns are completely randomized to provide a sense of generalization.

Table 1 First 10 rows of the classified dataset

	WTT	PTI	EQW	SBI	LQE	QWG	FDJ	PJF	HQE	NXJ	T.C
0	0.91	1.16	0.57	0.76	0.78	0.35	0.76	0.64	0.88	1.23	1
1	0.64	1.00	0.54	0.83	0.92	0.65	0.68	1.01	0.62	1.49	0
2	0.72	1.20	0.92	0.86	1.53	0.72	1.63	1.15	0.96	1.29	0
3	1.23	1.39	0.65	0.83	1.14	0.88	1.41	1.38	1.52	1.15	1
4	1.28	0.95	0.63	0.67	1.23	0.70	1.12	0.65	1.46	1.42	1

The eight analytical techniques used in this paper include Feature Selection (Logistic Regression), *K* Means Clustering, Quadratic Discriminant Analysis, CatBoost, Extreme Gradient Boosting, Boosting with Bayesian Optimization, Random Forest and Artificial Neural Networks.

Contributions of this paper are as follows:

- Create a machine learning algorithm, which can predict if a candidate is shortlisted or not.
- This is a classification problem with two classes: not selected and selected, represented by 0 and 1 s.
- After creating different ML models, a comparative analysis on the models is performed to figure out which one gives the best prediction results.

2 Literature Survey

The following are some papers in literature which deal with machine learning prediction.

Linsey [3] conducted a study at Ohio University and created a machine learning model to predict the influence of employment [4] at graduation. The machine learning models were then tested to identify the predictor with the highest accuracy. Major, GPA, Co-curricular activities and internships were the employability signals used for the prediction. The accuracy of the models is Logistic Regression—72.17%, Discriminant Analysis—69.81%, Decision Tree—71.70%, Artificial Neural Network—73.11%, SVM—87.26%.

Casuat and Festijo [5] created a machine learning approach for predicting the employability of the students [6]. 27,000 information (3000 observations and 9 features) pertaining to students based on three features namely, mock placement interviews, on-site training and GPA over a span of three years (2015–2018). The learning algorithms used were Decision Trees (DT), Support vector machine (SVM) and Random Forest (RF). Accuracy measures, f1-score and support measures, precision and recall measures were used for evaluation of the models. The study reached the conclusion that the Support Vector machine (SVM) [7] acquired 91.22% in accuracy measures which was significantly better than the other learning algorithms, DT 85%, RF 84%.

Nur et al. [8] performed a comparative study on Graduates' Employment in Malaysia by using Data Mining. This study conducted by the aforementioned authors uses the Ministry of Higher Education (MoHE) approved graduates' employment dataset from the year 2017. This paper compared 3 Machine Learning Models, namely Decision Tree, Logistic Regression (LR) and Artificial Neural Networks (ANN). The ANN model [9] achieved the highest accuracy 81.52%, with Decision Tree just trailing behind with an accuracy of 81.34%. An accuracy of 80.82% also placed the Logistic Regression model as a strong contender.

Nancy and Vineet [10] utilized a proficient data mining [11] technique to enhance students' employability prediction [12]. An alarming issue for many students (either at high-school level or graduate level) is the early prediction of employability [13] of students so as to provide them with a reassurance of continuing to work on this path or focus on higher studies. This paper primarily focuses on the HMM-SVM. Some existing methods such as *k*-Nearest Neighbor (*k*NN), Support Vector Machine (SVM) [1], Hidden Markov Model (HMM) and Artificial Neural Network (ANN) are used for comparison with the experimental outcomes that showed 93.4% accuracy which was perceived by the HMM-SVM classifier.

Othman et al. [14] utilized seven years of information (2011–2017) gathered through Malaysia's Ministry of Education tracer study for comprehension and implementation of classification techniques with respect to anticipating graduate employability [15]. Decision Tree, Support Vector Machines and Artificial Neural Networks were the three classification algorithms used, to work on a dataset of size 43,863. Through this study, the researchers were able to arrive at the conclusion that decision tree J48 produced a higher accuracy in contrast to other techniques with classification accuracy of 66.0651% and showed a scope of improvement to 66.1824% after the process of parameter tuning.

Aravind et al. [16] performed an extensive comparison of various regression models on the student placement prediction problem. A 2-phase methodology was implemented in which a simple dataset was considered in Phase 1, and an extended dataset was used in the following phase. A comparative study of seven models, namely—linear regression, decision tree regression, *K*-neighbor regression, random tree classifier, gradient boost regression [17], XGBoost regression and light GBM regression model. Root mean square error (RMSE) and prediction accuracy were the metrics considered for this study.

This work also used *K* Means Clustering and XGBoost with Bayesian Optimization for acquiring fairly high accuracy values, methods not implemented by the aforementioned studies.

3 Methodology

After the selection of the dataset, the next step is to segregate the input and output variables. The common issue when working with numerical data is the difference in magnitude wherein some features with high numerical values may alter the prediction, thus reducing the accuracy. To resolve this, the process of feature scaling (also known as Standardization) is performed. Feature scaling is a method used to normalize the range of independent variables (features of data), which is the process of transforming data into a standard scale. In the Pycaret package, however, an interesting notion is the process of implicit preprocessing of data. The development and comparison of the different models in this paper are discussed below. Testing is then done on a dataset the model has never seen before, as this is what we use to refer to the overall accuracy. Figure 1 represents the structural flow of the methodology.

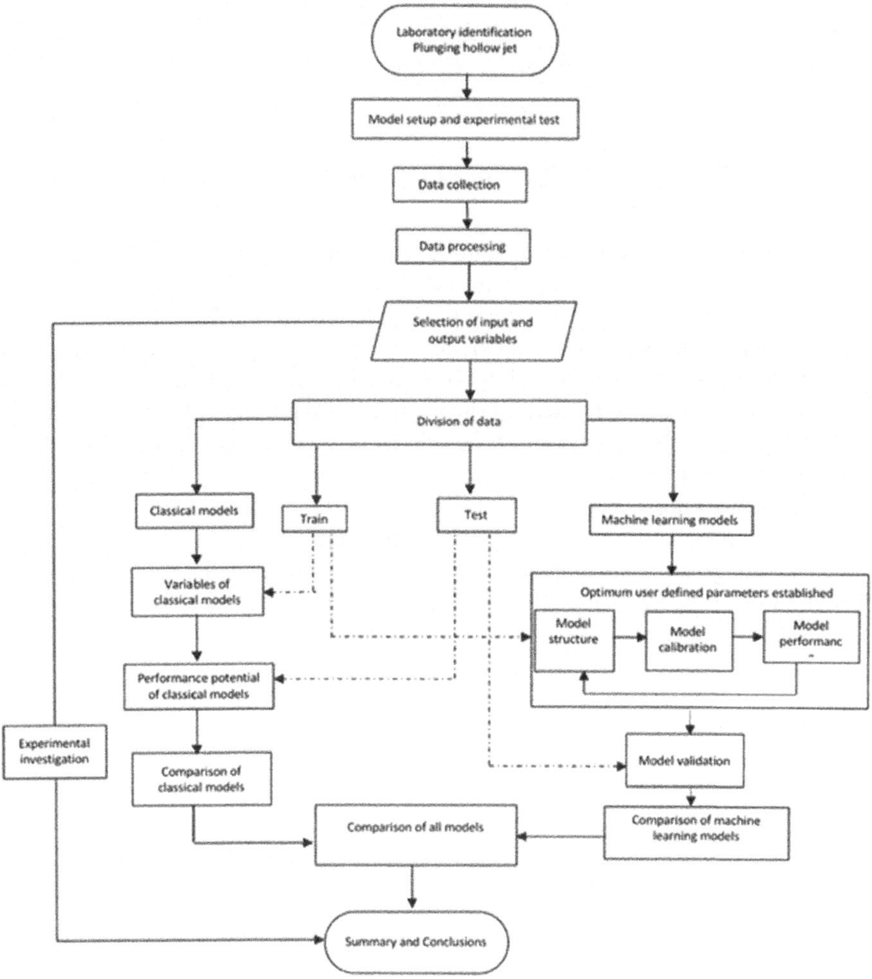

Fig. 1 The structural flow of the methodology

Table 2 Features and their corresponding *p*-values

Features	*P*-value
Const.	5.324075e−03
WTT	8.729204e−13
PTI	2.966677e−07
EQW	2.367872e−17
SBI	5.529602e−01
LQE	7.264250e−02
QWG	7.596567e−05
FDJ	7.388687e−04
PJF	2.776397e−12
HQE	2.449769e−19
NXJ	4.454545e−01

3.1 Dimensionality Reduction (Logistic Regression)

Feature Selection is a process that simplifies models, improves speed and prevents a series of unwanted issues arising from having too many features. *P*-value is the smallest level of significance at which we can reject the null hypothesis, given the observed sample statistic. Most statistical software calculates *p*-values for each test. These values are rounded off to 3 digits after the decimal. If the *P*-value is small, then there is stronger evidence in favor of the alternative hypothesis, that is, the closer to 0.000 the *p*-value, the better or more significant is the result you have obtained.

If a variable has a *p*-value > 0.05, we can disregard it. Table 2 corresponds to the features and their corresponding *p*-values.

From the above table, we can see that the "*p*-values" of the variables 'SBI','LQE' and 'NXJ' are greater than 0.05, thus concluding that these are not significant for the prediction. Thus, there is no harm in dropping these features.

3.2 K Means Clustering

Math Prerequisites

1. *Euclidean Distance:* This refers to the distance between 2 quantitative points (numerical values) expressed as a set of coordinates. It is described as [18]

$$d(xi, xj) = \surd\left(|xi1 - xj1|^2 + |xi2 - xj2|^2 + \cdots + |xip - xjp|^2\right) \quad (1)$$

2. *Centroid:* Mean position of a group of points (aka center of mass).

***K* Means Technique**
The steps involved are:

1. Select the desired cardinality. "*K*" stands for the number of clusters.
2. Determine the cluster seeds. A seed is essentially an elementary centroid picked indiscriminately or specified by the researcher in light of prior comprehension of the data.
3. On the basis of proximity, with respect to the Euclidean distance from the seeds, allocate each point on the graph to a centroid.
4. Adjust the centroids.
5. Then repeat steps 3 and 4 until we can no longer reassign points which completes the clustering.

K Means is an iterative process.

To figure out the optimum number of clusters, the Elbow Method is used.

Elbow Method: This method calculates the best *k* value by considering the percentage of variance explained by each cluster [18].

K-means is all about the analysis-of-variance paradigm. ANOVA—both uni- and multivariate—is based on the fact that the sum of squared deviations about the grand centroid is comprised of such scatter about the group centroids and the scatter of those centroids about the grand one:

$$\text{SStotal} = \text{SSwithin} + \text{SSbetween} \tag{2}$$

So, if SSwithin is minimized then SSbetween is maximized.

3.3 *Using PyCaret Package*

The following models are compared in PyCaret model.

- Logistic Regression Model.
- Quadratic Discriminant Analysis.
- Catboost Model.
- Extreme Gradient Boosting.
- Random Forest Classifier.
- Boosting with the help of Bayesian Optimization.

For all of the aforementioned models except the last one, a standard set of rules have been followed:

1. Create the Model.
2. Tune the model (through means of tuning, bagging and boosting).
3. Select the best performing model.
4. Predict and assess the accuracy of the most optimum model.
5. Compile the results.

Table 3 corresponds to the training accuracy of models during 10 Folds.

In the case of Boosting with the help of Bayesian Optimization, the following iterations are performed to acquire the best parameters by means of fine-tuning the model. The results are published in Table 4.

The parameters with their values, for the model with the highest accuracy are 'gamma': 0.21264372336401588, 'learning_rate': 0.10454114195929931, 'max_depth': 3.305746799396676, 'n_estimators': 107.48063146513937. The training accuracy achieved is 99%.

The training accuracy of Boosting with Bayesian Optimization was the highest out of all the models implemented in the PyCaret package.

Table 3 Training accuracy of models during 10 folds

	LR	QDA	CB	XGBoost	RF
0	0.9657	0.9714	0.9714	0.9571	0.9714
1	0.9571	0.9571	0.9286	0.9286	0.9429
2	0.9714	0.9714	0.9714	0.9429	0.9571
3	0.9429	0.9286	0.9143	0.9143	0.9143
4	0.9714	0.9429	0.9571	0.9714	0.9571
5	0.9714	0.9714	0.9429	0.9143	0.9429
6	0.9429	0.9429	0.9571	0.9571	0.9571
7	0.9571	0.9714	0.9429	0.9429	0.9429
8	0.9429	0.9571	0.9429	0.9143	0.9286
9	0.971	0.9565	0.942	0.9275	0.942

Table 4 Fine-tuning process

Iter	Target	Gamma	Learning rate	Max_depth	n_estimators
1	−0.2559	0.3057	0.3689	4.248	109.1
2	−0.3009	0.1443	0.7519	6.099	111.6
3	−0.2383	0.1279	0.1331	3.798	106.6
4	−0.263	0.0398	0.2965	8.396	113.1
5	−0.3202	0.04533	0.8946	7.586	109.5
6	−0.2963	0.2036	0.755	3.431	113.8
7	−0.2767	0.3942	0.5845	5.856	107.4
8	−0.2906	0.8052	0.6434	8.845	120.0
9	−0.2365	0.2126	0.1045	3.306	107.5
10	−0.3024	1.0	0.9369	3.0	106.7
11	−0.5	0.01675	0.0	3.684	107.2
12	−0.3027	0.1158	0.6639	9.706	106.1
13	−0.2661	0.9486	0.43	9.083	102.4

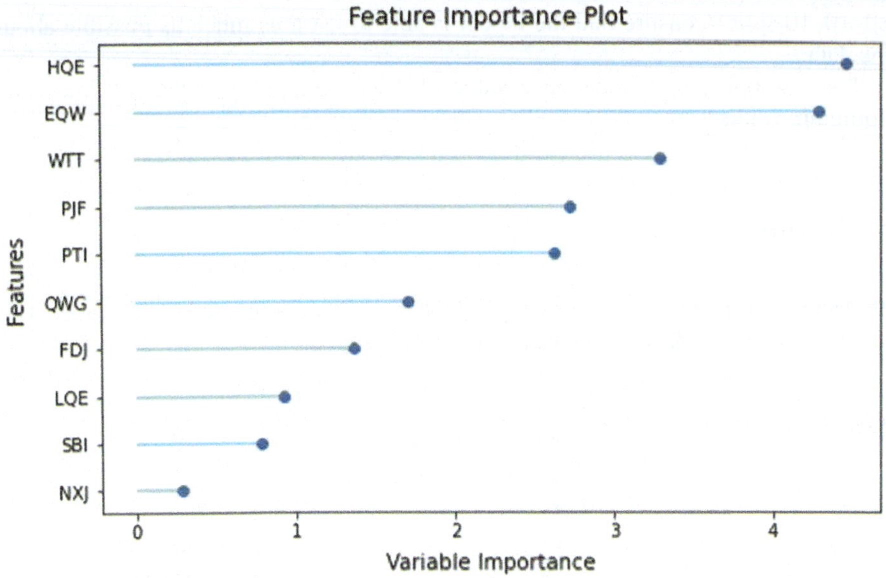

Fig. 2 Feature importance plot

This paper also describes the significance of various features in predicting the output, shown by Fig. 2.

3.4 *Deep Learning Model*

Artificial Neural Networks (ANNs) are a new, revolutionary approach (state-of-the-art machine learning) that take inspiration from the biological neural system in the human mind. Neural Networks are fundamentally different from other approaches in the sense that an ANN is organized in layers, and each layer consists of a set of nodes. The input layer communicates with one or more hidden layers, which in turn end up communicating with the output layer and each of these layers are connected by weighted links. ANNs have a broad practical scope of application due to their adaptive and fault tolerant nature.

The NN model uses state-of-the-art Deep Learning frameworks such as Google's TensorFlow2.0 to develop a business intuition on solving this problem. The input layer comprises of the 10 features (aka placeholders) while the output layer is made up of 2 nodes (as the output is one-hot encoded).

After increasing the number of hidden layers from 5 to 10 and keeping a uniform size of 50 nodes per hidden layer as well as tuning the hyperparameters, namely the batch_size and max_epochs gave a high training accuracy. All the fine-tuning of hyperparameters was done bearing in mind not to overfit. The split applied is an

80–10–10 split to ensure that the model is able to learn as much as possible about the data.

Early stopping with a patience value of 2 was used to be able to achieve the optimum results.

4 Results

Table 5 represents the test accuracy of all the models implemented in this paper. The same is pictorially depicted using a bar graph in Fig. 3.

Table 5 Accuracy of the models

Feature selection (LR)	93%
K means	95%
QDA	91.69%
CatBoost	92.69%
XGBoost	92.36%
Bayesian optimized XGB	98.34%
Random forest	91.03%
Neural network	97%

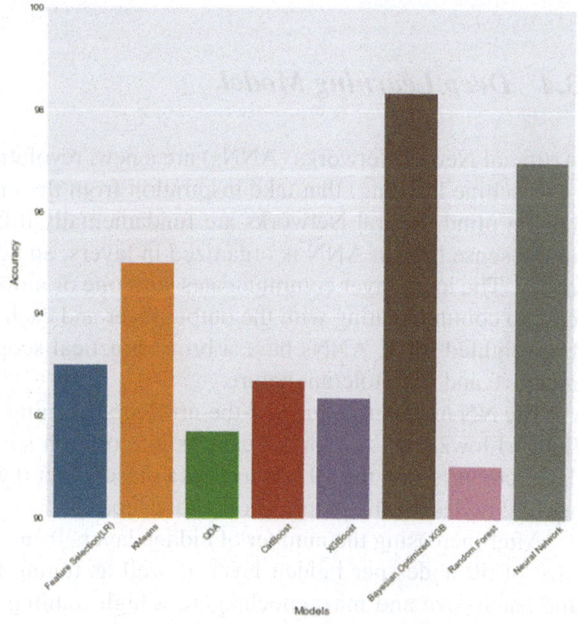

Fig. 3 Comparison of machine learning models

5 Conclusion

In conclusion, this work shows that though all models have an accuracy greater than 90%, the Bayesian Optimized Model with an accuracy of 98.34% outperforms all the other models with the Deep Learning Model trailing behind with an accuracy of 97%. Both K Means and Logistic Regression Models can be recommended with accuracies of 95% and 93%, respectively. CatBoost, though on the lower half, with 92.69% is also a fair choice along with Extreme Gradient Boosting (92.36%), QDA (91.69%) and finally Random Forest (91.03%).

References

1. Liu, Z., Chen, H.: A predictive performance comparison of machine learning models for judicial cases. In: 2017 IEEE Symposium Series on Computational Intelligence (SSCI), pp. 1–6. Honolulu, HI (2017). https://doi.org/10.1109/SSCI.2017.8285436
2. Mishra, A.: Classified Company Data (Interview Style Data) (2020). Accessed at: https://www.kaggle.com/aayushmishra1512/classifieddata
3. NACE—Predicting Employment Through Machine Learning (2019). https://www.naceweb.org/career-development/trends-and-predictions/predicting-employment-through-machine-learning/. Accessed 10 Nov 2020
4. Nemanick, R.C., Clark, E.M.: The differential effects of extracurricular activities on attributions in resume evaluation. Int. J. Sel. Assess. **10**, 206–217 (2002)
5. Casuat, C.D., Festijo, E.D.: Predicting students' employability using machine learning approach. In: 6th IEEE International Conference on Engineering Technologies and Applied Sciences (ICETAS), pp. 1–5. Kuala Lumpur, Malaysia (2019). https://doi.org/10.1109/ICETAS48360.2019.9117338
6. Casuat, C.D., Festijo, E.D.: Identifying the most predictive attributes among employability signals of undergraduate students. In: 16th IEEE International Colloquium on Signal Processing and Its Applications (CSPA), pp. 203–206. Langkawi, Malaysia (2020). https://doi.org/10.1109/CSPA48992.2020.9068681
7. Dhanya, N.M., Harish, U.C.: Sentiment analysis of twitter data on demonetization using machine learning techniques. In: Lecture Notes in Computational Vision and Biomechanics, vol. 28, pp. 227–237. Springer Netherlands (2018)
8. Nur, A., Daud, N., Athirah, R., Nur, S.: A comparative study on graduates' employment in Malaysia by using data mining. J. Phys: Conf. Ser. **1366**, 012120 (2019). https://doi.org/10.1088/1742-6596/1366/1/012120
9. Viswanathan, S., Kumar, M., Soman, K.P.: A comparative analysis of machine comprehension using deep learning models in code-mixed hindi language. Stud. Comput. Intell. **823**, 315–339 (2019)
10. Kansal, N., Kansal, V.: An efficient data mining approach to improve students' employability prediction. Int. J. Comput. Appl. **178**, 29–35 (2019). https://doi.org/10.5120/ijca2019919372
11. Mishra, T., Kumar, D., Gupta, S.: Students' performance and employability prediction through data mining: a survey. Indian J. Sci. Technol. **10**, 1–6 (2017). https://doi.org/10.17485/ijst/2017/v10i24/110791
12. Azziaty, N., Tan, K., Lim, C.: Predictive analysis and data mining among the employment of fresh graduate students in HEI. AIP Conf. Proc. **1891**, 020007 (2017). https://doi.org/10.1063/1.5005340
13. Chronicle of Higher Education: The Role of Higher Education in Career Development: Employer Perceptions. Chronicle of Higher Education, Washington (2012). http://chronicle.com/items/biz/pdf/Employers%20Survey.pdf. Viewed 18 Dec 2020

14. Othman, Z., Shan, S., Yusoff, I., Kee, C.P.: Classification techniques for predicting graduate employability. Int. J. Adv. Sci. Eng. Inf. Technol. **8**, 1712 (2018). https://doi.org/10.18517/ijaseit.8.4-2.6832
15. Yuzhuo, C.: Graduate employability: a conceptual framework for understanding employers' perceptions. High. Educ. **65**, 457–469 (2013). https://doi.org/10.1007/s10734-012-9556-x
16. Tadi, A., Reddy, S., Sai, A., Gurusamy, J.: A comparative study on machine learning algorithms for predicting the placement information of under graduate students. In: 2019 Third International Conference on I-SMAC (IoT in Social, Mobile, Analytics and Cloud) (I-SMAC), Palladam, India, pp. 542–546 (2019). https://doi.org/10.1109/I-SMAC47947.2019.9032654
17. Veerakumar, S., Dhanya, N.M.: Performance analysis of various regression algorithms for time series temperature prediction. J. Adv. Res. Dyn. Control Syst. **10**(3), 996–1000 (2018)
18. Hackerearth info web site (n.d.). https://www.hackerearth.com/practice/machine-learning/machine-learning-algorithms/clusteringalgorithms-evaluation-r/tutorial/

Speech Recognition for Medical Conversations Health Record (MCHR)

Nivedita Singh, M. Balasubramaniam, and Jitendra Singh

Abstract This paper provides an overview of major technical perspectives and recognizes the fundamental progress of the conversion of speech to text and also provides an overview of the techniques developed at each point of the speech-to-text conversion classification, a system that automatically transcribes conversations between doctor and patient. MCHR systems are known to store clinical data to capture the patient's electronic medical health record (EMHR) across time (Rajkomar et al. in NPJ Digit Med 1, 18 (2018) [1]). It eradicates the need to manually monitor the previous paper medical records of a patient and helps ensure consistency and standardization of data. When extracting medical data, electronic medical health records are more efficient for analyzing potential patterns and long-term improvements in a patient's clinical condition. MCHR systems manage doctor–patient conversational data recording as both an audio and text for the use of doctors in health data analysis. These systems help in analyzing clinical data in later stages or match patterns for other patients with similar problems or symptoms. Additional usage is from the conversation we can extract diagnosis information, drug prescriptions and dosage details which can be presented to doctors for verification and approval. This will reduce doctors' time; nowadays in large hospitals, patient loads are high and doctors are unable to perform prescription entry using text input in hospital management systems, as it's a time-consuming process. This paper presents the state-of-the-art in voice recognition and discusses how to build a clinical prescribing system for healthcare domain.

Keywords Medical transcription · Health conversation · End-to-end attention models · Patient–doctor communication · Qualitative research · Deep speech · Speech to text

N. Singh (✉) · M. Balasubramaniam · J. Singh
BDPM Group, Center for Development of Advanced Computing (C-DAC), Noida, India
e-mail: niveditasingh@cdac.in

M. Balasubramaniam
e-mail: mbalasubramaniam@cdac.in

J. Singh
e-mail: jitendrasingh@cdac.in

A. Noor et al. (eds.), *Proceedings of Emerging Trends and Technologies on Intelligent Systems*, Advances in Intelligent Systems and Computing 1371,
https://doi.org/10.1007/978-981-16-3097-2_16

1 Introduction

Recording consultations with the patient provides supportive information to recollect the treatment data given by doctors. It is useful for children and elderly patients who find it difficult to follow the directions of the doctor. The unsupervised clustering of conversation characteristics described those types correlated with higher communication scores in "speech words". The doctor–patient relationship will be rated high or low based on the performance and accuracy of the test collection (Fig. 1).

The recording of patient medical data will act as a digital health record for further analysis in the future [1]. In short, in the care of a patient, these recordings play a significant role. On the other way, the recording between the doctor and the patient may be a reason for stress and conflicts. Consultations related to personal behavioral problems may lead to legal problems if the information is passed on to voluntary individuals by mistake. If the doctor offers a different opinion compared to another doctor and the patient may not be able to obey the instructions, complications may occur [2]. When the doctor's audio is captured in a MCHR system, a doctor would be able to administer his medication to the patient. Data will be discreetly captured for all keywords. The doctor will be able to edit or delete or hand over any entry that has been generated using voice recognition. This review paper aims to present a new method of enumerating and matching different speech recognition systems using deep speech recognition techniques alongside speech to text conversion and applications that are at the forefront of this exciting and challenging field.

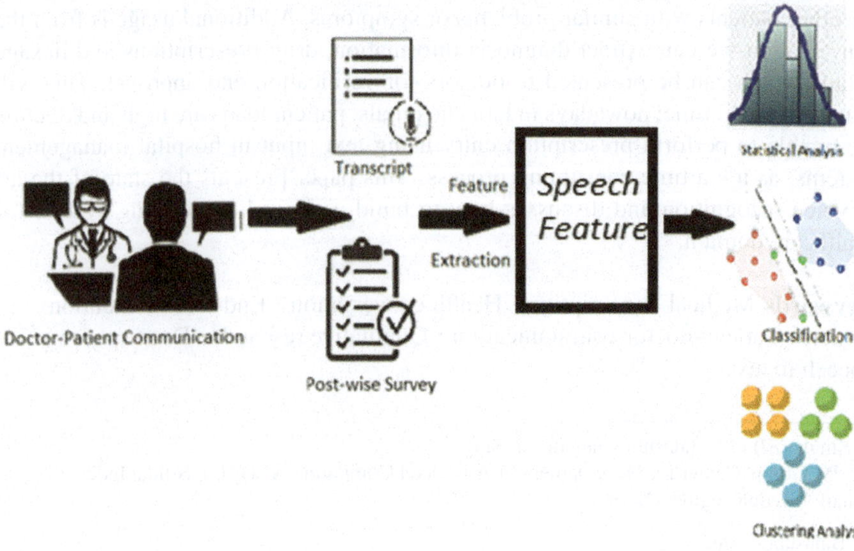

Fig. 1 Overview of communication analysis

2 Proposed Idea

The proposed system functions in two direction. The first is an audio file that will be the recorded conversation between doctor and patient, and the second is the speech-to-text conversion module which handles the speech-to-text conversion of the audio file.

2.1 *Module 1 (Speech to Text)*

The first part of the module will be a speech-to-text module. When the doctor and patient have their first talk, the audio will be recorded and simultaneously will the speech to text module [3]. Deep speech algorithm will be used to convert the audio file to text form. There is also a translator attached along with the speech-to-text module which will use NLP to translate the speech Doctor's comfortable language [4]. The text file obtained after the conversion will be saved in separate file.

2.2 *Module 2 (Speech Analysis)*

The second module will perform the analysis for the audio file recorded. These audio files will be saved in a separate folder for future reference. The primary research is to derive valuable details from the conversation. The extracted data can be categorized as major symptoms, the medication of the doctor, into more categories. This analysis will also be used for categorizing the patient based on the criticality of the disease [5]. This saved audio files can be considered as a digital health record for patients. The doctors will be able use these files as reference when similar cases come in the future.

3 Analysis

Patients should support their clinicians in the normal flow of communication which will help in a specific diagnosis, and the care can be administered accordingly. Patients can help attain these objectives. Being on time for all the appointments will help in stating the symptoms clearly. History of their health issue patterns such as when symptoms started, how symptoms interfere with life or any associated symptoms that could indicate a coexisting condition can be studied more efficiently [6]. Recording the doctor–patient conversation as both audio and text can be used for the health data analysis. The collected data will include the initial or beginning stage of the treatment till the prescription stage. With these data, we can analyze further in later stages or can match patterns for other patients with similar problems or symptoms. This paper is a continuation of a previous work, which focused on recording patient speech conversation data. The key emphasis of the proposed paper is on data collection for further analysis and on involving doctors in data collection without burdening them.

3.1 Purpose of Recording

There are many purposes of recording the conversation. Firstly, for categorizing the patients based on the disease diagnosed. Secondly, the prescription allotted by the doctor can be extracted out and a copy can be shared with the patient. Thirdly, the data may be used when similar cases are recorded for future reference. And finally, for health data analysis, this information can be used.

4 Technology Stack

Deep Speech API: In order to translate expression to text. The Deep Speech API allows you to transcribe voice to text from audio files in 80+ languages [7].

PYTHON: Python, developed by Guido van Rossum on 20 February 1991, is a popular programming language and is used for:

- Applications for Mobile (especially Android apps)
- Applications on desktops
- Applications on the internet
- Web servers and application servers
- Games
- Database connection
- And much, much more!

XML: Design of the User Interface. The Extensible Markup Language (XML) is a markup language that specifies a collection of rules for document encoding in a format that can be interpreted by humans and machines [8].

DATA BASE: Firebase Services for secure and remote access. It stores diagnostic, prescription, medication, and drug administration records.

5 Proposed System

The system generates speech at run time through windows or PC desktop applications or any input devices and processes the sampled speech data to identify the express text. Two files are assigned to store the recognize text [9]. It will support other larger systems, offering users a different option to access the database. The first file is saved in text form, and the other file is saved as an audio file.

Deep speech has been very helpful in developing IoT devices that require voice recognition [7]. Deep speech is an open-source offline voice recognition system from Mozilla. Based on Baidu's research paper, it uses a model trained by machine learning techniques.

Deep Speech uses TensorFlow lite to make the implementation part simpler, and it transforms the speech spectrogram into a text transcript using a neural network [10]. The speech recognition system is a good thing because the user can alter and change the system as per requirement and deep speech can be a stand-alone application to process speech recognition without a constant internet connection (Figs. 2 and 3).

5.1 Data Collection

To begin with, we have to collect the data from the web interface. Data generally include all those audio files. Firstly, A database has to be created for this application (Fig. 4).

5.2 Data Preprocessing

The audio files will be arranged based on the respective categories, the dataset preparation is mostly manual as it contains audio files, and most of the things in it will be irrelevant.

Steps to follow are:

- We have maintained the set of audio files.
- We have maintained set of respective folders.
- Set of respective levels (critical case, mild case, casual case).

Speech recognition, medical health record (MHR) logic:

MHR is made up of speeches, and NLTK and ML engines have the following components:

A user interface will be present which will have the list of patients based on their appointments. The doctor will be given a Login ID through which he/she can access the application Ones the patient is in the doctor's room; the doctor will check with the

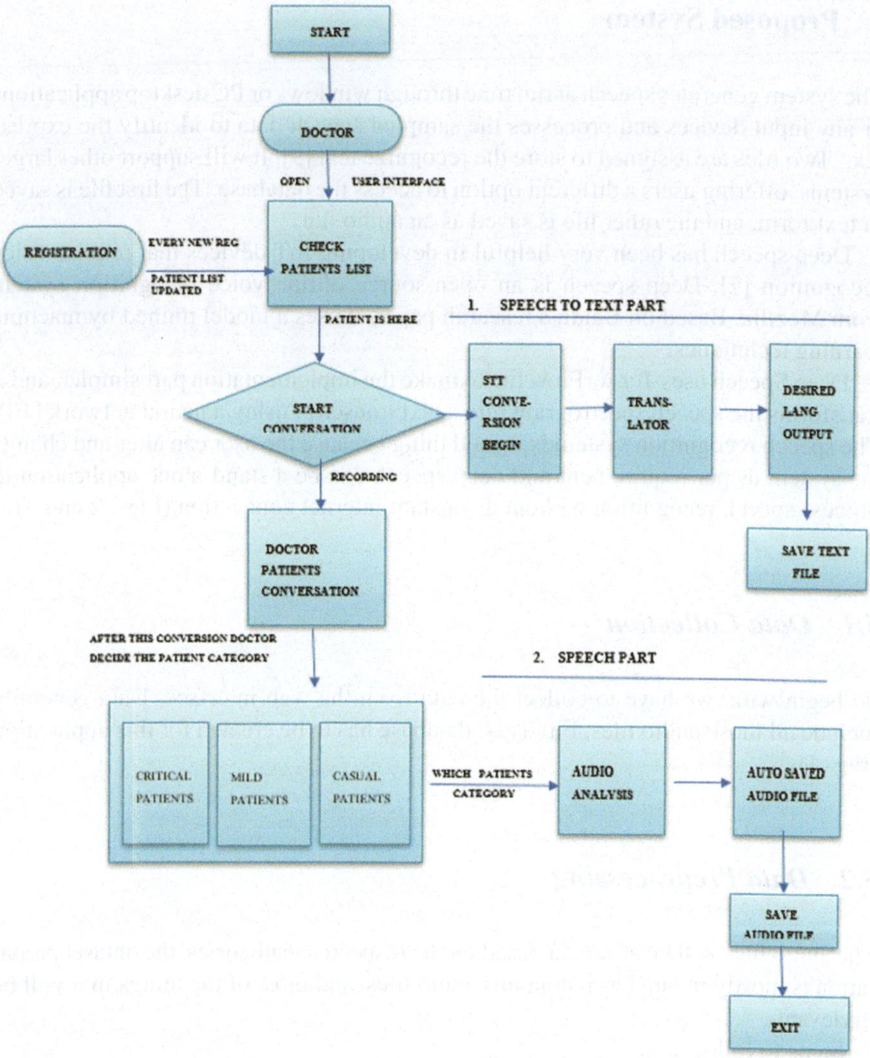

Fig. 2 An elaboration of the proposed idea

patient list and click on the "patient is here" radio button which will automatically start the conversation recording.

- After the conversation, the doctor will decide the patient category. The categories are further divided into
    1. Critical patients.
    2. Mild patient.
    3. Casual patients.

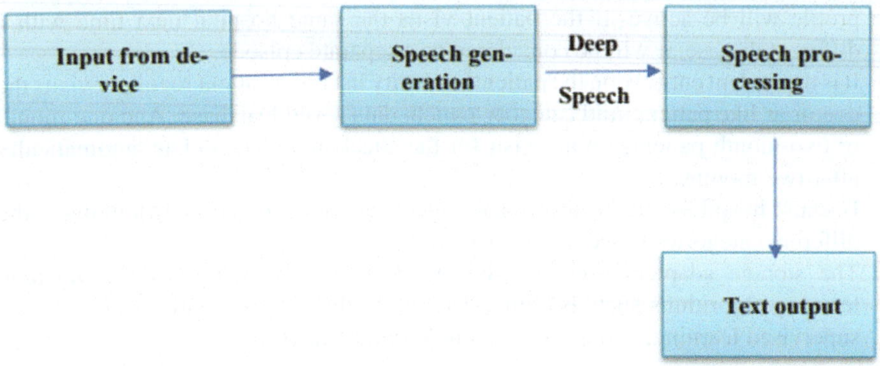

Fig. 3 Basic flowchart of speech to text system

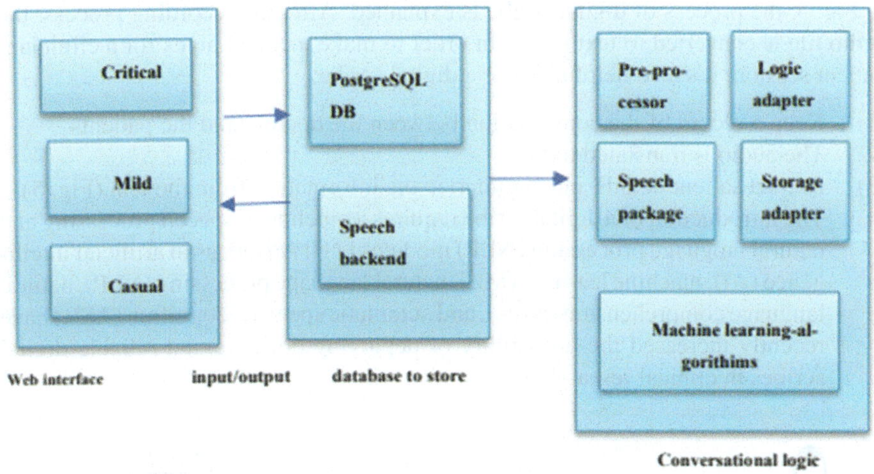

Fig. 4 Database flowchart of the proposed speech recognition medical conversation

- After this conversation, the audio file will be auto-saved.
- If the patient falls in the critical category, the data will be saved permanently in the database. For patients belonging to mild category, the data will be saved for 6 months and after 6 months the data will be automatically deleted from the system.
- If the patient belongs to a casual category, the data will be saved for one week, and after one-week data will be automatically deleted from the system, but will be saved in the backup. Casual cases normally include headache, normal cough, and fever, etc.
- An important feature of this application is that if the patient revisits, it will be flagged. Until the doctor has mentioned the patient as discharged, the patient

profile will be active. If the patient visits the same hospital next time with a different disease, it will be considered as a separate episode.

- It is dependent entirely on the patient's priority list if the patient regularly visits the doctor as like patient mild category patient data saved Database. And one-month or two-month patients do not visit for the checkup so data delete automatically after two months.
- Because hospitals will be allowed to collect patient information, belonging to the different categories based on the criticality.
- The storage adapters used are SQLite, PostgresDB, and MySQL. Machine learning algorithms such as comparison algorithms, Syncretistic, Naive Bayes, supervised learning, and reinforcement learning are used.

5.2.1 Digital Scribe

In Fig. 5, the process of digital scribe is explained. After the recording process, the audio file is converted to text using. In order to make medical notes for a clinician-patient session, it's also helpful to use a digital scribe:

(1) Keep a record of the conversation between the doctors and the patients.
(2) The audio is translated to text.
(3) Extract salient details and summarize the information from the text (Fig. 5).
(4) The introduction of a digital scribe requires a pipeline of speech processing and natural language processing (NLP) modules [11]. Advances in artificial intelligence (AI), machine learning (ML), natural language processing (NLP), natural language comprehension (NLC), and automatic speech recognition (ASR) have recently increased the probability of deploying effective and reliable digital scribes in clinical research.

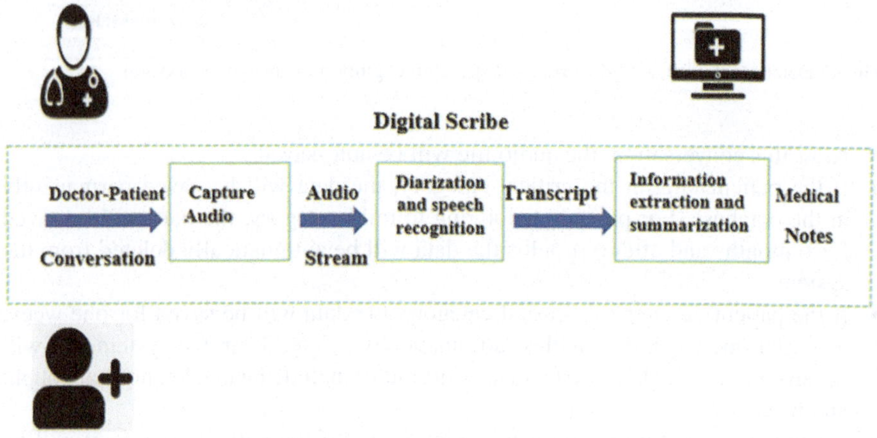

Fig. 5 Digital scribe

Figure 5, Pipeline Digital Scribe. A digital scribe captures the audio of a clinician–patient conversation, uses speech recognition to create a transcript, extracts information from the transcript, summarizes the data, and produces medical notes in the electronic medical health record (EMHR) that are important to the clinician-patient conversation [12]. Recognition of speech, information extraction, and summary are all based on AI and machine learning models that involve a lot of training and evaluation data. The information extracted will be categorized in four parts: the patient's detail, the symptoms expressed by the patient, information regarding the doctor's diagnosis as well as remarks, and lastly, the doctor's prescription.

Until now, research has concentrated on addressing fundamental issues in the creation of a digital scribe, like a clinical discussion ASR, which automatically populates the symptom analysis in a patient encounter, extracts symptoms from medical conversations, and produces dictation medical reports [13]. Although these developments are exciting, some issues prevent the complete implementation of a fully functioning digital scribe in a clinical setting and its evaluation (Table1).

6 Conclusion and Future Work

We explored developing deep speech recognition models for transcribing the conversation between doctor and patient in this paper. The quality of the system improves only with the quality and quantity of dataset we collect. The deep voice is more robust to noisy data and needs more clean-up. To understand the success of clinical activities, a thorough review is given. Nowadays, a user has to pay for using the app for speech recognition service, and users need to pay with a range from 4 US Dollars to 150 US Dollars for every 4 thousand applications. For different applications, Amazon Alexa is led by Google Assistant on the ASR system. In addition, another element that often needs to be linked to the Internet in order to process voice recognition. The proposed idea will completely work on deep speech for speech to text conversion with high accuracy.

In the future, the system can be modified by using a multilingual voice interaction system to interact with the user at deeper levels. The desired speech is created by the translating speech synthesis approach. In the future, this application implementation could be greatly improved by adding support for more data features, such as a user's demographic information. We also intend to add user-level language knowledge to the feedback feature of users. We are implementing the speech engine in python and using speech recognition libraries. Our voice recognition medical conversation application is in a first-level code and design phase right now.

Table 1 The challenges associated with the various tasks a digital scribe must perform

Task	Challenges
Recording audio	• Ambient noise levels are high • Authenticity of the microphone • Several speakers are present • Microphone positioning depending on the clinician's and the patient's preferences
Speech recognition	• Audio of varying quality • Several speakers are expected to speak • Non-lexical gaps, contradictions, false starts, interruptions • Ambiguity in Medical Terms • Owing to microphone distance and relative positioning, the volume of the speakers varies • Using multiple speakers to distinguish audio (speaker diarization)
Topic segmentation	• Structured and unstructured conversations • During a medical conversation, the progression of topics is nonlinear
Medical concept extraction	• Noisy output of program which map text to UMLS • Tool parameters for mapping text to UMLS are being fine-tuned • Inference dependent on meaning (understanding the appropriate meaning of a word or phrase given the context) • Zero anaphora, talk aloud, theme drift, and other spontaneous speech phenomenon
Overview of the summary	• Nonverbal communications that are unstructured are summed up • Integrating medical information for the recognition of specific data • Contextual inference • Solving the patient's contradictory details • Updating hypotheses as more information is released by the patient • Creating summaries to train a machine learning model for summarization
Data collection	• Privacy issues among clinicians and patients • Data collection and marking are both time-consuming and costly • Consent to audio recording and use of information for research purposes from the patient • Data anonymization and de-identification • Datasets that are expensive • Data are stored on a private server as a form of intellectual property • Clinicians are hesitant to be reported for fear of legal consequences and increased workload

Acknowledgements The idea behind this research paper was articulated by Joint Director Mr. Ajay Gupta. We are also grateful for the insightful comments offered by Mrs. Himani Goel and Dr. Sumit Soman. This research has been enhanced in countless ways by the kindness and understanding of one and all.

Special thanks go to our team members, Ms. Merrin Mary Solomon and Ms. Savita Kumari, who helped us with this research work and gave suggestion about the research paper. We are extremely grateful to C-DAC, Noida, for giving us this opportunity to work on this project.

References

1. Rajkomar, A., Oren, E., Chen, K., et al.: Scalable and accurate deep learning with electronic health records. NPJ Digit Med **1,** 18 (2018). https://doi.org/10.1038/s41746-018-0029-1
2. Walker, M., Littman, D., Kam, C., Abela, A.: Paradise: a framework for evaluating spoken dialogue agents. In: Proceedings of the 35th Annual Meeting of the Association of Computational Linguistics. ACL 97 (1997). Sorensen, J., Allouez, C.: Unary data structures for language models. In: Interspeech (2011)
3. Das, S.: Speech recognition technique: a review. Int. J. Eng. Res. Appl. (IJERA) **2**(3) (2012). ISSN: 2248-9622
4. Redye, B.R., Mahender, E.: Speech to text conversion using android platform. Int. J. Eng. Res. Appl. (IJERA) **3**(1) (2013). ISSN: 2248-9622
5. Dahl, G.E., Yu, D., Deng, L., Aero, A.: Context-dependent pre-trained deep neural networks for large vocabulary speech recognition. IEEE Trans. Audio Speech Lang. Process. (2011)
6. Azeta, A.A., Ikhu-Omoregbe, N.A., Ayo, C.K., Atayero, A.A.: Development and deployment of VoiceXML-based banking applications. J. Comput. Sci. Appl. Int. J. Niger. Comput. Soc. (NCS) **15**(1), 59–72 (2008)
7. Khaini, I.: Myanmar continuous speech recognition system based on DTW and HMM. Int. J. Innov. Eng.Technol. (IJIET) **2**(1), 78–83 (2013)
8. Goss, F., Blackley, S., Ortega, C., Kowalski, L., Landman, A., Lin, C., Meter, M., Bakes, S., Gradwohl, S., Bates, D., Zhou, L.: A clinician survey of using speech recognition for clinical documentation in the electronic health record. Int. J. Med. Inf. **130** (2019). https://doi.org/10.1016/j.ijmedinf.2019.07.017
9. Vimala, C., Radha, V.: A review on speech recognition challenges and approaches. World Comput. Sci. Inf. Technol. J. (WCSIT) **2**(1), 1–7 (2012). ISSN: 2221-0741
10. Miao, Y., Glowered, M., Metz, F.: EESEN: end-to-end speech recognition using deep models and wefts-based decoding. In: ASRU, (2015). Bandana, D., Choroski, J., Serdyuk, D., Brakes, P. Benjie, J.: End-to-end attention-based large vocabulary speech recognition. abs/1508.04395 (2015). http://arxiv.org/abs/1508.04395
11. Kain, A., Hossam, J. P., Ferguson, S.H., Bush, B.: Creating a speech corpus with semi-spontaneous, parallel conversational and clear speech. Tech Report: CSLU-11-003, Aug 2011
12. Kumah-Crystal, Y.A., Pirtle, C.J., Whyte, H.M., Goode, E.S., Anders, S.H., Lehmann, C.U.: Electronic health record interactions through voice: a review. Appl. Clin. Inform. **9**(3), 541–552. Published online 18 July 2018. https://doi.org/10.1055/s-0038-1666844
13. Lin, Y.-S., Ji, C.-P.: Research on improved algorithm of DTW in speech recognition. In: International Conference on Computer Application and System Modeling (ICCASM 2010), pp. 418–421. IEEE (2010)

Albatross Optimization Algorithm: A Novel Nature Inspired Search Algorithm

Keertan Krishnan, Akshara Subramaniasivam, Kaushik Ravichandran, and Natarajan Subramanyam

Abstract The venture of meta-heuristic optimization algorithms into the field of biology has only accelerated growth in swarm intelligence and evolutionary algorithms. We present one such novel approach, inspired by several avian algorithms, mimicking the lifestyle and breeding pattern of the Laysan Albatross. The proposed algorithm is suitable for any nonlinear continuous function optimization task and proves to perform better for training neural networks, compared to gradient-based approaches such as back-propagation and other bio-inspired approaches such as the Whale Optimization Algorithm. We also show that it is also not as susceptible to overfitting. We tested our algorithm on a total of 12 datasets and obtained an average train accuracy of 0.785, average training standard deviation of 0.037, average test accuracy of 0.793 and an average test standard deviation of 0.035.

Keywords Albatross optimization algorithm (AOA) · Bio-inspired algorithm · Neural networks · Meta-heuristic · Breeding Threshold Distance(btd) · Evolutionary algorithms (EA)

1 Introduction

A search algorithm involves finding a structure in a search space which could be a combinatorial optimization or mathematical problem of finding a discrete or continuous value or finding the best solution that solves a problem. The solution to an optimization problem is often a metonymy from nature. A primary source of this inspiration is due to the vastness of nature and the infinitely many possible meanings and implications it has. Nature-inspired algorithms are often preferred to solve complex optimization tasks due to various reasons. Imitation of the natural world leads to solutions that have already been tried and tested. Not requiring any gradient information ensures that issues such as local minima are avoided. Use of randomness

K. Krishnan · A. Subramaniasivam · K. Ravichandran (✉) · N. Subramanyam
PES University, Bengaluru, India
e-mail: natarajan@pes.edu

A. Noor et al. (eds.), *Proceedings of Emerging Trends and Technologies on Intelligent Systems*, Advances in Intelligent Systems and Computing 1371,
https://doi.org/10.1007/978-981-16-3097-2_17

to depict unpredictability of nature allows the possibility for exploration of the entire search space. Binitha and Sathya [1] provide an extensive survey of this taxonomy and classification of algorithms.

Genetic Algorithms are based on Darwin's theory of evolution and its inception laid the foundation for nature-inspired algorithms. Nature-inspired algorithms can be further classified depending on their foundations into environment-based algorithms, bio-inspired algorithms and physics/chemistry-based algorithms. **Environment-based algorithms** draw inspiration from phenomenon occurring in the natural environment as a basis. The lightning search algorithm(LSA) [2] uses the principles of lightning strikes to search through solution spaces effectively. Several novel approaches for different problems, such as the ant colony system(ACS) [3], imperialistic competitive algorithm(ICA) [4] have achieved results using such environment-based techniques. **Physics-based algorithms** use notions of the physical world to achieve optimizations over solution spaces such as the gravitational search algorithm(GSA) [5] uses the concept of gravitation, whereas chemistry-based algorithms implement their solutions using structures and notions of chemistry to achieve their optimization targets such as simulated annealing(SA) [6], which uses the principles of annealing in metallurgy as its inspiration. **Bio-inspired algorithms** use biological and natural processes as their inspiration and can be further classified into Swarm-based algorithms, evolution-based algorithms and ecology-based algorithms. **Evolution-based algorithms** use the principles of evolution and natural selection to achieve the best solutions over multiple iterations. **Ecology-based algorithms** find inspirations from ecological sources as wide-ranging as symbiosis (PS20), biographical (BBO) and weed-colonies(AWC). **Swarm-based algorithms** are designed to replicate a mathematical model of the intelligent behaviour of groups of animals in social situations as they gather information about the solution space.

Swarm-based algorithms such as the Whale Optimization Algorithm [7] mimic the bubbling and spiralling hunting technique of humpback whales to explore solution spaces effectively. The Cuckoo Optimization Algorithm [8] mimics the reproductive behaviour of cuckoos, which consist of laying eggs in another healthy birds' nest and the Squirrel Optimization Algorithm [9] mimics squirrels' movement towards the most food-bearing tree. These algorithms outperform most other nature-inspired algorithms in searching the solution space effectively and are well known for their contributions to this field.

Our contributions in this paper are as follows: We propose a novel nature-inspired algorithm with inspiration drawn from the breeding and lifestyle of the Laysan Albatross. The reproductive process of the Albatross bird was studied extensively and modelled mathematically. The proposed algorithm was validated on 12 well-known classification datasets obtained from the University of California at Irvine (UCI) machine learning repository [10] and DELVE repository [11] to train feed-forward neural networks. A comparison with familiar nature-inspired algorithms for training neural networks shows that our algorithm outperforms the others in terms of accuracy and minimizes overfitting.

The rest of the paper is structured as follows, Sect. 2 presents a discussion of the biological background and inspiration of this work. Section 3 presents the algorithm, definition of terminologies and relevant formulae. Section 4 covers the implementation details of this algorithm when applied to train neural networks. Section 5 presents the evaluation methodologies used to evaluate the performance. Section 6 presents the results obtained, with comparisons made to back-propagation and the Whale Optimization Algorithm. Section 7 evaluates the three algorithms, with a discussion of the results. This paper is the first look into the proposed algorithm for which we propose possible explorations in Sect. 8.

2 Bio-inspiration

Table 1 shows the terminologies used in the rest of this paper and the relation between their real-world occurrence and their implication in the algorithm. Albatrosses are part of the Avian class and the Diomedeidae biological family in Linnaean taxonomy. They are some of the largest extant birds, with their wingspan over 12 feet from tip to tip. They spend most of their lives close to, or over the oceans. Albatrosses are true globetrotters, with individuals often having been documented covering more than 120,000 km in the total distance over a single year. Similar to many other birds, albatrosses possess excellent vision. However, albatrosses are part of only a handful of birds which have yellow or red droplets of oil in their colour receptors, which helps to see through the haze and improves distance vision.

To find the optimal parameters to the optimization function, a population of albatrosses is initialized. Each candidate solution has the potential to carry only a single egg. However, not every individual lays an egg in every iteration, similar to how an Albatross does not lay an egg every year. Additionally, the nest is made in the ground with damp moss and other foliage the albatross parents find in their immediate surroundings, corresponding to the Breeding Threshold Distance. When albatrosses do

Table 1 Terminologies

Term	Meaning	Implication
Location	A location in the natural habitat	Any point in the solution space
Albatross	A bird capable of reproducing in a location near it and moving around	A feasible solution which is capable of either giving rise to a new solution or moving towards the best solution
Egg	An offspring of the Albatross laid at a distance from the Albatross' location	A new feasible solution after random movement, not capable of immediate reproduction
Breeding threshold distance	Maximum distance up to which an egg can be laid	The maximum distance from the initial solution up to which a new solution can be made

give birth, only a single egg is laid. This egg hatches and in a year is ready to join the total population of albatrosses. The entire population of albatrosses then drifts in the direction of the best location for survival and breeding, in a typical curvilinear fashion. This randomness allows for more exploration of the solution space than would have been possible in the case of following undeviating paths.

Additionally, albatrosses being birds with such long lifespans, we did not consider it apt to dynamically alter the solutions' fitness based on age. However, as albatrosses get older, the radius within which they lay their eggs diminishes, as they try to offset their dotage by laying and keeping the eggs closer, for added protection from predators.

3 Algorithm

A random population of albatrosses is first initialized, each one corresponding to a candidate solution. The Breeding Threshold Distance (BTD) as shown in Fig. 1 is the greatest distance in a particular direction at which an albatross can spawn a new solution. This BTD and maximum birth limit are then randomly initialized for every Albatross. The maximum birth limit corresponds to the maximum number of eggs an Albatross can lay in its lifetime, which ranges from 5 to 21. The formula to calculate Breeding Threshold Distance is given in the given Eq. (1), where γ is the rate at which the BTD decreases over iterations.

$$\text{BTD}_i = \frac{\gamma * \text{eggs\_left}_i * (u - l)}{\text{max \_eggs}_i} \tag{1}$$

Fig. 1 Breeding Threshold Distance

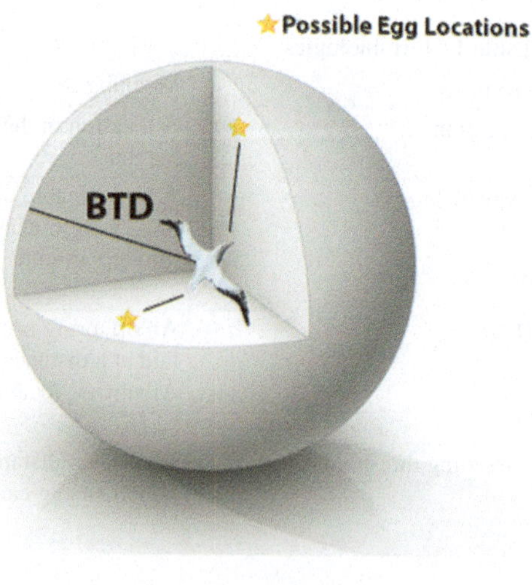

With a certain probability per iteration, each Albatross lays an egg at a maximum of Breeding Threshold Distance from its initial position. Thus, each albatross does not lay an egg every iteration, although it has a 50% chance of doing so. This is akin to the natural world, where albatrosses lay one egg every 2 years. Thus, the probability of an albatross carrying an egg depends on the following formula:

$$f(x) = \begin{Bmatrix} 1 \text{ if } r_2 < 1 - p_e \\ 0 \text{ if } r_2 > 1 - p_e \end{Bmatrix} \tag{2}$$

Constraining each solution vector (Albatross) to only lay one egg keeps the solution space manageable, ensuring that the maximum population is not reached too quickly. In such a situation, there would be pockets of the solution space that have become saturated with solutions, while other regions of the solution space would not be explored at all. Not all of the eggs laid survive; however, a certain percentage of the eggs that were laid are forcefully killed, depicting the loss of eggs due to natural factors and predation. The remaining eggs are added to the population of adult albatrosses, and there is no distinction found between the adult albatrosses and the juveniles that have just joined the population.

As in the wild, with age, the BTD of the albatross reduces as it lays an increasing number of eggs. The BTD of every Albatross reduces linearly over its lifetime depending on the number of eggs already laid. This means that if an albatross is alive over many iterations and has its BTD linearly decreasing, it implies that the albatross was never part of the least fit albatrosses. Thus, to be alive over many generations is a pointer to the fact that a good solution is either one that is currently being occupied by the albatross, or is somewhere near the albatross itself, thus allowing the spawning of eggs closer to the parent.

If the total population of albatrosses is greater than the maximum permissible population, the least fit albatrosses are killed. This is simply natural selection at play and is a central feature of all bio-inspired algorithms. All the mature albatrosses now move towards the healthiest albatross as they desire a healthy environment for the growth of their offspring, corresponding to the best solution. To achieve this, the albatrosses travel a certain percentage of the distance between itself and the direction of the best albatross, as shown in Fig. 2. However, choosing to explore rather than exploit is a very important trade-off. Exploitation refers to making decisions based on the knowledge at a certain point in time, to maximize gain. Exploration refers to making decisions that have previously been explored, in the hope of finding solutions better than the ones found with the current amount of knowledge. In complex optimization problems, it is not advisable to exploit prematurely, as the entire solution space has not been explored. To facilitate both, a random vector is added to the vector pointing in the ideal direction of travel towards the best albatross.

Thus, the final direction in which the albatross ends up travelling in is a combination of this random direction and the ideal direction. The travelling is shown in Fig. 2. As we know the vector from an albatross to the fittest albatross, we compute the new location of the corresponding albatross as the vector sum of this vector and another randomly chosen vector. This ensures that all the albatrosses always move towards

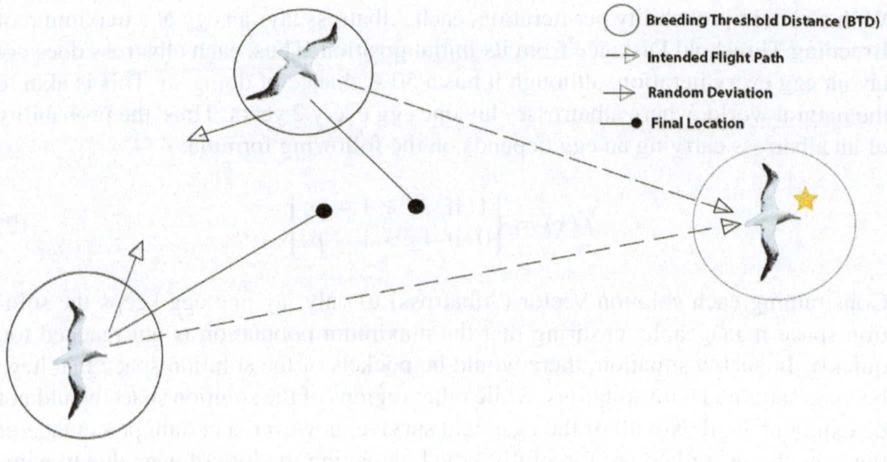

Fig. 2 Albatross travelling

the fittest albatross along with a good amount of exploration. Once the albatrosses have reached the final locations for that iteration, a certain percentage of the worst Albatrosses are killed, to allow room for potentially healthier eggs to mature and grow. The number of remaining albatrosses is given by Eq. (3).

$$N = n * (1 - p_a) \tag{3}$$

Finally, we evaluate the remaining albatrosses for fitness and stop if required; otherwise, the life cycle is continued for the remaining Albatrosses, starting from laying of eggs.

4 Implementation

AOA theoretically is capable of solving all optimization problems in any form. It also supersedes the performance of state-of-the-art optimization algorithms. We train a neural network using this algorithm to evaluate its performance. Depending on the datasets being trained on, they are prone to a large number of local minima and inflexion points. Not only are neural networks challenging to learn and model mathematically, but they are also of great significance to the field of machine learning and deep learning, which rely heavily on the existing training power of gradient-based algorithms such as back-propagation. Thus, any improvements over these prevalent methods are not only of theoretical interest but also have widespread implications due to the extensive employment of neural networks. This algorithm was implemented for a vanilla ANN in Python. The pseudocode for the algorithm implemented is given in Algorithm 1.

Algorithm 1: Albatross Optimization Algorithm

```
[ht]
vlen = no.ofweightsandbiases
max_albatross = 200
ncand_sol = 100
tot_egg_mat = [0] * ncand_sol
for i = 0toncand_sol do
  | v_mat[i] = [random(−2, 2)] * vlen
end
max_egg_limit = [random(5, 21)] * ncand_sol
while fitness(best_sol) ≯ threshold do
    new_egg_mat, tot_egg_mat = give_birth_with_prob(v_mat, egg_mat, btd_mat)
    eggs_left = max_egg_limit − tot_egg_mat
    btd = update_egg_mat(eggs_left, max_egg_limit)
    update_btd(v_mat, btd_mat)  eggs = generate_eggs(new_egg_mat, v_mat, btd)
    v_mat.append(eggs)  total_egg_mat.append([0] * length(eggs))
    max_egg_mat.append([random(5, 21)] * length(eggs))
    if v_mat.shape[0] ≥ max_albatross then
      | kill_worst(v_mat, v_mat.shape[0] − max_albatross)
    end
    travel(v_mat)  kill_worst(v_mat, 0.2 * v_mat.shape[0])
end
```

$$m = 2n + 1 \tag{4}$$

$$\dim = n * m + m * o + m + o \tag{5}$$

To utilize the algorithm to train multi-layer perceptrons, a random population of albatrosses is first initialized, each one corresponding to a set of weights and biases of the neural network. Each albatross is a potential neural network, whose dimensions are given by Eq. (4). Initially, this set of solutions is randomly initialized. The maximum birth limit corresponds to the maximum number of eggs (solution vectors) an Albatross can lay in its lifetime. Thus, additional solutions are solution vectors spawned from existing albatrosses. This maximum birth limit along with the Breeding Threshold Distance(BTD) for each Albatross is then randomly initialized. Not every Albatross lays an egg at a maximum of Breeding Threshold Distance from its initial position. Additionally, 10% of the eggs are killed randomly. The actual position of the eggs is given by Eq. (6).

$$p_e^{ij} = p^{ij} + \mathrm{BTD}_i + \mathrm{factor}_{ij} \tag{6}$$

The eggs are then added to the solution space following which 200 of the fittest solutions (albatrosses) are maintained, discarding the remainder. All the solution vectors then travel a certain percentage of the distance in the direction towards the best solution having the least cost. Once again, 20% of the costliest Albatrosses(solutions) are discarded, to make space for additional candidate neural networks. We finally evaluate the current population of albatrosses(neural networks) for fitness using binary cross-entropy. The algorithm then stops if a certain fitness has been reached; otherwise, the process is continued.

5 Evaluation

In this section, we present the evaluation metrics, with details of the datasets and the multi-layer perceptron (MLP) neural network used.

5.1 Dataset

To test the algorithm and its suitability for neural network training, the algorithm is used for training over 12 datasets, which were selected from the University of California, Irvine (UCI) machine learning repository [10] and the DELVE repository [11]. The datasets all present a classification task where the class of the target variable is aimed to be determined. The performance of our model, naturally, depends on the dataset used. The details of the datasets used are presented in Table 2, along with the MLP structure used for each dataset. To incorporate variation into the datasets, the chosen datasets vary in terms of several features, training examples, trainability, classes and testing examples. This provides an additional challenge to the algorithm. Each dataset is divided as 67% of the instances for training and 33% for testing purpose using random sampling.

Table 2 Dataset metrics

Dataset	#Classes	#Features	#Training samples	#Testing samples	MLP structure
Breast cancer	2	8	461	238	8-17-1
Parkinson	2	22	128	67	22-43-1
Credit	2	61	670	330	61-123-1
Blood	2	4	493	255	4-9-1
Vertebral	2	6	204	106	6-13-1
Australian	2	14	455	235	14-29-1
Chess	2	36	2109	1087	36-73-1
Hepatitis	2	10	102	53	10-21-1
Tic-tac-toe	2	9	632	326	9-19-1
Monk	2	6	285	147	6-13-1
Diabetes	2	19	771	379	19-39-1
Liver	2	10	387	191	10-21-1

5.2 Neural Network

The neural network used for training for all datasets was an MLP with a single hidden layer (having $2 * N + 1$ neurons) and possible activation units being one among softmax, reLU and sigmoid as show in Fig. 3. Three aspects of interest are the architecture of the multi-layer perceptron, representations of the Albatross search agents and the fitness function chosen to evaluate the fitness of the Albatrosses. The proposed neural network architecture for each dataset presented in Table 2. To eliminate the variations due to scale, each of the features is max-min normalized as given by Eq. (7).

$$v' = \frac{v_i - \min_A}{\max_A - \min_A} \tag{7}$$

Each albatross search agent is a single-dimensional vector containing the weights and biases which is fit to the network to obtain results through feed-forward of inputs. The output is then evaluated and the cost of each potential solution vector is found. The fitness is inversely proportional to the cost, where the least costly solution is assumed to be the best albatross. We use binary cross-entropy to evaluate this cost. Each solution vector consists of a set of weights and biases which is spliced for use as shown in Fig. 3.

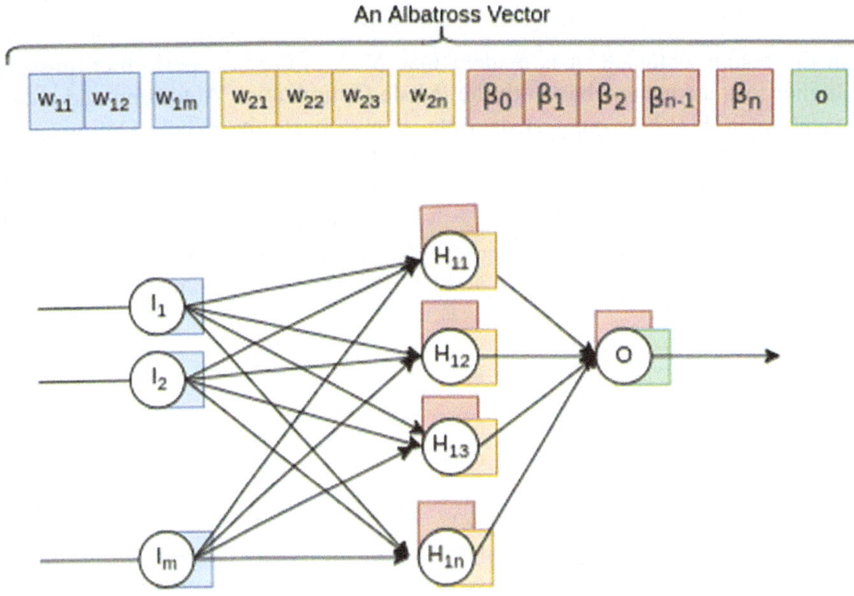

Fig. 3 Neural network architecture

6 Results

This section presents the results of the proposed algorithm. The results have been obtained from an Apple MacBook Pro 2017 Intel(R) Core(TM) i5-7360U CPU @ 2.30GHz, 8 GB RAM using Python v3.7.3. The benchmark algorithms for training neural networks in the literature are back-propagation and the Whale Optimization Algorithm. Our results have been compared against these algorithms and summarized in Tables 3 and 4. The results have been obtained through a total of 10 runs for each dataset. The test accuracy, training accuracy and the standard deviations for both the mean test accuracy and the training accuracy have been presented. In Figs. 4 and 5, the obtained results over the 10 runs are presented in the form of boxplots. Training result boxplots are shown in red, whereas testing result boxplots are shown in blue. It can be seen that the spread of test results obtained through using Whale Optimization Algorithm is much higher than either back-propagation or AOA. It is also seen in dataset results of TicTacToe, Monk, Vertebrae and Parkinsons that the Whale Optimization Algorithm presents much higher training accuracies than those achieved by its testing. Based on these results, several findings come to the fore. These are presented in the following section.

7 Comparison

From the results obtained, it is evident that AOA is the best performer, providing higher test accuracies than both the Whale Optimization Algorithm and back-propagation in 11/12 of our test datasets. It also beats the Whale Optimization

Table 3 Results

Dataset	Whale Optimization Algorithm				Albatross Optimization Algorithm			
	TrainAVG	TrainSTD	TestAVG	TestSTD	TrainAVG	TrainSTD	TestAVG	TestSTD
Australian	0.871	0.009	0.782	0.126	0.854	0.023	0.853	0.024
Blood	0.767	0.009	0.791	0.01	0.756	0.02	0.793	0.014
Breast Cancer	0.971	0.003	0.837	0.16	0.947	0.031	0.957	0.021
Chess	0.669	0.146	0.612	0.139	0.9	0.029	0.896	0.034
Credit	0.87	0.007	0.817	0.065	0.861	0.029	0.855	0.036
Diabetes	0.535	0.014	0.491	0.064	0.532	0.036	0.538	0.048
Hepatitis	0.911	0.046	0.767	0.036	0.834	0.046	0.791	0.026
Liver	0.72	0.015	0.718	0.022	0.722	0.022	0.76	0.024
Monk	0.814	0.05	0.693	0.108	0.76	0.051	0.787	0.05
Parkinsons	0.873	0.036	0.687	0.23	0.854	0.039	0.834	0.041
TicTacToe	0.699	0.019	0.677	0.017	0.679	0.038	0.702	0.027
Vertebrae	0.817	0.014	0.676	0.102	0.729	0.089	0.753	0.081

Table 4 Results

Dataset	Back-propagation			
	TrainAVG	TrainSTD	TestAVG	TestSTD
Australian	0.623	0.094	0.628	0.102
Blood	0.755	0.038	0.733	0.041
Breast cancer	0.88	0.014	0.879	0.014
Chess	0.775	0.017	0.786	0.013
Credit	0.833	0.037	0.841	0.034
Diabetes	0.527	0.014	0.542	0.027
Hepatitis	0.864	0.028	0.727	0.044
Liver	0.739	0.029	0.696	0.04
Monk	0.653	0.061	0.602	0.054
Parkinsons	0.752	0.02	0.731	0.014
TicTacToe	0.649	0.029	0.643	0.021
Vertebrae	0.727	0.012	0.636	0.022

Algorithm on the additional metric and shows that it is empirically resistant to overfitting, providing better and more stable results, as seen by the lower STD values for several datasets.

7.1 Comparison with Back-propagation

Back-propagation is a gradient-based approach and hence is prone to issues such as convergence to local minima and saddle point issues. Overall, the test results using the back-propagation proves to be the worst, on all but two(chess, diabetes) datasets. However, in these cases, back-propagation only shows higher accuracies than the Whale Optimization Algorithm, but not AOA. The issues with gradient-based approaches are present and back-propagation often converges to local minima, without powerful optimization tools such as momentum, which also have their drawbacks.

7.2 Comparison with Whale Optimization Algorithm

The Whale Optimization Algorithm proves to be more competitive than the back-propagation algorithm and performs better than back-propagation in 7/12 dataset results. However, a characteristic of the Whale Optimization Algorithm is its spiralling behaviour. While this spiralling behaviour is beneficial for exploration, it also causes the algorithm to tend to overfit severely, as seen in the high STD values of

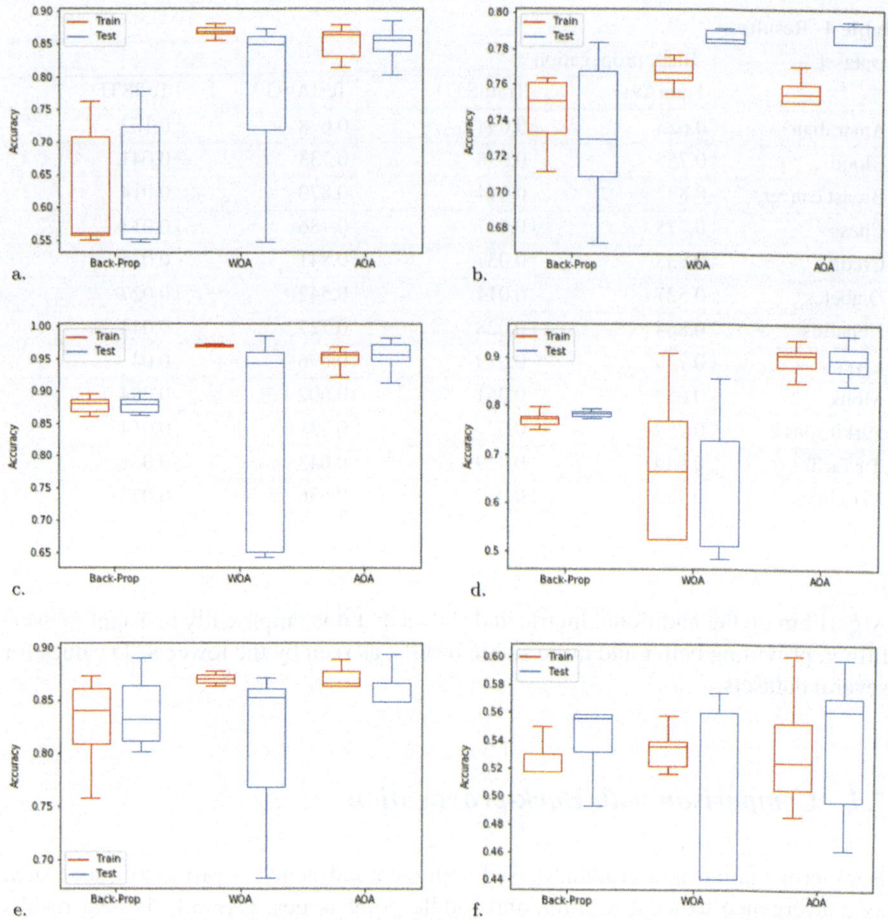

Fig. 4 **a** Australian,**b** blood, **c** breastcancer, **d** chess, **e** credit, **f** diabetes

the Parkinson's and breast cancer dataset. Due to this occasional overfitting, it renders eccentric results, as seen in the unexpected dip in test set performance of the Hepatitis, Monk, Parkinsons, Breast Cancer, Vertebrae and Australian datasets.

8 Conclusion and Future Work

A novel nature-inspired Albatross Optimization Algorithm(AOA) was proposed in this paper. This was done by observing the living style and reproductive patterns of the Laysan Albatross. The algorithm was put to test in scenarios where it was used to find optimal neural network functions for their training using 12 well-known

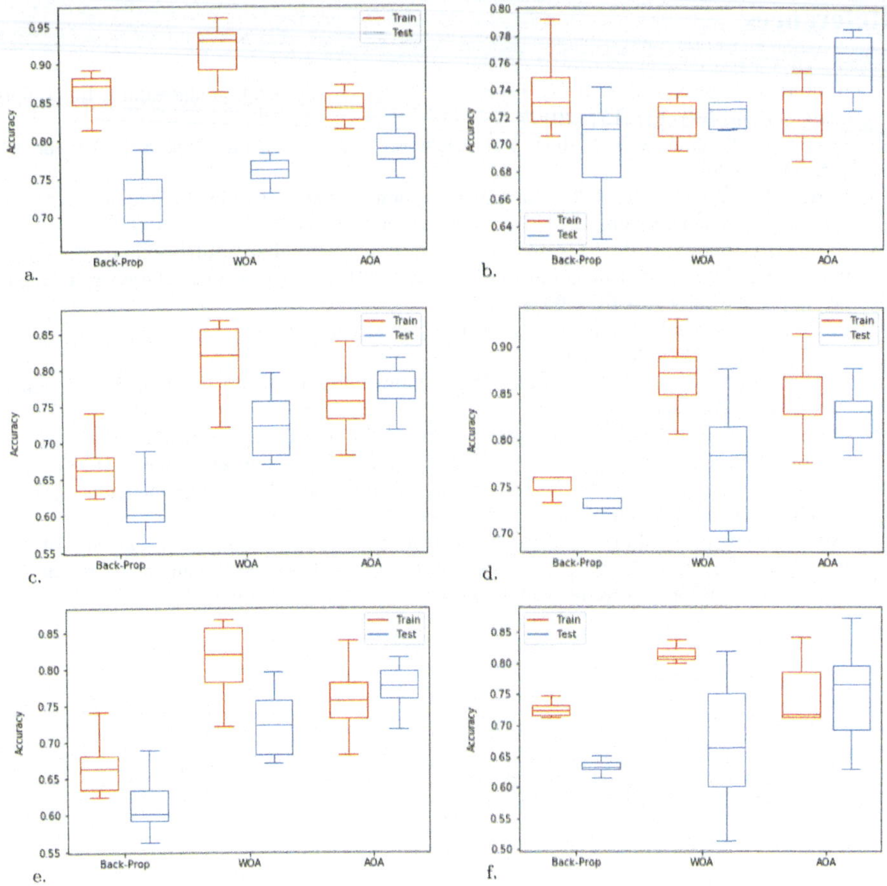

Fig. 5 **a** Hepatitis, **b** liver, **c** monk, **d** Parkinsons, **e** TicTacToe, **f** vertebrae

classification datasets. A rigorous comparative study was also performed with the most popular gradient-based algorithm (back-propagation) and bio-inspired algorithm (Whale Optimization Algorithm) which are currently employed for this task. The results that were obtained proved to be competitive to the state-of-the-art algorithms for the task.

The present work is a first look into the proposed Albatross Optimization Algorithm, which in the future may be extended to more constrained optimization problems of a larger scale. An evaluation of our proposed algorithm on the CEC benchmark is under works, which would give a good idea on the soundness of the algorithm. The applicability of this algorithm concerning non-stationary optimization problems, where the optimization objective is temporally dependant is of high interest. Multi-objective optimization problems and NP-hard problems may also be subject to further exploration using this algorithm, which has a high level of real-world applicability.

References

1. Binitha, S., Sathya, S.S., et al.: A survey of bio inspired optimization algorithms. Int. J. Soft Comput. Eng. **2**(2), 137–151 (2012)
2. Shareef, H., Ibrahim, A.A., Mutlag, A.H.: Lightning search algorithm. Appl. Soft Comput. **36**, 315–333 (2015)
3. Dorigo, M., Gambardella, L.M.: Ant colony system: a cooperative learning approach to the traveling salesman problem. IEEE Trans. Evolution. Comput. **1**(1), 53–66 (1997)
4. Atashpaz-Gargari, E., Lucas, C.: Imperialist competitive algorithm: an algorithm for optimization inspired by imperialistic competition. In: 2007 IEEE Congress on Evolutionary Computation, pp. 4661–4667. IEEE (2007)
5. Rashedi, E., Nezamabadi-Pour, H., Saryazdi, S.: GSA: a gravitational search algorithm. Inform. Sci. **179**(13), 2232–2248 (2009)
6. Kirkpatrick, S., Gelatt, C.D., Vecchi, M.P.: Optimization by simulated annealing. Science **220**(4598), 671–680 (1983)
7. Mirjalili, S., Lewis, A.: The whale optimization algorithm. Adv. Eng. Softw. **95**, 51–67 (2016)
8. Rajabioun, R.: Cuckoo optimization algorithm. Appl. Soft Comput. **11**(8), 5508–5518 (2011)
9. Jain, M., Singh, V., Rani, A.: A novel nature-inspired algorithm for optimization: squirrel search algorithm. Swarm Eolution. Comput. **44**, 148–175 (2019)
10. Dua, D., Graff, C.: UCI machine learning repository (2017). http://archive.ics.uci.edu/ml
11. Rasmussen, C.E., Neal, R.M., Hinton, G.E., van Camp, D., Revow, M., Ghahramani, Z., Kustra, R., Tibshirani, R.: The delve manual. http://www.cs.toronto.edu/delve (1996)

NLP-Based Tools for Decoding the Language of Life

Aparna Chauhan and Yasha Hasija

Abstract As the scientific know-how of the people around the world is expanding, the requirement of new technologies is also growing rapidly. This is evident by the number of papers being published and the new discoveries of scientists that are changing the definition of impossibility day by day. This paper explains one such technology which has made possible not only the recognition of natural language (i.e., human language) by computers but generation of speech and text which is natural language processing (NLP). When machine learning came into picture for assaying large amount data (statistical), deriving meaning from data became easy. Statistical prediction could be made for data containing millions of data points. However, analysis and prediction from textual data still remained a challenge. In 1950s, Alan Turing's publication—Computing Machinery and Intelligence, introduced NLP in computational field which dealt with conversion of human language to machine-readable form and generated written or spoken output. NLP can be further be applied in bioinformatics for deducing the structure and function of a protein from its primary chain sequence or deriving end products of functional genes from their basal sequences as many researchers have found the sequences to be similar to human language calling it the 'language of life'. Current studies are based upon using the rules of NLP in analyzing gene and protein sequences. This article is aimed at exploring the various applications of natural language processing in the field of bioinformatics and medical informatics.

Keyword Natural language processing · Medical informatics · Bioinformatics · Language of life · Protein structure prediction

---

A. Chauhan · Y. Hasija (✉)
Department of Biotechnology, Delhi Technological University, Bawana Rd, Shahbad Daulatpur Village, Rohini, New Delhi 110042, India

A. Noor et al. (eds.), *Proceedings of Emerging Trends and Technologies on Intelligent Systems*, Advances in Intelligent Systems and Computing 1371,
https://doi.org/10.1007/978-981-16-3097-2_18

1 Introduction

The concept of NLP which is subdomain of computer science and artificial intelligence was developed in the early 1950s. It basically involves programming computers so as to identify and analyze human language (natural language), thereby achieving tasks such as speech recognition [1], machine translation [2], analysis of sentiments [3] and emotions, classification of text (e.g., spam filter), development of chatbots and many more. Although initially the rules for natural language processing were complex and hard-coded, during late 1980s a revolution was brought about by the use of machine learning algorithms for language processing which focused on statistical models (like hidden Markov model) to make probabilistic decisions based on the text input by assigning weights to features that constitute the data. It includes natural language interpretation (NLI—understanding of natural language by a machine) as well as natural language generation (NLG—conversion of structured data into plain-text of human language as in chatbots) as shown in Fig. 1. NLI and NLG are the two major challenges of NLP besides speech recognition. The analysis of literal meaning of a word is comparatively easier than identifying the meaning behind a word (opinion mining). So, besides solving problem related to grammar correction and syntax analysis, NLP also aims to understand semantics and with deep

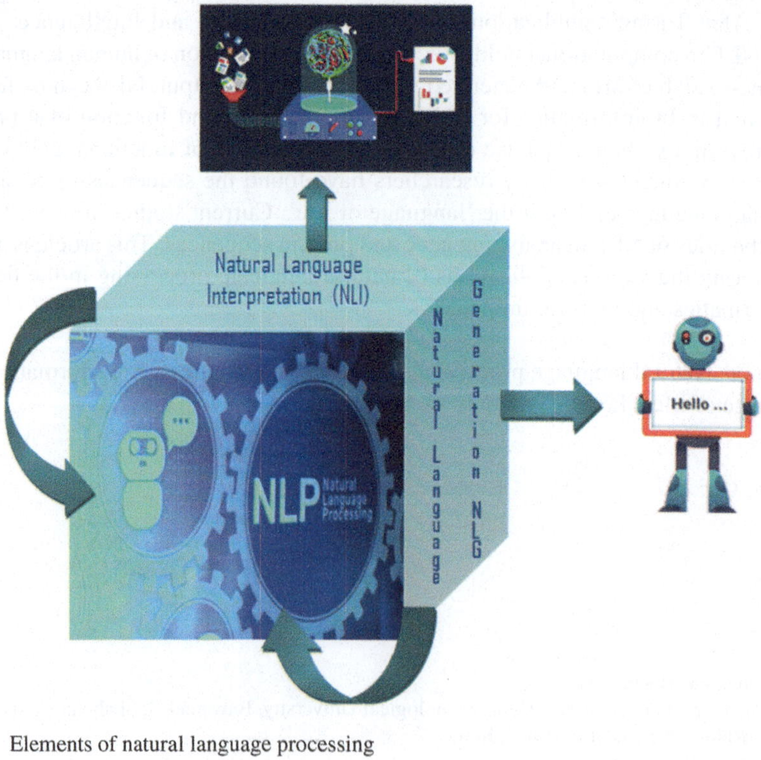

Fig. 1 Elements of natural language processing

learning and artificial intelligence the approach has been examination of patterns for improving the understanding of a program. NLP uses the following techniques for semantic analysis.

- Word sense disambiguation: It uses the context of a sentence to derive meaning behind a word.
- Named entity recognition: Artificial neural network is used to categorize words into groups such as person, animal, place and time for interpretation of a word.
- Natural language generation: It derives tone of emotion behind a text based on database.

Before the concept of NLP came into existence, only numerical data could be interpreted using a machine. In fact, all of the machine learning algorithms required and input of numerical vector to perform the task of classification or regression. This implies that pre-processing of textual data is necessary to convert a corpus of text into vector format. One of the solutions is using bag of words model which represents the textual data as multiset based on the word count without paying any regards to the grammar or order. Refining the data produces better results; thus, pre-processing not only involves conversion of text into vector form, but also removes the redundancy in corpuses, which may include elimination of very common words (like 'a', 'and', 'the', etc.), calculating term frequency and inverse document frequency, etc. When the data are refined, a model is trained on it. In this way a workflow is formed, which can be stored in the form of pipeline for further use. Most of the NLP systems have pre-defined pipeline which works best for corpuses relating to a particular field. It is clear that the improvement of algorithms has enhanced interpretation of vague components that are frequently used by people while speaking or while searching the web, and thus, the rigid rule-based NLP has been almost washed off by the wave of artificial intelligence that has struck the computational industry. With approximately 6,500 languages being spoken across the world, and variation in the context of same language spoken in two different regions, natural language processing has a long way to go in order to interpret the meaning behind overabundance of textual data (comprising of various human languages) which is everyday being generated online. Besides human language, NLP is also entering the field of biology with the same principles being used for interpreting language of life [4], i.e., genetic language.

2 NLP Real-World Applications

Applications of NLP (Fig. 2).

Fig. 2 Applications of natural language processing (NLP)

2.1 *Search Engines*

The textual data need to be classified as it is one of the major requirements for web browsing, filtering data, sentiment analysis and translation of one language to another by identifying and categorizing words according to the language. Natural language processing makes use of convolutional neural networks and other artificial intelligence algorithms for the task of categorizing words, and the results are displayed on the browser based on this classification. Smart searches are possible, in which the user writes the text string and the machine does the rest of the work of analyzing the meaning behind the text and give the best matching result. The ambiguity of human language leads to complexity in string match algorithms and may require the user to select multiple options for searching a domain on a search engine, the use of natural language processing tools help to refine the search and provide more accurate results for a text field, and thus, NLP has become an integral part of search engines which help user to find data online from databases on servers. Another advantage of NLP in search engines is the feature of search auto complete which was instigated by Google and is now used by companies for improving the user experience. It is very helpful in cases where user might not know the full term, instead just a few keywords related to the search. Also, it speeds up the process of searching and NLP-based search and recommendation systems have become common [5].

2.2 *Spell Check*

With the world moving at a faster pace, it becomes difficult for people to manually check all grammatical and spelling errors in a document or mail. The spell check feature makes the task easier and faster without obstructing the work of a user. It makes use of natural language processing to achieve grammatically correct text for different languages [6] in very less time. NLP is integrated into word processing documents and email composers for enhancing the experience of users and saving time and frustration that may be caused by misinterpretation of incorrectly written text. Some of the application softwares that use NLP are Microsoft Word, WordPress and Gmail compose block.

2.3 *Spam Filter*

It is one of the first applications that used NLP and machine learning to segregate our spam and ham mails. As email grew as a medium for communication over large distances and an intermediate for sending official and non-official letters, many companies saw it as an advantage to promote their products and services. Soon the accounts of users were clogged by unimportant and most of the time irritating mails, and it became a tedious task to dig out the important mails from the overflowing mail boxes. Although initially machine learning was limited to numerical data, NLP made it possible to categorize the mails using classification algorithms (SVM, Naïve Bayes, ensemble classifiers, K-nearest neighbor) and separate out spam emails. The state-of-the-art classifiers utilize natural language processing and apply filter on the email messages as spam mails have specific features which are mined by computer algorithms in anti-spam filters [7].

2.4 *Diagnosis of Diseases*

A faster and reliable method for diagnosis of diseases through unstructured records of a physician is use of natural language processing software. Extraction and analysis of data or images is done to predict the occurrence of a disease. Deep learning algorithms are applied on images of skin or on mammogram for determination of the risk of skin or breast cancer respectively. Researchers have been able to predict skin cancer with accuracy of about 94% [8]. Although the technique has become very popular, questions are raised as the features based on which the deep learning algorithm identifies the presence of a disease is not known. One of the best examples is software which analyzed over 500 mammographic image charts in few hours which helped in saving 500 h of work of physicians and also diagnosed breast cancer earlier, thereby facilitating faster treatment.

2.5 *Virtual Assistants*

Virtual agents are software applications or platforms which assist humans as they understand human language and respond to the user, answering queries and performing tasks that the user asks. The technology has escalated significantly since the development of voice activated toy—Radio Rex in 2011, and the first smart virtual assistant as we know it today was launched in 2011 (Siri—VDA installed in iPhone 4S Smartphone). There has been no looking back since then, and a large number of virtual agents have begun to dominate the IT industry, e.g., Google Assistant, Alexa, etc. Besides these, there are also chatbots which have become indispensable to firms providing services online. Chatbots as the name suggests chat with the customer and try to sort common problems related to a service so that every time a customer has a query which can be resolved without a customer care executive, it can be solved quickly at any time as chatbots are available 24 h a day. As these platforms are completely dependent on interpretation and generation of human language, natural language processing is the core of these programs.

2.6 *Machine Translation*

As people are connecting more and more, virtual distance is diminishing between individuals and the world has become a global city over internet where people interact through various social media websites and other platforms. However, with over thousands of languages being spoken across the globe, translation of one language to another has become an urgent necessity to allow smooth communication. Machine translation which is heavily dependent upon natural language processing comes as a relief for people as they are able to communicate in their native language while also enabling multilingual web search [9]. It also helps avid travelers to converse with local people of a region whose language they don't know. Some examples of machine translation that applies artificial intelligence to achieve accurate translations are Google translate and Skype translator. Skype translator allows on-the-fly interpretation and conversion of spoken words in real time, thereby facilitating conversation between people speaking distinct languages.

3 NLP in Bioinformatics

When bioinformatics emerged as an interdisciplinary field after successful completion of Human Genome Project and discovery of genetic code, natural language processing entered the field of biotechnology for unraveling the structures and functions of protein using the protein/DNA sequences as many researchers found them to be closely related to human language when it came to the composition of the sequence. As researchers have successfully predicted protein structures and functions based on preexisting language processing techniques, text mining and natural language processing technologies are being largely employed in bioinformatics research. Plethora of biological data is available across web servers like DisProt [10] (for prediction of protein disorder), SPPIDER [11] (for protein interaction site prediction) and InterPreTS [12] (for prediction of protein–protein interaction), extraction of data related to protein interactions from these servers and developing models through NLP to produce meaningful results have attracted the attention of researchers making it a research hotspot in bioinformatics. Analysis methods such as context-free grammar analysis [13] and ontology analysis [14] have been applied along with machine learning algorithms like ensemble learning and Bayesian network for recognition of protein name and identifying interaction between the proteins.

Different NLP-assisted text mining systems are used to derive variety of information about biological systems and interactions between its components. Some are used to identify names of proteins, genes and other factors essential for metabolic pathways which are in turn necessary for further study of interaction between these entities. The NLP systems used for these purposes iterate through the biological databases (like MEDLINE) and automatically capture the required information (such as genes related to signal transduction pathway) while iterating through the text (based on the query). Besides recognition of words and abstracts, recognition and linking of concepts has also been achieved through processing of written language by the process of concept normalization. It has further enhanced the search through databases by extracting publications and papers based not only on entity recognition but also on concept recognition.

Also, NLP entered the field of biotechnology with analysis of medical (clinical) records which were available in abundance in scientific publications. Extraction of the records and deriving meaningful results was done through natural language processing systems with evaluation being done on biomedical text corpuses available online like CRAFT (Colorado Richly Annotated Full Text) [15] and many more which gained popularity as text mining was introduced in biological literature analysis. This was an effort for the mining of literature relevant to the research topic from the exponentially growing literature online, with PubMed (https://www.ncbi.nlm.nih.gov/pubmed) being the first breakthrough for bioscientists for finding relevant literature with computational tools being used for classification of documents. PubMed (by NCBI) uses NLP rules for searching through MEDLINE database (which stores publications related to life sciences, health, medicine and other relevant topic (Fig. 3). The domain of medicines had been of great interest to computational linguistics for

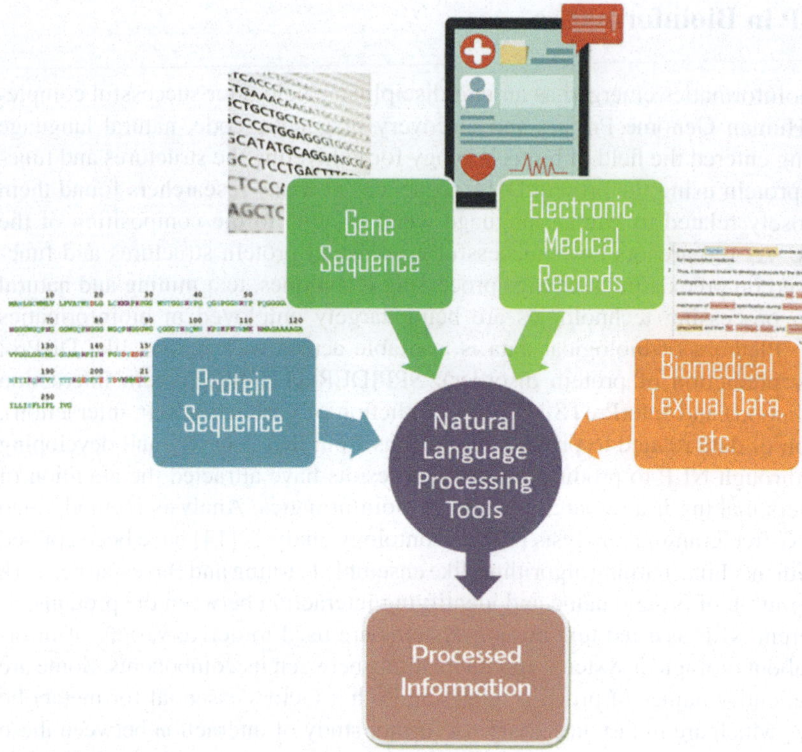

Fig. 3 Natural language processing in bioinformatics

a long time since early 1990s. MedLEE, a natural language processing system developed in 1994, has been successfully applied in medicine since its development. Clinical NLP is advantageous as it makes possible the transformation of plain medical text into organized data which can be processed by computers for treatment of patients and advancement in medical science. Another tool which is used for part of speech tagging for research papers in MEDLINE database.

4 Research Work

Table 1 lists some of the research work done for applying NLP in biology and other algorithms (in which NLP is applicable) as well as some tools used for mining of biological text.

Table 1 Development of NLP concepts in bioinformatics

Topic	Year	Description
Natural language parsers for NLP problems in Biology [16]	January 2007	As the interest grew in developing syntactic parsers for solving natural language processing-related problems in the field of biology, Andrew B Clegg and Adrian J Shepherd in 2009 used dependency graphs to evaluate the performance of various parsers available as open source and which were previously used on bioinformatics projects based upon dependency graphs. Tests were further performed upon the two most commonly used tools—Bikel and Charniak—Lease (released in 2005) parsers, as they turned out to have the best accuracy among the others used by community of computational linguistics Part of speech (POS) tagging is an important part for NLP, so biomedical vocabulary for the purpose of POS tagging which Charniak parser could access along with Bikel 0.9.8 gave the best results. The high scores were achieved due to good POS (part of speech) tagging with most of the errors being associated with POS errors
Use of NLP for generation of networks associated with transcription regulation in *Escherichia coli* K-12 [17]	August 2007	With the huge amount of information and databases available online, manually curating biological databases are not only a laborious but also a very expensive process. Due to the amount of manual work required, it is also prone to errors. This study aimed at use of state-of-the-art NLP systems which were implemented on corpuses of abstracts as well as complete papers. These automatically created regulatory interaction networks in machine-readable form. Due to heavy dependence of systems biology on organized data (used by bioinformatics tools to study biological processes), curation of textual publications which are end results of the vast researches being carried out to uncover the complex interactions between proteins, genes, transcription factors and other biological compounds involved in gene expression, became a necessity. But manual curation was required to ensure dependability of the data and resources As the literature was growing day by day, manual maintenance became a near to impossible task. So to generate the transcriptional regulatory network of species *E. coli* K-12, RegulonDB and EcoCyc databases were mined as well as text was collected from NCBI PubMed through very selective search strategies and curation of the collected data was done. It helped not only in verifying the information which was already curated manually, but also in discovering new information that may have escaped manual curation. Although NLP turned out to be very useful tool for analysis of literature, it could not be used as a fully automated system in place of a trained professional who can uncover all the information in a text and also interpret graphs and figures

(continued)

Table 1 (continued)

Topic	Year	Description
Natural language parsers for extraction of protein–protein interaction [18]	2008	Protein–protein interaction or PPI extraction involves mining of information related to pairs of proteins which are supposed to interact according to biomedical texts. Since the data on biomedical text were growing at a very rapid pace with new papers being published almost every day, researchers faced the challenging task of digging out papers related to their research topic. Fully automated extraction of PPI became need of the hour. This paper aimed at evaluating different framework based state-of-art syntactic parser in extracting protein–protein interactions. First framework is dependency parsing which makes use of tree structure. In this dependency tree format the nodes represent the words while the edges refer to the association between those words. In this experiment, MST [19] and KSDEP [20] were used as representative parsers for dependency parsing framework. The second framework used was Phrase Structure Parsing, which have output is a phrase structure tree of Penn Treebank-style. The four parsers included under this framework were—NO-RERANK [21], RERANK [22], BERKELEY, STANFORD. Deep parsing was the last framework based on which parsers were evaluated. Two variants of Enju parser were taken for the evaluation—ENJU and ENJU-GENIA. The output of the parsers was fed as input feature (statistical) to machine learning classifiers. The classifier was run with features extracted from different combinations of all the parsers as well as different parse representations (like CoNLL, PTB, HD, SD, PAS) which were available for a parser. The PPI extraction method was based on SVM with SubSet Tree Kernels [23]. The two types of features which were incorporated were—BoW (Bag of Words) and parser output features

(continued)

Table 1 (continued)

Topic	Year	Description
Concept normalization of biomedical text making use of natural language processing [24]	2013	Concept normalization systems are used by computers for extraction of suitable data related to a concept from large unstructured text by linking concepts in text to sources containing information that relates to those concepts. The objective of the paper was to study the effectiveness of using NLP for rule-based (as the tools used in normalization of context of biomedical nature are dictionary-based.) normalization of concepts. The two concept normalization systems MetaMap and Peregrine were used upon the Arizona disease corpus. Their performance was analyzed first without using NLP and afterwards using rule-based NLP. Assessment of performance was done for exact as well as inexact boundary matching Due to the voluptuous textual data present, extraction of useful information requires processing by computer systems. This further involves recognizing and linking a particular concept to sources that contain more information relevant to the concept. Abundant researches have been focused on concept recognition (named entity recognition) with lesser studies being done for the more cumbersome task of concept normalization. This study investigates the application of rules (based on natural language processing) for improvement of performance of concept normalization systems for biomedical text. AZD corpus which contained disease concept annotations (along with Unified Medical Language System codes), concept names as well as initiation and end positions of disease when they occurred in middle of sentences was used as data for training (one-third) and testing (two-third) NLP module Combination of annotations of normalization systems with part of speech tagging and chunking information developed the rules for the NLP module making use of OpenNLP tool suite (which is based upon concept of maximum entropy) NLP rules were divided into five sub modules—coordination, abbreviation, term variation, boundary correction and filtering. Only the annotations of Peregrine were utilized for the development of NLP module, MetaMap annotations were not used. The results showed an improvement of 12.3% (in case of exact boundary matching) and 11.1% (in case of concept identifier matching) for Peregrine, while for MetaMap there was an improvement of 14.1% (in exact boundary matching) and 12.9% (in concept identifier matching). In error analyzing, the major sources of error were term variation and error in boundary correction. Filtering errors were the result of incorrect annotation by gold standard. Coordination and abbreviation errors turned out to be few relatively. Although there were some errors, the use of NLP module was definitely advantageous for normalization of disease concepts and the module developed can be utilized for any system that normalizes concepts. However, its use in normalization of concepts besides diseases needs to be studied further for wider application

(continued)

Table 1 (continued)

Topic	Year	Description
PubTator—A tool for curation of biological text online [25]	2013	Online assessment of knowledge became indispensible for current researches going on in biomedical field. The source of this knowledge were biological databases which contained data curated by experts. There was necessity for automatic computer tools that assist researchers in biocuration. PubTator not only showed the capability of improving the efficiency, but also increasing the accuracy of curating online literature manually. With an interface similar to PubMed, PubTator offers unique components which differentiate it from other tools used for annotating and searching literature. It offers options for highly specific search as it allows user to choose from the following six options—PubMed, Gene, Chemical Disease, Semantic Search and PMID List. It supports both searches for keywords as well as semantics. Text mining tools which showed superlative performance were used for recognition of biological entities. These included • GeneTUKit [26]—for mention of genes • GenNorm [27]—for gene normalization • DNorm—for disease normalization • SR4GN [28]—for species • tmVar [29]—for mutations Lookup approach based on dictionary for chemicals It differs from PubMed as biological entities which have been computed previously are highlighted with different colors for different entities. Genes are given purple color, green means chemical, orange for diseases and so on. It simplifies the task of looking up for annotations. For annotation purpose, three tasks are supported • Document Triage—Curatable articles are prioritized based on their reading. Identification of curatable articles can be done by either ticking the check box present on the left of the article or by indicating whether the article on the top of PubTator annotations page • Entity annotation—PubTator also allows for annotation of bio-entities the steps for which are given in the online tutorial • Relationship annotation—It further allows the user to specify what kind of relation they want to extract from the literature, like PPI, relations between a gene and disease PubTator was developed to cater the need for automated computer based systems that curate and annotate biological concepts. An integral component of the text mining techniques utilized for development of the tool is NLP, which is necessary for analyzing and deriving valuable information from textual data. Existing mining techniques and better algorithms are the future goals which need to be accomplished automated biocuration in order to assist researchers in their studies

(continued)

Table 1 (continued)

Topic	Year	Description
Applying NLP for studying molecular interaction that may lead to NSCLC (non-small cell lung cancer) [30]	2014	With 80–85% of lung cancer patients belonging to the category of NSCLC and a very low survival rate has made it compulsory to dig deeper into the causes and treatment methods. The pathogenesis of NSCLC at molecular level is unknown till now; thus, investigation of pathways connected to it and the genes involved in its generation need to be analyzed for development of a remedy This paper aimed at deriving those molecular interactions (and hence the NSCLC related phenotypes) by using NLP tool of GeneWays on mouse genome database along with tool for annotating mammalian phenotype. Results of this study suggested that TG, NF1, NF2 and E2F1 genes may be an integral part in advancement of NSCLC. However due to the restriction on amount of literature available on the pathways involved in progression of NSCLC, further study needs to be done for knowing the role of above mentioned genes as well as uncovering the role of other genes that may have an important function in development of NSCLC
Structural modeling of protein–protein interactions by using natural language processing [31]	2018	NLP methods are progressively being applied to biomedical texts for retrieving information through matching of concepts (normalization) and entities (recognition). Biomedical researches provide results that restrict the binding mode, which can further be utilized in preparation of docking tools. This paper is based on a text mining (TM) tool which extracts information related to binding sites and orientation of proteins from abstracts published on PubMed and successfully applied it for docking of protein molecules. Extension of the TM tool was prepared making use of natural language processing methods for deriving the residue context from biomedical papers. The dictionaries which existed already for identification of protein interactions were not sufficient for accurately predicting their binding. The modified TM tool that was developed in this research was able to distinguish word that related to binding sites and other interactions for improving the performance of TM

5 Recent Advances

Natural language processing has proved its prominence in computational biology with NLP-based algorithms being applied to not only summarize enormous medical and clinical data, but also uncovering the biological interactions such as protein–protein interaction (PPI) networks from published works. A recent progress has been the use of NLP-based search engines specific for clinical data processing, indexing and searching [32] and discovering mutations which effect PPI from curated literature [33]. NLP has solved the long unresolved problem of organizing relatively unstructured data especially the electronic health record (ETH) which include clinical notes and preliminary reports generated by medical professionals. Another interesting application has been the use of virtual assistant in laboratories [34] providing a reliable support especially to researchers with physical impairments enabling to carry out daily routine in laboratory with ease. Development of independent NLP pipelines for clinical and medical information [35] depicts the power that this technology holds in bioinformatics research as it has displayed its usefulness in the most complex problems of computational biology ranging from protein structure prediction [36] to defining relations of genes with particular diseases [37].

6 Future Prospects

Natural language processing is a very strong tool for using computer systems in analyzing textual data and deriving results besides supplementing text mining tasks. With the ever-growing health problems and an urgent need for their treatment, the application of NLP for extracting useful information from research papers and articles can help in prognosis of diseases which are almost impossible to cure in last stages and may not have any symptoms in early stages. Also, as the interaction between proteins and other biological elements in our body have major effect on our health, using NLP to predict structure and function of proteins and their interactions with other proteins or metabolic products can further help in determining the pathways involved in pathogenesis and other malfunctioning in the processes in our body. NLP has a long way to go in order to predict accurately the occurrence of a disorder by using the medical records of a patient. Better tools and algorithms that can make use of NLP are bound to generate better tools especially text mining tools for extracting valuable information from the humongous amount of medical and biological text present online.

References

1. Chapelle, C.A., Chung, Y.-R.: The promise of NLP and speech processing technologies in language assessment. Lang. Test. **27**, 301–315 (2010). https://doi.org/10.1177/0265532210364405
2. Khan, N.S., Abid, A., Abid, K.: A Novel Natural Language Processing (NLP)–based machine translation model for English to Pakistan sign language translation. Cognit. Comput. **12**, 748–765 (2020). https://doi.org/10.1007/s12559-020-09731-7
3. Velupillai, S., Mowery, D., South, B.R., Kvist, M., Dalianis, H.: Recent advances in clinical natural language processing in support of semantic analysis. Yearb. Med. Inform. **24**, 183–193 (2015). https://doi.org/10.15265/IY-2015-009
4. Nambiar, A., Heflin, M., Liu, S., Maslov, S., Hopkins, M., Ritz, A.: Transforming the language of life. In: Proceedings of the 11th ACM International Conference on Bioinformatics, Computational Biology and Health Informatics, pp. 1–8. ACM, New York, NY, USA (2020). https://doi.org/10.1145/3388440.3412467
5. Guo, W., Gao, H., Shi, J., Long, B., Zhang, L., Chen, B.-C., Agarwal, D.: Deep natural language processing for search and recommender systems. In: Proceedings of the 25th ACM SIGKDD International Conference on Knowledge Discovery & Data Mining, pp. 3199–3200. ACM, New York, NY, USA (2019). https://doi.org/10.1145/3292500.3332290.
6. Zaky, D., Romadhony, A.: An LSTM-based Spell Checker for Indonesian Text. In: 2019 International Conference of Advanced Informatics: Concepts, Theory and Applications (ICAICTA), pp. 1–6. IEEE (2019). https://doi.org/10.1109/ICAICTA.2019.8904218
7. Srinivasan, S., Ravi, V., Alazab, M., Ketha, S., Al-Zoubi, A.M., Kotti Padannayil, S.: Spam Emails detection based on distributed word embedding with deep learning. Presented at the (2021). https://doi.org/10.1007/978-3-030-57024-8_7
8. Brinker, T.J., Hekler, A., Utikal, J.S., Grabe, N., Schadendorf, D., Klode, J., Berking, C., Steeb, T., Enk, A.H., Von Kalle, C.: Skin cancer classification using convolutional neural networks: systematic review. J. Med. Internet Res. **20**, 1–8 (2018). https://doi.org/10.2196/11936
9. Madankar, M., Chandak, M.B., Chavhan, N.: Information retrieval system and machine translation: a review. Procedia Comput. Sci. **78**, 845–850 (2016). https://doi.org/10.1016/j.procs.2016.02.071
10. Vucetic, S., Obradovic, Z., Vacic, V., Radivojac, P., Peng, K., Iakoucheva, L.M., Cortese, M.S., Lawson, J.D., Brown, C.J., Sikes, J.G., Newton, C.D., Dunker, A.K.: DisProt: a database of protein disorder. Bioinformatics **21**, 137–140 (2005). https://doi.org/10.1093/bioinformatics/bth476
11. Langdon, Q.K., Peris, D., Kyle, B., Hittinger, C.T.: Sppider: A species identification tool to investigate hybrid genomes with high-throughput sequencing. Mol. Biol. Evol. **35**, 2835–2849 (2018). https://doi.org/10.1093/molbev/msy166
12. Russell, R.B., Aloy, P.: InterPreTS: protein interaction prediction through tertiary structure. Bioinformatics **19**, 161–162 (2003). https://doi.org/10.1093/bioinformatics/19.1.161
13. Temkin, J.M., Gilder, M.R.: Extraction of protein interaction information from unstructured text using a context-free grammar. Bioinformatics **19**, 2046–2053 (2003). https://doi.org/10.1093/bioinformatics/btg279
14. Skusa, A., Rüegg, A., Köhler, J.: Extraction of biological interaction networks from scientific literature. Brief. Bioinform. **6**, 263–276 (2005). https://doi.org/10.1093/bib/6.3.263
15. Verspoor, K., Cohen, K.B., Lanfranchi, A., Warner, C., Johnson, H.L., Roeder, C., Choi, J.D., Funk, C., Malenkiy, Y., Eckert, M., Xue, N., Baumgartner, W.A., Bada, M., Palmer, M., Hunter, L.E.: A corpus of full-text journal articles is a robust evaluation tool for revealing differences in performance of biomedical natural language processing tools. BMC Bioinform. **13** (2012). https://doi.org/10.1186/1471-2105-13-207
16. Clegg, A.B., Shepherd, A.J.: Benchmarking natural-language parsers for biological applications using dependency graphs. BMC Bioinform. **8**, 1–17 (2007). https://doi.org/10.1186/1471-2105-8-24

17. Rodríguez-Penagos, C., Salgado, H., Martínez-Flores, I., Collado-Vides, J.: Automatic reconstruction of a bacterial regulatory network using Natural Language Processing. BMC Bioinform. **8**, 1–11 (2007). https://doi.org/10.1186/1471-2105-8-293
18. Miyao, Y., Sagae, K., Sætre, R., Matsuzaki, T., Tsujii, J.: Evaluating contributions of natural language parsers to protein-protein interaction extraction. Bioinformatics **25**, 394–400 (2009). https://doi.org/10.1093/bioinformatics/btn631
19. McDonald, R., Lerman, K., Pereira, F.: Multilingual dependency analysis with a two-stage discriminative parser, p. 216 (2006). https://doi.org/10.3115/1596276.1596317
20. Sagae, K., Tsujii, J.: Shift-reduce dependency DAG parsing, pp. 753–760 (2008). https://doi.org/10.3115/1599081.1599176
21. Chiang, D.: Statistical parsing with an automatically-extracted tree adjoining grammar, pp. 456–463 (2000). https://doi.org/10.3115/1075218.1075276
22. McClosky, D., Charniak, E., Johnson, M.: Reranking and self-training for parser adaptation, pp. 337–344 (2006). https://doi.org/10.3115/1220175.1220218
23. Sætre, R., Sagae, K., Tsujii, J.: Syntactic features for protein-protein interaction extraction. In: CEUR Workshop Proceedings, p. 319 (2007)
24. Kang, N., Singh, B., Afzal, Z., van Mulligen, E.M., Kors, J.A.: Using rule-based natural language processing to improve disease normalization in biomedical text. J. Am. Med. Informatics Assoc. **20**, 876–881 (2013). https://doi.org/10.1136/amiajnl-2012-001173
25. Wei, C.H., Kao, H.Y., Lu, Z.: PubTator: a web-based text mining tool for assisting biocuration. Nucleic Acids Res. **41**, 518–522 (2013). https://doi.org/10.1093/nar/gkt441
26. Huang, M., Liu, J., Zhu, X.: GeneTUKit: a software for document-level gene normalization. Bioinformatics **27**, 1032–1033 (2011). https://doi.org/10.1093/bioinformatics/btr042
27. Wei, C.H., Kao, H.Y.: Cross-species gene normalization by species inference. BMC Bioinform. **12** (2011). https://doi.org/10.1186/1471-2105-12-S8-S5
28. Wei, C.H., Kao, H.Y., Lu, Z.: SR4GN: a species recognition software tool for gene normalization. PLoS ONE **7**, 7–11 (2012). https://doi.org/10.1371/journal.pone.0038460
29. Wei, C.H., Harris, B.R., Kao, H.Y., Lu, Z.: TmVar: a text mining approach for extracting sequence variants in biomedical literature. Bioinformatics **29**, 1433–1439 (2013). https://doi.org/10.1093/bioinformatics/btt156
30. Li, J., Bi, L., Sun, Y., Lu, Z., Lin, Y., Bai, O., Shao, H.: Text mining and network analysis of molecular interaction in non-small cell lung cancer by using natural language processing. Mol. Biol. Rep. **41**, 8071–8079 (2014). https://doi.org/10.1007/s11033-014-3705-5
31. Badal, V.D., Kundrotas, P.J., Vakser, I.A.: Natural language processing in text mining for structural modeling of protein complexes. BMC Bioinform. **19**, 1–10 (2018). https://doi.org/10.1186/s12859-018-2079-4
32. McEwan, R., Melton, G.B., Knoll, B.C., Wang, Y., Hultman, G., Dale, J.L., Meyer, T., Pakhomov, S.V: NLP-PIER: a scalable natural language processing, indexing, and searching architecture for clinical notes. AMIA Jt. Summits Transl. Sci. Proceedings. AMIA Jt. Summits Transl. Sci. **2016**, 150–159 (2016)
33. Qu, J., Steppi, A., Zhong, D., Hao, J., Wang, J., Lung, P.-Y., Zhao, T., He, Z., Zhang, J.: Triage of documents containing protein interactions affected by mutations using an NLP based machine learning approach. BMC Genomics **21**, 773 (2020). https://doi.org/10.1186/s12864-020-07185-7
34. Austerjost, J., Porr, M., Riedel, N., Geier, D., Becker, T., Scheper, T., Marquard, D., Lindner, P., Beutel, S.: Introducing a virtual assistant to the lab: a voice user interface for the intuitive control of laboratory instruments. SLAS Technol. Transl. Life Sci. Innov. **23**, 476–482 (2018). https://doi.org/10.1177/2472630318788040
35. Jin, Y., Li, F., Yu, H.: BENTO: A visual platform for building clinical NLP pipelines based on CodaLab. In: Proceedings of the 58th Annual Meeting of the Association for Computational Linguistics: System Demonstrations, pp. 95–100. Association for Computational Linguistics, Stroudsburg, PA, USA (2020). https://doi.org/10.18653/v1/2020.acl-demos.13
36. Liu, B., Zhang, D., Xu, R., Xu, J., Wang, X., Chen, Q., Dong, Q., Chou, K.-C.: Combining evolutionary information extracted from frequency profiles with sequence-based kernels for

protein remote homology detection. Bioinformatics **30**, 472–479 (2014). https://doi.org/10.1093/bioinformatics/btt709
37. Zou, Q., Li, J., Wang, C., Zeng, X.: Approaches for recognizing disease genes based on network. Biomed Res. Int. **2014**, 1–10 (2014). https://doi.org/10.1155/2014/416323

Zomato Review Analysis Using NLP

Puppala Bala Bhavana, Hari Kishan Kondaveeti, and Asish Kumar Dalai

Abstract Natural language processing is generally used to understand the interactions between humans and machine, and it is also a subpart of artificial intelligence. Sentimental analysis is a subpart of AI and used to differentiate the words by the patterns that are given by people like good, bad, or average. Sentiment analysis is used to know about user's likes and dislikes toward a product. Zomato is an app which is used for ordering foods from different restaurants and allows user to drop their reviews and ratings of respective food that they ordered. These reviews given by the customers will be taken for sentimental analysis. The buying decision of a customer is also depending on the reviews and ratings. So, reviews have a lot more important. It is estimated that 79% of unstructured data is producing every day. A lot of textual data are generating through social media, healthcare, e-commerce applications. We as humans cannot analyze these huge volumes of data. Sentiment analysis will help us to do this task very easily. In this project, we can train with millions of data and predict new data as positive, negative, and neutral. We use artificial neural networks (ANNs) to train the model and to predict the reviews as positive, negative, or neutral. Artificial neural networks use feedforward networks and work as biological neural networks.

Keywords Natural language processing · Artificial neural networks · Sentimental analysis · Feedforward network

P. B. Bhavana · H. K. Kondaveeti · A. K. Dalai (✉)
School of Computer Science Engineering, VIT-AP University, Beside Secretariat, Near Vijayawada, Amaravati, Andhra Pradesh, India
e-mail: asish.d@vitap.ac.in

P. B. Bhavana
e-mail: balabhavana.puppala@vitap.ac.in

H. K. Kondaveeti
e-mail: kishan.kondaveeti@vitap.ac.in

A. Noor et al. (eds.), *Proceedings of Emerging Trends and Technologies on Intelligent Systems*, Advances in Intelligent Systems and Computing 1371,
https://doi.org/10.1007/978-981-16-3097-2_19

1 Introduction

In today's world, everything is digitalized. Whenever we buy any products, food items online, our decision to buy the product is based on the reviews and ratings given by the customers. Reviews play a vital role in the buying decision of a customer. Companies also look into the reviews given by the customers from different locations to know and analyze whether the customers are satisfied by their products and services. "**Zomato Reviews Analysis**" project is based on the data collected from the Zomato Bangalore branch. Bangalore is the capital of Karnataka. One can find all the different restaurants across the world in Bangalore. There are an estimated twelve thousand plus restaurants located in the city cuisines, reviews, and ratings given by customers. Through this project, we analyze the sentiment of the reviews given by the customers using **natural language processing (NLP)** and predict them as positive, negative, and neutral by using **artificial neural networks (ANN)** [1].

1.1 Overview

Now everywhere and everything are data, but a person who can wisely use the data will be in the highest position; nowadays we humans are nothing without data. But the data we find are unstructured and it is in huge GB's. It would become really difficult for a person to manually clean it and analyze data which are categorical or in document formats that require more human work, higher cost, and more time. Instead of this, if we train a computer to analyze the data, it does perform accurately though it depends on how you trained the computer but what methods you have to use to train a computer. This is where NLP comes into the picture.

Now NLP will convert the data that a computer can be learned easily, and later ANN trains it to identify the mood or sentiment of that particular word.

Zomato is an online food delivery application where a customer can order food from different restaurants and can give reviews and ratings about the food ordered from restaurants. The reviews of the food of respective restaurants can be used for sentiment analysis and will be predicted as positive, negative, and neutral. This will help the restaurants to know the customer satisfaction level and can improve their services and product quality.

1.2 Existing Methods and Recent Works

In today's digital world, many people are buying products online. After customers bought the products, they share their opinion about the particular product as ratings and reviews. These reviews have a major impact on the products, and further decisions took by the customers. Reviews will help the companies to know whether the customers are satisfied with their products and services or not. By analyzing reviews, companies can better understand the customers and can improve their product quality, sales decisions, introduce new schemes, offers, etc. But it is very difficult and time-taking for humans to analyze every review given by the customers as they are huge. By using the sentiment analysis concept, we can analyze the reviews easily within a fraction of seconds.

Recently, they are using NLP to train the language to the system and sentimental analysis to detect the sentiment of the word, and later using ANN they are training the model and predicting the new reviews as positive, negative, or neutral.

2 Proposed Method and Solution

2.1 Method

An artificial neural network is a computing system inspired by the analogy of biological neurons, which is used to process the information as a human brain does. ANN uses a feedforward network that consists of different layers, namely input, hidden, and output layers. Artificial neuron is also called "perceptron".

A biological neuron contains dendrites, cell body, axon, nucleus, synapse, etc. The information will be given to the soma by dendrites. Then, it will be transferred through the axon, and finally, through synapse which connects the axon terminals of two neurons, the information will be transferred to another neuron. This is how a biological neuron works and shares the information among millions of neurons present in the brain.

By the inspiration of biological neurons, artificial neuron was implemented. In artificial neuron (perceptron), the information is given by the inputs which have some weights. Then, we apply some activation function to get the output.

2.2 Solution

In this, we applied sentiment analysis steps like sentence segmentation, word tokenization, stemming, removing stopwords, count vectorization to the better understanding of a text to the system, and later we randomly initialized weights to the inputs. Later, it goes into the hidden layers where all the calculations happen; using activation functions we define the output of the given input.

Forward Propagation: Forward propagation happens from the left to right where every neuron gets active, and this results in the activation of all the neurons in the node and with a weight limit.

Then, it compares the output value which is predicted one and the input which we were given called actual value and calculates the error or cost function [2].

Later, this cost function or error value is reduced using gradient descent [3].

Backpropagation: Backpropagation is nothing but vice versa of forward propagation; it happens from left to right where the cost function value is backpropagated and it updates the weight of the nodes to reduce the error. This continues until we get the predicted value.

This repeats certain times by changing the weights of inputs after each observation. Later we deployed this model in a web framework that was done using flask.

3 Experimental Results

We have done Zomato review analysis using NLP and ANN where we tried to acknowledge the system using NLP later using ANN we trained the model where it calculates the moods of the texts given then differentiates them as positive, negative, or neutral, and then, we deployed our model in a flask web framework.

Now our model predicts whether the given new review is negative, positive, or neutral.

3.1 Theoretical Analysis

See Figs. 1, 2, 3 and 4.

3.2 Natural Language Processing (NLP)

The data we find are unstructured, and it is in huge GB's. It would become really difficult for a person to manually clean it and analyze data which is categorical or in document formats that require more human work, higher cost, and more time. Instead of this, if we train a computer in a pattern to analyze the data, it does perform accurately though it depends on how you trained the computer but what methods you have to use to train a computer. This is where NLP comes into the picture. Now NLP will convert the data that a computer can be learned easily.

Natural language processing is a part of AI which trains the system with all the human languages. NLP is used to work with semantics to do applications used in the real world with different structures of the language with suitable efficient algorithms we also tried to make it to understand all the text.

Sentimental analysis [4] plays a major role in NLP. There are some steps followed by it.

- Sentence segmentation
- Word tokenization

Fig. 1 Block diagram

- Stemming
- Lemmatization
- Removing stopwords.

3.3 Artificial Neural Network (ANN)

ANN is similar to our biological nervous system. It calculates the moods of the cleaned text or content by giving them weights and adding activation function [5] to

```
t="Great Food.Some dishes are really good but some are very bad sounds like sour....."
s=re.sub("(n\'t)|(\w+nt)"," not",t)
s=re.sub("(\'ve)"," have",s)
s=re.sub("(\'s)"," is",s)
s=re.sub("(\'m)"," am",s)
s=re.sub("\\\\n"," ",s)
s=re.sub(".\\\\."," ",s)
s=re.sub("\\\\."," ",s)
s=re.sub("\\\\x"," ",s)
s=re.sub("x8"," ",s)
s=re.sub('[^a-zA-Z]',' ',s)
s=s.lower()
s=s.split()
s= [word for word in s if not word in set(stopwords.words('english'))]
ps = PorterStemmer()
s = [ps.stem(word) for word in s if not word in set(stopwords.words('english'))]

s=' '.join(s)
b=[]
b.append(s)
x1=cv.fit_transform(b).toarray()
x1
x1 = sequence.pad_sequences(x1, maxlen=1500)
c1=model.predict(x1)
#lb.inverse_transform(c1)

[nltk_data] Downloading package stopwords to
[nltk_data]     C:\Users\Lenovo\AppData\Roaming\nltk_data...
[nltk_data]   Package stopwords is already up-to-date!

[5]: c1

:[5]: array([[2.2473617e-38, 1.0000000e+00, 5.2446408e-24]], dtype=float32)

[6]: prediction=np.argmax(c1)

[7]: prediction

:[7]: 1
```

0 -> Negative 1 -> Neutral 2-> Positive

Fig. 2 Predicted—neutral

it. Later it performs the loss functions and uses gradient descent functions to reduce the loss.

4 Conclusion and Future Scope

By using the "Zomato Reviews Analysis" project, we can predict the reviews of different restaurants as positive, negative, or neutral successfully. This will help the restaurants to know the customers' satisfaction and in improving food quality and in bringing new food items with different flavors. It gave the accuracy as 97.88% with a loss of 0.054 which says the model is performing well.

These can be used in the healthcare department where you train the model with medical terminology and use searching or retrieving the information relevantly from all the sources.

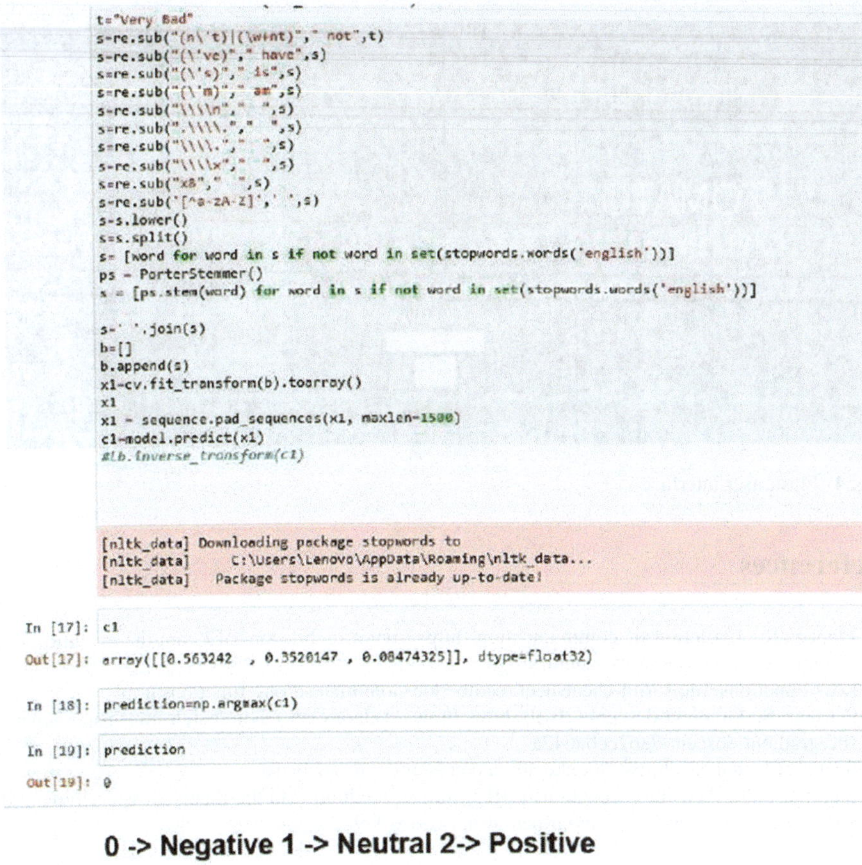

Fig. 3 Predicted—positive

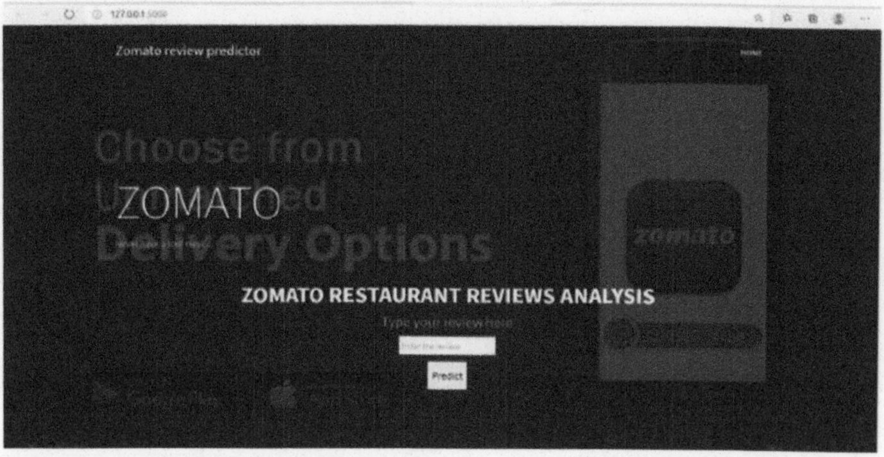

Fig. 4 Flask user interface

References

1. Munoz, R.: Evolution of communication. https://www.mobilecon2012.com/the-evolution-of-communication-through-the-centuries/
2. Loss Functions: https://ml-cheatsheet.readthedocs.io/en/latest/loss_functions.html
3. Patrikar, S.: Gradient descent (2019). https://towardsdatascience.com/batch-mini-batch-stochastic-gradient-descent-7a62ecba642a
4. Sentiment Analysis: https://monkeylearn.com/sentiment-analysis/
5. Sharma, A.V.: Activation functions (2017). https://medium.com/the-theory-of-everything/understanding-activation-functions-in-neural-networks-9491262884e0

LSB Technique-Based Dual-Image Steganography Using COS Function

Aiman Jan, Shabir A. Parah, Muzamil Hussan, and Bilal A. Malik

Abstract Due to the advancement in technology, the exchange of information via a network has been taking such a great peak that each and every person is highly dependent on the Internet, like in the current Covid-19 pandemic situation. But sharing information/secret medical data by means of the Internet has created a risk to the information loss. Thus, data exchange should be done in such a way that its presence cannot be perceived by the third party with bad intentions and should not be able to modify or alter the important message. For the safety of important data, many algorithms like cryptography, steganography, watermarking, etc. have been used by researchers to safeguard data sharing via an insecure Internet. Cryptography and steganography together can provide better security to the data as one encrypts the data and the other hides its presence in multimedia, like image, audio, video, etc. In this paper, data encryption and image steganography have been presented to provide security to the message that is being transferred through the Internet. The experimental outcomes show that the proposed technique is able to transfer important data securely without coming into the eyes of a hacker. The PSNR value for the proposed system is about 51.14 with 1 bit per pixel (bpp) capacity. As per the security view, the presented method has a 7.99 entropy value, which means the system is secure against attacks.

Keywords Covid-19 pandemic · Cryptography · Encryption · Image steganography · Medical data · Multimedia · Perceive · Security

A. Jan (✉) · S. A. Parah · M. Hussan
Department of Electronics and Instrumentation Technology, University of Kashmir, Srinagar, India

B. A. Malik
Institute of Technology, University of Kashmir, Srinagar, India
e-mail: bilalmalik@kashmiruniversity.ac.in

A. Noor et al. (eds.), *Proceedings of Emerging Trends and Technologies on Intelligent Systems*, Advances in Intelligent Systems and Computing 1371,
https://doi.org/10.1007/978-981-16-3097-2_20

1 Introduction

In today's digital world, most of our daily activities are dependent on the Internet. Many forms of communication like financial, health, business-related tasks are done online. Thus, constantly tons of data are being partaken through the Internet. Like in the present scenario of the Covid-19 pandemic, technological innovations are playing a great role where almost each and every work is done online, whether it is related to the education task, financial work, or sharing patient records for diagnosing remotely. But, sharing information over the Internet has created a risk of either data loss or data modification that may result in wrong information, economic failure, or wrong diagnosis [1]. Many methods have been put forth by the researchers for the safeguard of data, like cryptography, steganography, watermarking techniques, and are looking for the different ways by which more security can be provided to the transferring information.

Cryptography is one of the best methods used for data security. It converts the secret data into an unreadable form. It modifies/encrypts data so that the intruder should not be able to read the actual message or information or data. But, if the eavesdropper one way or another way is successful in decrypting the information, he can alter it and may transfer the wrong information to the receiver. Hence, transferring encrypted information is not the proper solution to the safety of data as it attracts third-party attention by its form. So, the data that can be either the important message, personal data, medical information, medical records or a medical image must be transmitted in such a form that the intruder should not be able to find the presence of secret data during transmission. Steganography is a good option for hiding the information. It embeds the data into an unperceivable form. On cover object classification, steganography can be categorized into different techniques [2] shown in Fig. 1. Of them, image steganography is the commonly used technique. Data in the form of bits are embedded into any of the multimedia, like image audio, video, etc. But, sharing

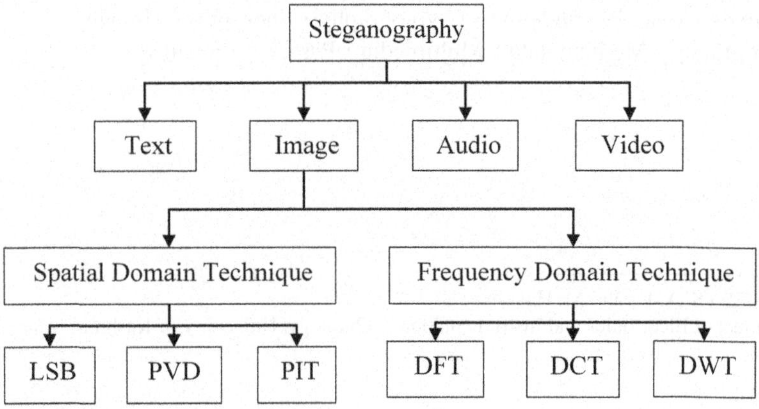

Fig. 1 Steganography classification

that multimedia which is the need of the hour by means of insecure Internet alone cannot withstand the attacks. Therefore, dual-image steganography (cryptography along with image steganography) is the better way to secure data and transmit it in an unperceivable manner. It not only conceals the information but encrypts the data prior to embedding, thus, providing security to the data being transferred to the receiver. So, it gives security to the message in both manners combinedly.

Steganography can embed data into a cover object by two domains: spatial domain and frequency domain [3]. In the spatial domain, data are directly embedded into the cover object, whereas in the frequency domain cover object is first converted into its frequency coefficients, and then, the data are embedded into frequency coefficients. The spatial domain technique is the simplest and fastest technique, whereas frequency domain technique is the complex method. There are many ways of embedding data into an image with the spatial domain, like least significant bit (LSB), pixel value differencing (PVD), etc. Of them, LSB embedding steganographic technique is the basic, fastest, and simplest one. It directly replaces the LSB bit of image pixels with the secret message/information bit. However, embedding data into an image can deteriorate the image quality. Hence, along with security, the algorithm should balance the quality, imperceptivity, and robustness conflicts.

Enormous research based on image steganography has been done for providing security to the data being transferred to the receiver. Like in [5], the enhanced image steganography method has been reported. The scheme has used an odd–even substitution method for embedding purpose. Thus, the method has tried to improve payload but needs to further improve security with good PSNR value. Another image steganographic scheme has been presented in [6]. The framework has used reversible data hiding and chaos encryption scheme based on wavelet transform. Though the method is good in providing security and a good amount of payload, the use of transformation techniques has made the system complex. One more information hiding technique has been proposed [7] in which patient record has been embedded inside an image for secure transmission. The method provides security to the data that are being transferred through an insecure network and, however, can be improved to achieve better quality analysis values. In [8], the watermarking method has been used to hide data in the cover image for security reasons. However, it can be improved in terms of PSNR value. The discussed schemes/techniques, however, have improved some security to the data by concealing its existence, but need further improvement in the payload with a good PSNR value, so that a good amount of information gets transferred in one go without coming into the eyes of an intruder.

In this paper, the dual-image steganography technique is proposed to enhance security to the data being transmitted through the Internet. Important data in the form of an image are encrypted before embedding with the cryptography method and are then embedded into LSB pixels of the cover image to form a stego image by LSB techniques. The analysis shows the technique is capable of transferring data securely via the Internet by balancing the conflict triangle parameters.

2 Proposed Work

Initially, at the sender side, the secret medical data are encrypted and are concealed into a cover image to form a stego image with combined cryptography and steganography techniques. The formed encrypted image is then communicated through the wireless network. At the receiver side, the received image is extracted and the message is decrypted to get the complete and correct information. The encryption, embedding, extraction, and decryption are shown in Fig. 2 and are explained as follows:

2.1 *Embedding and Encryption Method*

(i) Choose secret medical data in the form of an image.
(ii) Generate a sequence using the cos function according to the size of the secret data.

$$\text{Random sequence} = \cos(p * \cos^{-1}(k(z)))$$

where 'p' and 'k' are the initial values to be set. 'z' is the number of iterations.
(iii) Apply circular shift to the generated bits as per the desired shifts.
(iv) Generate encrypted message by performing logical XOR operations between shifted bits and secret data bits.
(v) Choose the cover image.
(vi) Embed the encrypted data bits into LSB pixels of an image.
(vii) Repeat the process until the data bits get embedded into the cover image and generate a stego image.

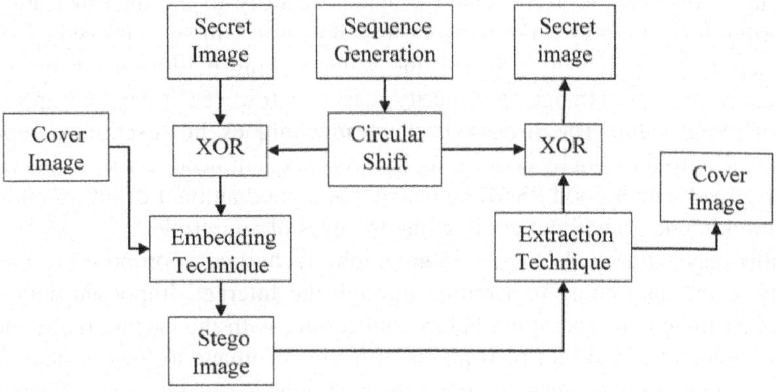

Fig. 2 Proposed method

2.2 *Extraction and Decryption Method*

(i) Take the stego image generated above.
(iI) Extract data bits for LSB of pixels.
(iII) Repeat the process till all bits get extracted from the cover image.
(iv) Perform XOR operation on extracted bits and generated shift bits and get the message back into its original form.

3 Simulation Analysis

This technique has been verified using MATLAB. Here, two images have been taken for analysis to validate the proposed scheme. One image is used as a cover image with 256 × 256 and the other as a secret medical image of 256 × 256 size.

In order to check the quality, security, and capacity of an image, various analyses like peak signal-to-noise ratio (PSNR), normalized cross-correlation (NCC), normalized absolute error (NAE), structural similarity index (SSIM), number of changes per rate (NPCR), unified average changed intensity (UACI), and entropy [4] have been applied on the cover image, stego image, and retrieved image.

Table 1 clearly shows that the proposed method can transmit data securely without being deteriorated through the wireless medium. Different parametric value shows that the stego image is not much different from the original image, thus making difficult for the eavesdropper to detect the message. The scheme has been tested for security results. The proposed scheme has about 99.19% NPCR value and 27.04% UACI value. The proposed method has also been tested for randomness, and the results show a 7.9922 entropy value. Thus, the technique along with quality is able to provide security to the data.

Table 1 PSNR, NCC, NAE, and SSIM values for different cover images

Image	PSNR	NCC	NAE	SSIM
Lena	51.1306	1.0000	0.0020	0.9966
Baboon	51.1465	1.0000	0.0019	0.9987
Pepper	51.1287	1.0000	0.0021	0.9968
Plane	51.1471	1.0000	0.0014	0.9965
Girl	51.1365	1.0000	0.0043	0.9958
Boats	51.1159	1.0000	0.0019	0.9974
Milkdrop	51.1426	1.0000	0.0024	0.9956
Sailboat	51.1570	1.0000	0.0020	0.9975
Man	51.1526	1.0000	0.0032	0.9976
Bridge	51.1452	1.0000	0.0022	0.9989
Average	**51.1403**	**1.0000**	**0.0023**	**0.9971**

Table 2 Comparison with already existing techniques

Algorithm	Payload	PSNR
Brar and Sharma [5]	42B	42.4302
Raju and Prabha [6]	0.2 per image	52.1374
Parah et al. [7]	1 bpp	39.13
Parah et al. [8]	1.5 bpp	45.37
Proposed	**1 bpp**	**51.1403**

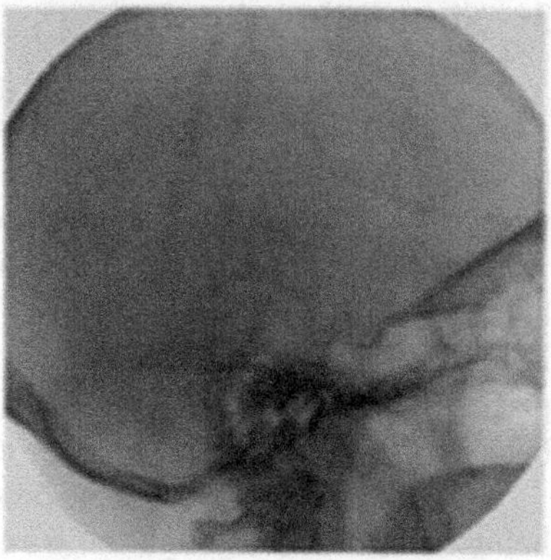

Fig. 3 Secret data image before embedding

From Table 2, one can find that the proposed system has a good PSNR value with a good amount of capacity in comparison with the existing techniques and thus is more secure than the already available methods as the quality of the stego image of the proposed method that is to be transferred is better. Hence, the presented technique can withstand attacks. Additionally, the data are encrypted to add more security to the data that are not done by the compared techniques.

It is clear from the analysis that the proposed method gives better PSNR value with a good amount of payload with respect to the already presenting frameworks. The proposed method along with good quality values also gives good NPCR, UACI, entropy values, and can retrieve the secret data back with negligible loss (refer to Figs. 3 and 4), which means the method is robust against attacks.

4 Conclusion

This paper proposes the dual-image steganography technique based on LSB substitution and encryption techniques. Information/secret medical data are first encrypted with the help of the COS function, whereas data are embedded into the cover image using the LSB technique. Experimental results show the authenticity of the technique by performing various analyses on the cover image and stego image. It is clear from

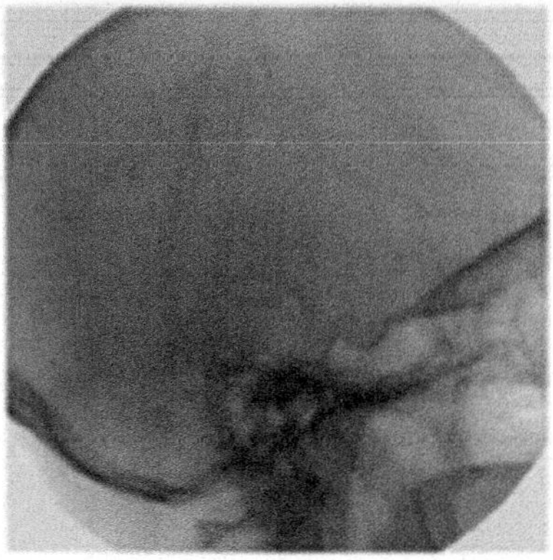

Fig. 4 Secret data image after extraction

the results that the technique is better in terms of quality and security and hence can be suitable for secure data transmission.

References

1. Jan, A., Parah, S.A., Malik, B.A.: A Novel Laplacian of Gaussian and Chaotic encryption based image steganography technique. In: 2020 International Conference for Emerging Technology (INCET), Belgaum, pp. 1–4 (2020). https://doi.org/10.1109/INCET49848.2020.9154173
2. Kundu, D., Upreti, A.: Survey of various steganography tools. In: 2018 International Conference on Automation and Computational Engineering (ICACE), Greater Noida, pp. 117–120 (2018). https://doi.org/10.1109/ICACE.2018.8687092
3. Dhawan, S., Gupta, R.: Analysis of various data security techniques of steganography: A survey. Inf. Secur. J. Glob. Perspect (Taylor & Francis) (2020). https://doi.org/10.1080/19393555.2020.1801911
4. Jan, A, Parah, S.A., Malik, B.A.: Logistic Map-Based Steganography using Edge Detection. Innovations in Computational Intelligence and Computer Vision. Springer, (2020). https://doi.org/10.1007/978-981-15-6067-5_50
5. Brar, R.K., Sharma, A.: Improved steganography using odd even substitution. Learn. Anal. Intell. Syst. **17**, 1–8 (2020). https://doi.org/10.1007/978-3-030-48118-6_2 (Springer)
6. Raju, K.U., Prabha, D.N.A.: Chaos Encryption Images on Lossless and Reversible Data Hiding using Wavelet Transform. Easychair Preprint (2345) (2020). https://easychair.org/publications/preprint/2tVH
7. Parah, S.A., Sheikh, J.A., Akhoon, J.A., Loan, N.A.: Electronic health record hiding in images for smart city applications: A computationally efficient and reversible information hiding technique for secure communication. Futur. Gener. Comput. Syst. (2018). https://doi.org/10.1016/j.future.2018.02.023 (Elsevier)
8. Parah, S.A., Loan, N.A., Shah, A.A., Sheikh, J.A., Bhat, G.M.: A new secure and robust watermarking technique based on logistic map and modification of DC coefficient. Nonlinear Dyn. (Science and Business Media B.V., Springer) (2018) https://doi.org/10.1007/s11071-018-4299-6

'NIS' a Network Investigation System

Himani Garg, M. Balasubramaniam, Ajay Kr Gupta, and Jitendra Singh

Abstract Project Implementation (or Project execution) is that part where vision, thoughts, and plans become reality. Usually, the muddles related to credential sharing, login by spiteful users, connecting system from totally different locations at odd times, and problems related to information security of projects claims regular audit and tracking. With these challenges, this paper introduces a smart solution 'NIS' to monitor, control, and manage site infrastructures, i.e., hardware and network. This provides a quick solution to help-desk and users to troubleshoot every possible glitch related to network, storage, and connectivity and also helps in circulating updates, alerts, and urgent notifications and at the same time share this information to the running project whenever the user logs in the system.

Keywords Network Investigation · Electron · CORS Whitelisting · Internet Speed · Data Analysis · Information Security

1 Introduction

An impeccable project headship and continual monitoring is important to know where you are going! Nothing is much significant than supervising the accomplishment of tasks and the evolution of the project to ensure that everything goes smoothly. With this to go in the right direction, one needs to take care of various factors that are different from project development and testing but a part of the implementation and post-implementation. Some major heads are discussed below to support the

H. Garg (✉) · M. Balasubramaniam · A. K. Gupta · J. Singh
BDPM Group, Center for Development of Advanced Computing (C-DAC), Noida, India
e-mail: himanigoel@cdac.in

M. Balasubramaniam
e-mail: mbalasubramaniam@cdac.in

A. K. Gupta
e-mail: ajaygupta@cdac.in

J. Singh
e-mail: jitendrasingh@cdac.in

A. Noor et al. (eds.), *Proceedings of Emerging Trends and Technologies on Intelligent Systems*, Advances in Intelligent Systems and Computing 1371,
https://doi.org/10.1007/978-981-16-3097-2_21

above statements and to forward attention to some hidden points which are typically prioritized low.

1.1 Resources

Projects rely on the effective employment of finite resources, whether these are people, equipment, or facilities. To focus on inappropriate hardware/software resources, it is usually difficult to find the increasing need/upgradation of resources on every site as the project and data grows.

1.2 People

They are probably the most important resource on any project and the challenge that an organization has to meet when using project management relies on the efficient utilization of these people. Correct people must be used; they should timely log-in in the system, not follow malicious practices, and solve the client issues uniformly.

1.3 Environment

A project does not exist in a vacuum; to deliver successful projects, an organization must be aware of those factors (both internal and external) that will have an impact on the project: uninterrupted network connectivity, all possible provision of WiFi/LAN connections, the accurate latitude and longitude details, list of users allowed to login via that site, security factors, etc.

As long as an organization is fully aware of these challenges and deals with them appropriately, they can easily be overcome. The benefits of using NIS, by far, outweigh these challenges.

Through NIS one can apply a structured methodology and conscientious data analysis which can then be applied to projects across an organization to privilege it with information security, streamlined, and structured post-implementation data, and a chance for meticulous study across site-wide incorrect practices without compromising security.

2 Problem Statement

With the implementation of any project across users at different locations, there is one common problem faced by any organization for using that system by users and that is internet connectivity issues, scanty bandwidths, low configurations computers, and first-generations learners. Incessant questions and knots that transpire while deploying and running any application are:

i. What is the internet speed and bandwidth over every site?
ii. What are the possible options for WiFi/LAN connectivity?
iii. Where is my application running geographically?
iv. Are they running the application on the best-viewed browser and its version?
v. Are login ids shared among users?
vi. Is every site installed with basic hardware requirements?
vii. Is there any login made at an odd time or from a different place?
viii. Is my system being used by any unauthorized user?

3 NIS and Its Methodology

3.1 Origin

To solve these kinds of issues either end-user need to be computer literate for support-related activities or there should be independent software installed on the remote computers where companies can access and analyze the root cause of software break-down. Considering a similar kind of approach, C-DAC, Noida, has designed and created a desktop application called "NIS" which will not only provide the overall system and network information but also help the software implementer to analyze the root issues faced by users across various sites.

3.2 Features of NIS

- A simple and portable executable file (exe) that captures all system details on which installed.
- Take account of latitude and longitude information from which the user has logged in to the application.
- Determine upload/download internet speed following Ookla.
- Version management of app and version up-gradation by auto-update feature.
- Auto-startup feature after the power is turned on.
- Serene integration with the main project.
- Follows CORS Whitelist concept.
- In-built user manual.

3.3 Technology Used

Electron is a framework that enables you to create desktop applications with JavaScript, HTML, and CSS. NIS has been developed using electron as its framework combined with chromium rendering engine and Node JS runtime [5].

Node JS is an open-source server-side runtime environment developed on Chrome's V8 JavaScript engine [1]. It provides an event-driven, non-blocking (asynchronous) I/O and cross-platform runtime environment for building highly scalable server-side applications using JavaScript [2].

jQuery is a JavaScript library designed to simplify HTML DOM tree traversal and manipulation, as well as event handling, CSS animation, and Ajax [3].

JSON is a lightweight format for storing and transporting data. All the NIS data, version management uses JSON format.

HTML and CSS has been used for designing a user interface with a better responsive experience.

3.4 Architecture of NIS

NIS follows a very elementary architecture, developed over electron framework which combines Chromium rendering engine and Node.js. Electron additionally provides a feature to receive connections from the outside and act as a server, i.e., HTTPS server (Refer Fig. 1).

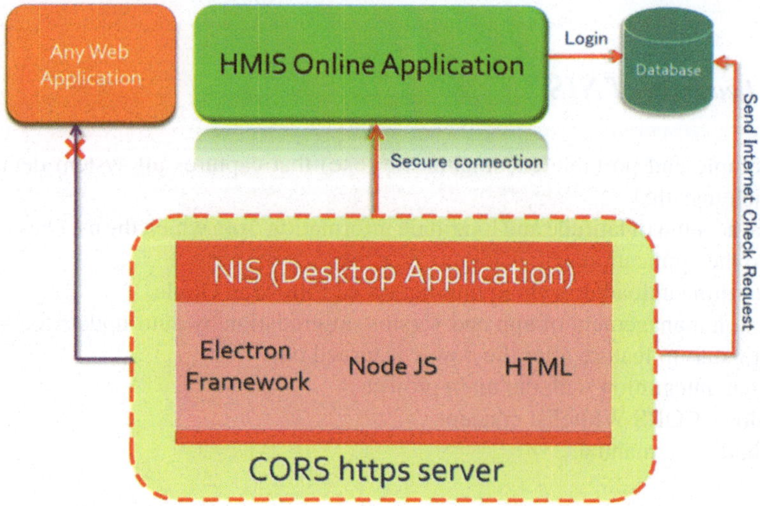

Fig. 1 Architecure of NIS

This server has been notably designed to act as CORS (Cross-origin resource sharing) server that allows restricted resources on a Web page to be requested from another domain outside the domain from which the first resource was served [4].

The same server forms a secure connection with HMIS (Hospital Management Information System) to lift all system details from NIS to HMIS (or any Web application) and parade the same as NIS System Data information; under the pattern of some benchmark values on which the instant scenario will be examined to rationalize the system and its network requirements up to mark to successfully use all the services of Web application without the crunch.

The user can, however, individually request to check the internet connection and know its speed with NIS and send the same information to cloud servers for further analysis and problem resolution.

3.5 Methodology and Workflow

On deploying the NIS application on the user system, some information can be collected, captured, and used for the analysis of the system along with the usage of the software at a particular location. The information gathered by the NIS application comprises of the version of Mac address, OS platform, architecture, amount of memory and system resources consuming, internet service provider, and internet bandwidth, geographical location of the place, and network interfaces.

Once the information is gathered, as soon as the user login into a Web application, it will be available to our cloud servers on various database tables with unique identification number to the system and location (Refer Fig. 2). This collection of data or information is presented in many analytical ways such as BIRT reports, Dashboards, Pie charts, and many more. The system is also helpful in the analysis of the system and figure out the configuration of different systems used at various locations, their internet and bandwidth connectivity, available browser; its version, and resource utilization.

The NIS application will work on the system as system resources similar to windows applications wherein the user need not have to be skilled or experienced in configuring the NIS application. The application will automatically start with the system login same as other applications and gather all required information such as OS type, platform, architecture, Mac address, network configurations, interfaces, CPU details, resources utilized, and bandwidth to the cloud servers. During the situation of low bandwidth/speed fluctuations or no internet, the application will generate an alert (in red) to the monitoring tool, i.e., Dashboard for information and immediate resolution.

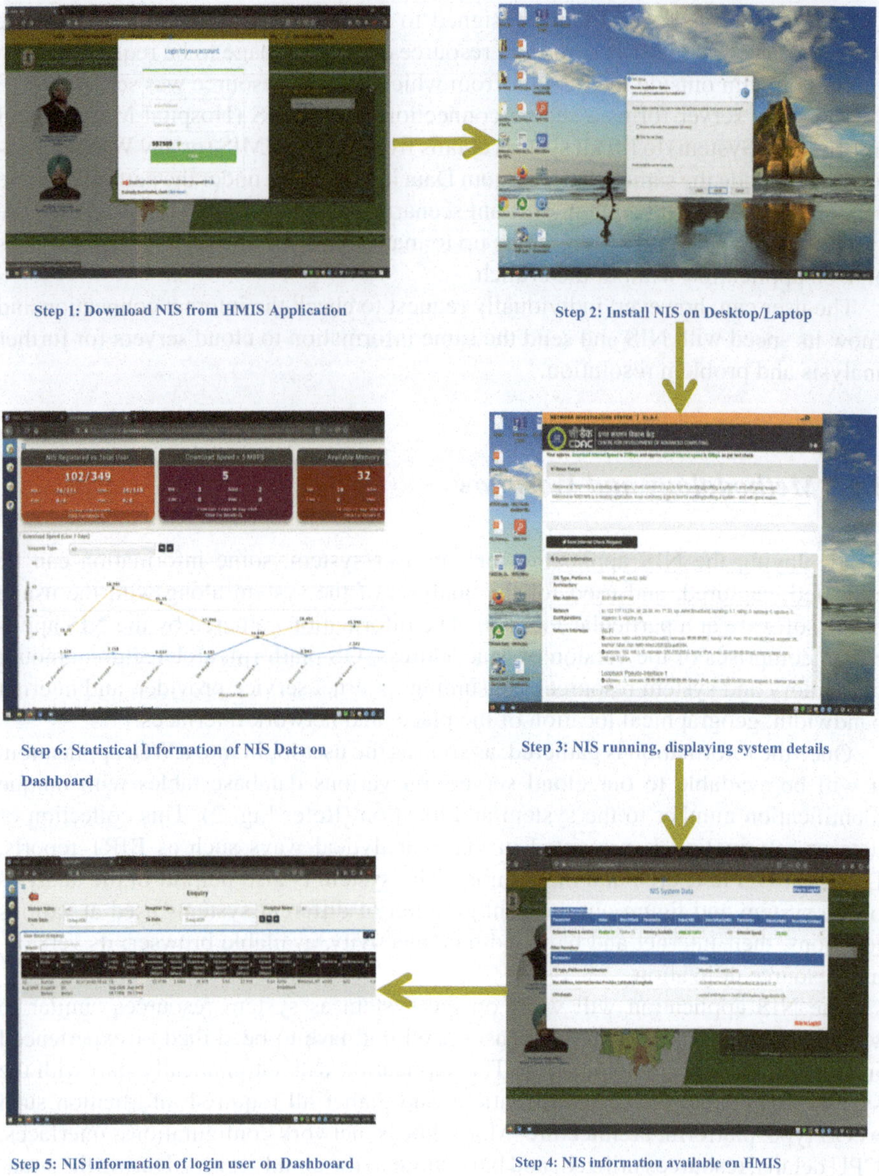

Fig. 2 Workflow of NIS

3.6 *Exceptional Elements of NIS*

There can be plenty of other tools and approaches that could be followed to obtain the information of any system and analyze its data for betterment. In this section, we highlight the specific features of NIS that make it useful for end users. This allows for continuous monitoring of the required parameters.

General Aspects

Effortless Handling: Once the application is deployed, the primary functionality of the application will initiate immediately and that would "act as system resources". The NIS application will work on the system as system resources similar to windows applications wherein the user need to be skilled or experienced in configuring the NIS application. Also, NIS will overlong run in the background, recalculating the details every 30 min without fail and post the same to the main application.

Auto-update functionality with Patch Management: Knowing the actual situation of the system and reachability issues of the user on remote locations where the system is configured, we designed the NIS application in such a way that users need not have to re-configure the application repeatedly. The NIS application has in-built auto-update functionality that works every time NIS starts (Refer Fig. 3). The application will continuously interact with the cloud servers to search for any regular updates of the application. In addition to this, if there is an update to the application, the system will automatically download the update on the system with each update while also maintaining a versioning system for tracking and information purpose.

Integration with HMIS application: The NIS application is designed and developed in a way that this application can be easily integrated with the any running application (currently with HMIS Punjab) through communication via HTTPS server. The information collected by the NIS application is stored on databases in high-end cloud

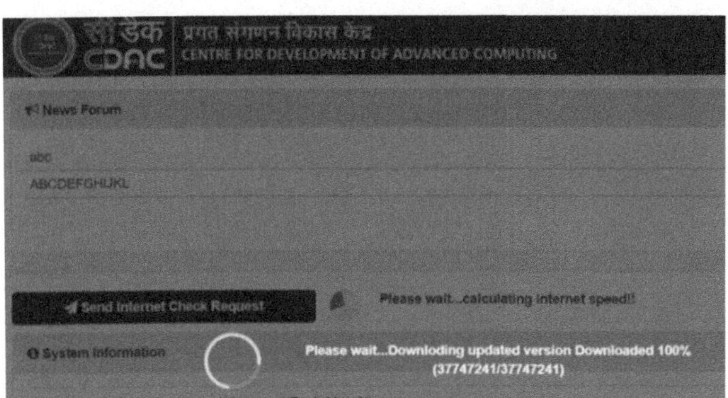

Fig. 3 Auto-update feature of NIS

servers, which can be accessible to the HMIS application and can be used by higher officials and technical teams for references and further actions (Refer Fig. 4).

Technical Aspects

CORS Whitelist: Typically, NIS does not allow pages outside its domain to access data. Only domains that are needed for cross-origin communication are whitelisted. In general terms, the Web site except the specific whitelisted application will not be able to read the NIS data and misuse it. NIS handles all its whitelisting with special login privileges and can be done by C-DAC Authorized users only (Refer Fig. 5).

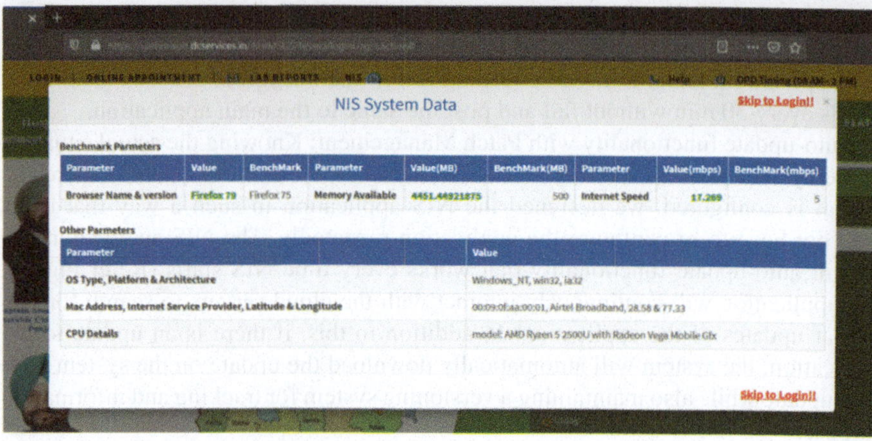

Fig. 4 NIS integration with HMIS

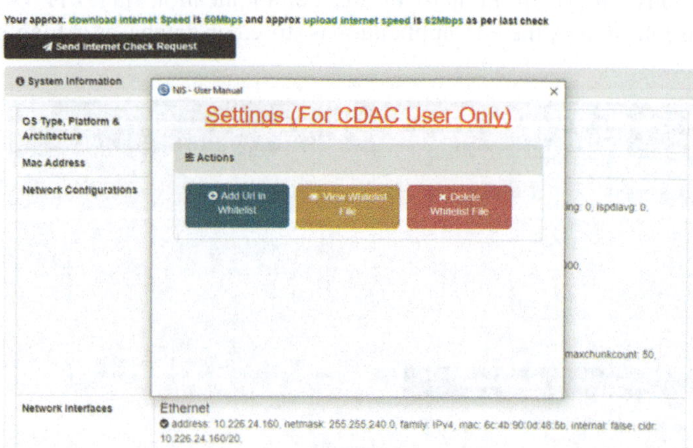

Fig. 5 CORS Whitelist concept

Data Analysis: NIS data has been statistically arranged into varied reports and graphs for every user. This data study will further be used to examine the infrastructure conditions, browser login variations, scanty bandwidth across locations, and inappropriate logins done on different systems.

4 Results

With the development of the NIS application along with integration with the HMIS Punjab application, accessibility of local systems at remote locations has become difficult for implementers. Some live examples from various locations (28 locations) are shown in below screenshots for information/references to indicate the usability of the NIS application where network connectivity is always an issue for implementers.

Initially, we are capturing three major KPIs in the NIS application which is primarily used to showcase the NIS system capabilities. These KPIs provide the actual data captured by the NIS application and highlighted on the dashboard with drill-down functionality (Refer Fig. 6).

NIS Registered Users Versus Total Users

In this KPI, the application is capturing the total number of users working on HMIS applications and systems where the NIS application is installed and configured (Refer Fig. 7).

Downloading Speed > 5 MBPS

In this KPI, the application is configured to capture the bandwidth in hospitals or locations with a speed less than 5 Mbps (where the Threshold value is set to 5 Mbps). This threshold value can be standardized in NIS application for different locations. Through this, authorities can analyze and take corrective measures to overcome issues on bandwidth or speed issues (Refer Fig. 8).

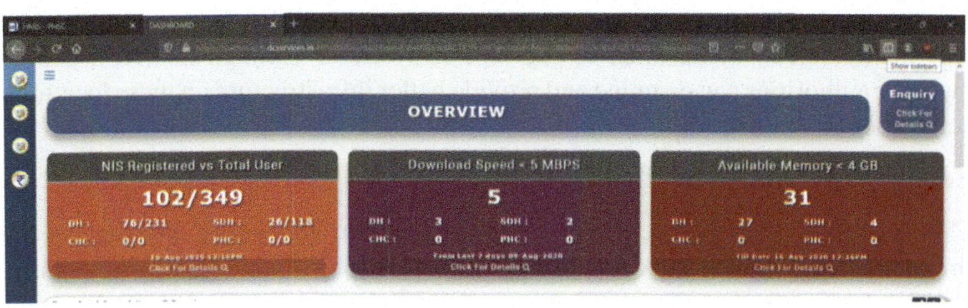

Fig. 6 NIS KPI's on dashboard

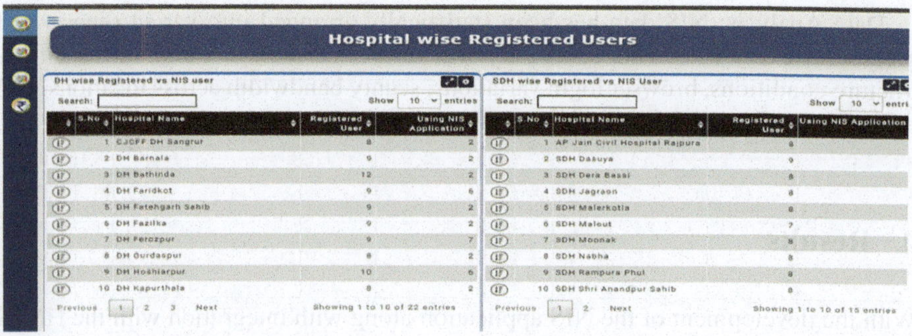

Fig. 7 NIS versus hospital users

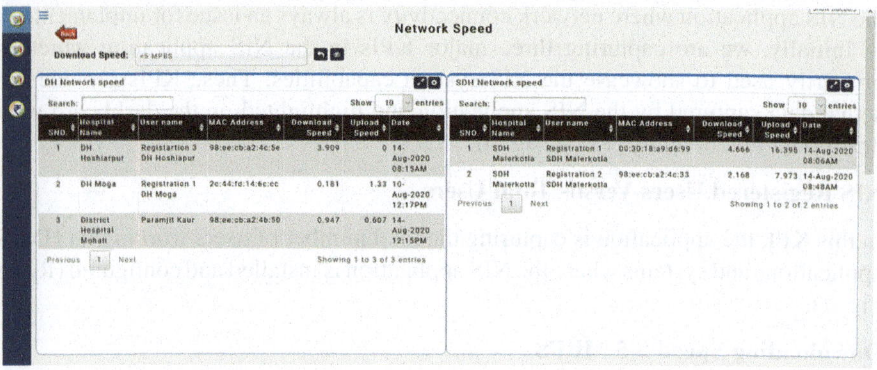

Fig. 8 Stores having speed < 5 MBPS

Available Memory < 4GB

In this KPI, the application is configured to capture the memory installed on the system and available memory for the time of using the software (HMIS). Here also, the threshold value can be attuned in NIS application with less than 4 GB available memory (RAM) installed on the remote systems. Through this KPI, departments or authorities can examine and provide better infrastructure on these locations/authorities (Refer Fig. 9).

Enquiry Tab

This space will provide the recent activity of each user logging into the system, first and last login date, MAC address, browser used, and its version, geographical coordinates, hardware and network information. As soon as the user requests to check internet speed, the details will again be logged in to the cloud-based server and can be viewed under Enquiry Tab (Refer Fig. 10).

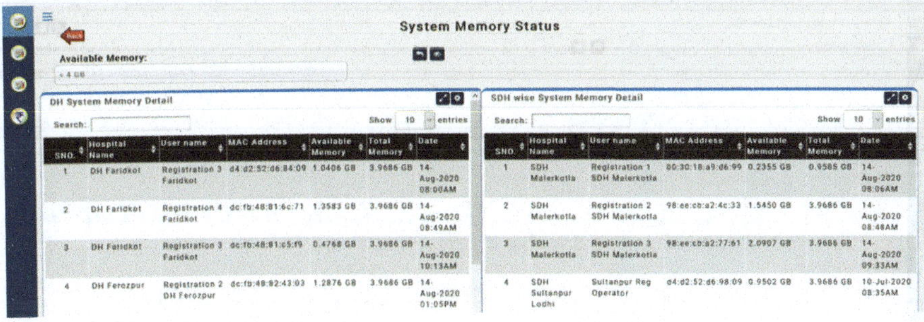

Fig. 9 Stores having RAM < 4 GB

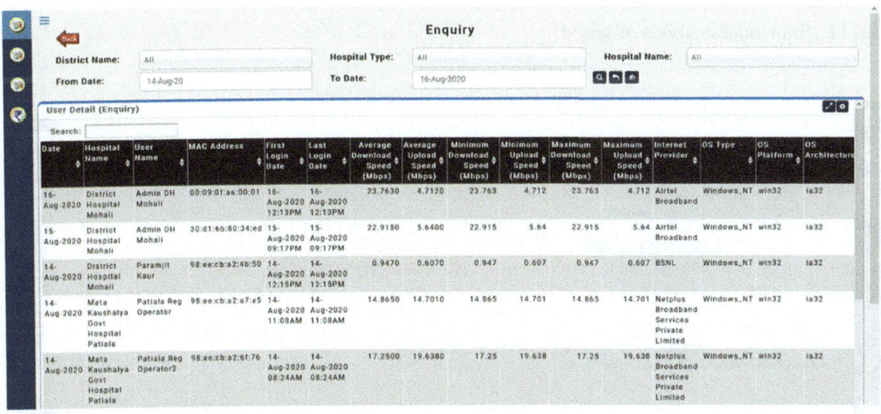

Fig. 10 Recent login entries of users

Other Dashboard Tab With Analytics

NIS application is also equipped with tons of new features wherein information is captured, analyzed, and shared with departments, companies, developers, and technical teams. Some of these new features contain a news portal, alerts on every time login, privacy of platform, bar charts pie charts for information captured using the NIS application (Refer Figs. 11 and 12).

Detailed Analysis and Comparative Study

Information security refers to some set of practices and methodologies which are designed and implemented to guard all sort of confidential and sensitive information or data from any unauthorized access.NIS being a infosec tool, attempts to cater information from disparate systems, which can further be analyzed based on few defined indicators:

a. Exposure of multiple user login on same mac address.
b. Revelation of odd time/odd place login for further necessary actions.

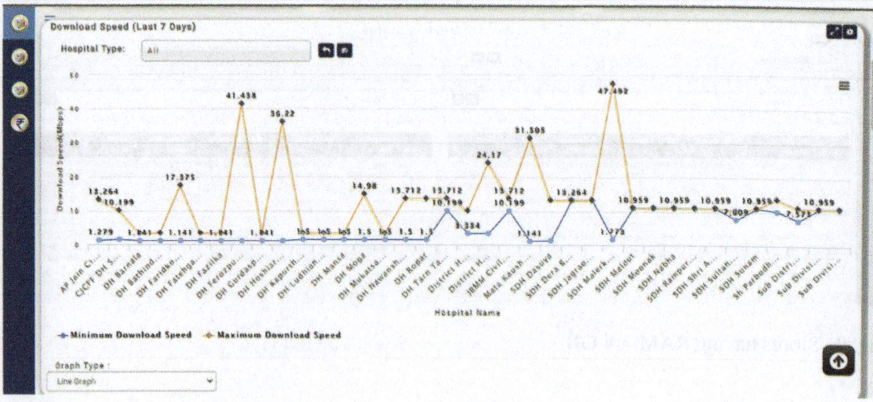

Fig. 11 Statistical analysis of speed

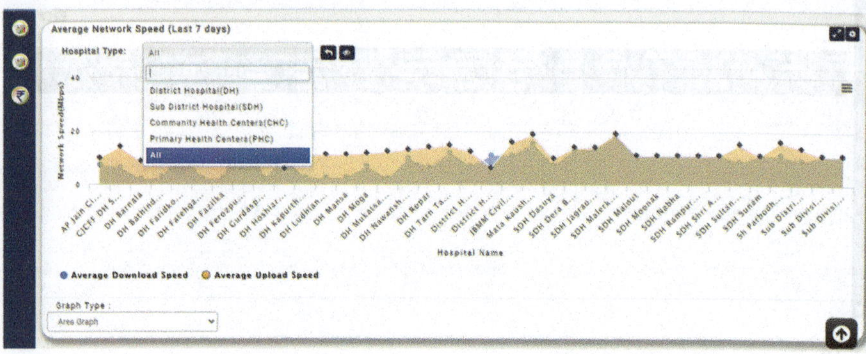

Fig. 12 Graphical analysis of speed

c. Detection of any VPN connected.
d. Anomaly detection by monitoring the internet speeds, so that when breach occurs, alert can be prompted.

This analysis and algorithms are under the future scope of NIS. The same can be done by merging AI concepts for automatically detecting the malicious actions and notify the user in real time. The below figure explains the proposed data analysis flow that will be the further step after gathering the information from all users (Refer Fig. 13).

5 Future Scope and Conclusion

The solution developed by C-DAC, Noida, called "NIS"—Network Investigation System—is envisioned to be the first step in solving the most common problems of

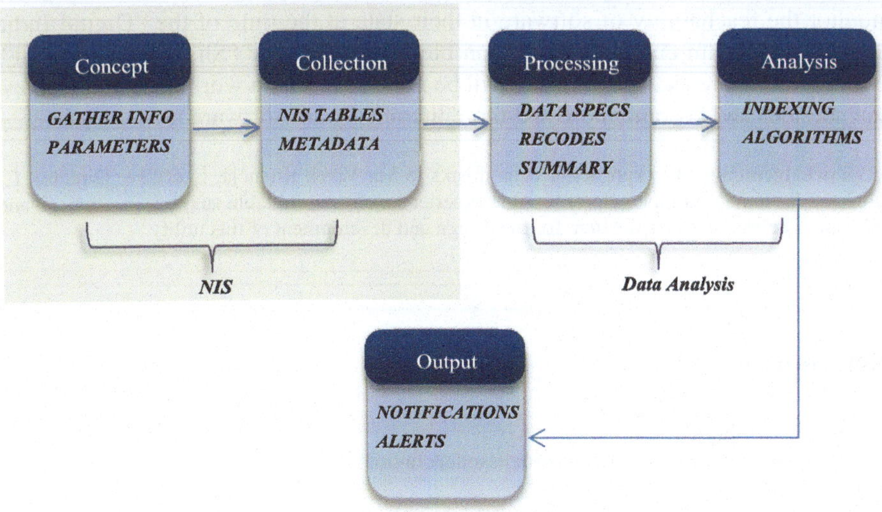

Fig. 13 Data analysis flow

the users working in remote locations where reachability to access the system and human participation is extremely difficult. The below comparison will sum up the current and future possibilities with NIS (Refer Table 1).

With this solution, implementers and developers are more focused on designing applications with better connectivity and low bandwidth consumption. Organizations with experienced software development can also implement or integrate such software with applications where usability, reliability, and consistency in software usage is a challenge. In addition to this, technical teams, head officials, and authorities can

Table 1 Comparison between Project with NIS and without NIS

Project without NIS	Project with NIS
1. Illegitimate login from a different computer but with known credentials 2. Delay in identification of abnormal functionality of application in case of browser issues 3. Slow page load and data retrieval affairs questions project performance even in the case of passive internet speeds 4. Slow performance due to depressed hardware/software environment 5. Suspicious login at odd times but with authorized users 6. A separate Alert Management System integration is compulsory for push notifications and alerts	1. Uncharted MAC Address alert feature will contrast in case of new device login attempted 2. The current benchmark shows up to the mark browser and its minimum version 3. Can be tracked with instant internet speed check request 4. Well-timed update about of crouched system configurations all over client-side systems 5. Login recording with odd/even time graphs 6. NIS can behave as an alert distribution system with notifications floating over screens

monitor the reachability of software in their state at the time of the "Digital India Initiative" program curated by our Hon'ble Prime Minister "Sh. Narender Modi". With this software aka "NIS," users will be more focused on working with the latest technologies and become more and more literate toward software technologies.

Acknowledgements We express our deep thanks to Mr. Vivek Khaneja, Executive Director; C-DAC Noida (India) with the help of his vast experience, efforts, valuable suggestion, support, and motivation helped us to great extent for the design and development of this utility.

References

1. https://nodejs.org/
2. https://www.tutorialsteacher.com/nodejs/nodejs-tutorials
3. https://en.wikipedia.org/wiki/JQuery
4. https://en.wikipedia.org/wiki/Cross-origin_resource_sharing
5. https://www.electronjs.org/docs

Improvement in SHA-3 Algorithm Using Different Internal Methods and Operations

Vanita Jain, Rishab Bansal, Mahima Swami, and Dharmender Saini

Abstract In this paper, we propose a novel approach to change the internal architecture of schoolbook implementation of SHA-3 algorithm. With this new architecture and internal operations, we aim at decreasing the time taken by the original algorithm. Specifically, we reduce the number of internal rounds of the algorithm, and to compensate for the loss in confusion and diffusion, we make SHA-3's internal operations more complex. We change the algorithm on its core bit level and make its internal operation more interdependent on the neighbouring bits to achieve more diffusion which results in better confusion.

Keywords Hashing functions · Information security · Cryptography · SHA-3 · Secure hashing · Sponge construction · KECCAK

1 Introduction

Cryptographic hash functions [1, 2], especially for authentication applications, such as message authentication codes, password protection and digital signature, are of major importance for a number of security applications. Cryptography is a form of testing data integrity. It is used to make sure no access or modification is made to the data transmitted in a message [3]. The intensity of SHA [4] mainly depends on a range of factors. These include the facility to calculate the hash value, the failure to generate a message that has a hash, the inability to modify a message without altering its hash. In addition to this, two separate messages of the same hash value cannot be identified. Safe hashing algorithms cannot only protect data from misuse

V. Jain (✉) · R. Bansal · M. Swami · D. Saini
Bharati Vidyapeeth's College of Engineering, New Delhi, India
e-mail: vanita.jain@bharatividyapeeth.edu

R. Bansal
e-mail: rishabbansal.it1@bvp.edu.in

M. Swami
e-mail: mahima.it1@bvp.edu.in

A. Noor et al. (eds.), *Proceedings of Emerging Trends and Technologies on Intelligent Systems*, Advances in Intelligent Systems and Computing 1371,
https://doi.org/10.1007/978-981-16-3097-2_22

or modification but also guarantee user authentication. SHA is used to sign content with a digital fingerprint and can be used for encrypting and decrypting passwords by many operating systems. Even in wireless communications, SHA-1 and SHA-2 are very important and are widely used for secure communication [5]. It shows, however, numerous weaknesses and limitations which cause an appropriate replacement to be found. The hash functions [6–8] can be classified into two types: Keyed and Unkeyed. Keyed hash functions can be used for message authentication code (MAC) [9]. It requires two inputs, a message and a secret key. Unkeyed hash functions can be classified into three categories: block ciphers, customised functions and arithmetic functions. Hash functions have one way property which means computing things in one direction is easy and the vice versa is hard. For example, for a message m, it is easy to calculate $H(m) = M$, but it is hard to obtain m from a given M. Collision attack [10] means finding a message n such that $H(m) = H(n)$.

2 Related Work

After some successful attacks on SHA-1 [11, 12], NIST launched a public competition to promote the development of a new cryptographic hash function. The winning algorithm of this competition will be named as SHA-3. Team Keccak won the competition and their algorithm is specified in Federal Information Processing Standard (FIPS) 180-3, Secure Hash Standard. Keccak is based on the sponge construction [13]. The benefit of using this type of architecture is to obtain a more secure and robust against generic and known attacks. Sponge construction also makes the use of the compression function more simple and flexible. Keccak has 2 phases: Absorption (input phase) and Squeezing(output phase). In the Absorption Phase, input blocks are XORed into a substate; this state is then converted as a whole using the permutation function. In Squeezing Phase, output blocks are obtained from the same subset and altered with the state transformation function. The size of this altered state is called rate (r). The size of the part which remains unaltered is called capacity (c). For a permutation function that computes b-bit blocks, capacity is equal to b-r. Security level of the scheme depends upon the capacity of the algorithm. Maximum security level of the scheme is usually half of capacity $2^{c/2}$. Padding makes sure that the length of the input is divisible by r to break input into blocks. A pattern of $10 * 1$ is used for padding in SHA-3. The size of the input block varies depending on the size of the output: Keccak-512 has a block size of 576 bits, Keccak-384 has 832 bits, Keccak-256 has 1088 bits and Keccak-224 has 1152 [14]. SHA-3 has five main functions through which the message blocks are passed 24 times, namely Iota, Chi, Pi, Rho, Theta. The input block is broken into a tensor of 1600 bits ($5 \times 5 \times 64$). Then these 1600 bits are passed through these functions 24 times (in original implementation). $P[u, w]$ is an array with $u, w = 0, 1, 2, 3, 4$.

2.1 *Theta (θ)*

In this step, each bit is XORed with ten other neighbouring bits [10].

$$Q[u] = \text{XOR}(P[u, 0], P[u, 1], P[u, 2], P[u, 3], P[u, 4]) \tag{1}$$

$$R[u] = \text{XOR}(Q[u - 1], \text{rotate}(Q[u + 1], 1)) \tag{2}$$

$$P[u, w] = \text{XOR}(P[u, w], R[u]) \tag{3}$$

$$[u, w = 0, 1, 2, 3, 4]$$

2.2 *Rho (ρ)*

In this step, each row of 64 bits is rotated cyclically according to the rotation table given in Table 1 [10].

2.3 *Pi (π)*

In this step, all the bits are changed according this equation [10]:

$$Q[u, 2u + 3w] = P[u, w] \tag{4}$$

$$[u, w = 0, 1, 2, 3, 4]$$

Table 1 Rotation constants (original)

$y\backslash x$	0	1	2	3	4
0	0	36	3	41	18
1	1	44	10	45	2
2	62	6	43	15	61
3	28	55	25	21	56
4	27	20	39	8	14

Table 2 Round constants (original)

I[01] = 1	I[02] = 32898
I[03] = 9223372036854808714	I[04] = 9223372039002292224
I[05] = 32907	I[06] = 2147483649
I[07] = 9223372039002292353	I[08] = 9223372036854808585
I[09] = 138	I[10] = 136
I[11] = 2147516425	I[12] = 2147483658
I[13] = 2147516555	I[14] = 9223372036854775947
I[15] = 9223372036854808713	I[16] = 9223372036854808579
I[17] = 9223372036854808578	I[18] = 9223372036854775936
I[19] = 32778	I[20] = 9223372039002259466
I[21] = 9223372039002292353	I[22] = 9223372036854808704
I[23] = 2147483649	I[24] = 9223372039002292232

2.4 *Chi* (χ)

In this step, binary bitwise operations like AND, OR, XOR, NOT are performed in the bits according to this equation [10]:

$$P[u, w] = \text{XOR}(Q[u, w, \text{NOT}(Q[u+1, w]) \text{ AND } Q[u+2, w])) \tag{5}$$

$$[u, w = 0, 1, 2, 3, 4]$$

2.5 *Iota* (ι)

To break the uniformity of the process, in this step, a unique constant is XORed with the $P[0, 0]$ row. This produces entropy in the each step. The constants are given in Table 2 [10].

3 Proposed Architecture

There are five main operations in KECCAK algorithm: theta, rho, pi, chi, iota. We changed theta, rho, chi, iota.

3.1 *Theta (θ)*

Original implementation This is the original implementation [10] of theta in SHA-3.

$$Q[u] = \text{XOR}(P[u, 0], P[u, 1], P[u, 2], P[u, 3], P[u, 4]) \tag{1}$$

$$R[u] = \text{XOR}(Q[u-1], \text{rotate}(Q[u+1], 1)) \tag{2}$$

$$P[u, w] = \text{XOR}(P[u, w], R[u]) \tag{3}$$

$$[u, w = 0, 1, 2, 3, 4]$$

Our implementation This is our implementation of theta method. These extra operations increase the algorithm's diffusion factor. Instead of 10, it now XORs each bit with 20 neighbouring bits.

$$Q[u] = \text{XOR}(P[u, 0], P[u, 1], P[u, 2], P[u, 3], P[u, 4]) \tag{6}$$

$$R[u] = \text{XOR}(Q[u-1], \text{rotate}(Q[u+1], 1)) \tag{7}$$

$$S[u] = \text{XOR}(R[u-2], \text{rotate}(R[u-2], 1)) \tag{8}$$

$$T[u] = \text{XOR}(S[u-2], \text{rotate}(S[u-3], 1)) \tag{9}$$

$$P[u, w] = \text{XOR}(P[u, w], T[u]) \tag{10}$$

$$[u, w = 0, 1, 2, 3, 4]$$

3.2 *Rho (ρ)*

Original implementation Table 3 [10] shows the original rotation constants.

Table 3 Rotation constants (original)

$y \backslash x$	0	1	2	3	4
0	0	36	3	41	18
1	1	44	10	45	2
2	62	6	43	15	61
3	28	55	25	21	56
4	27	20	39	8	14

Table 4 Rotation constants (modified)

$y\backslash x$	0	1	2	3	4
0	8	43	26	4	8
1	7	34	21	53	47
2	60	18	37	57	4
3	20	17	19	41	44
4	25	2	58	51	62

Our implementation Table 4 shows our modified rotation constants. We tested random values for these constants and selected those which provided maximum diffusion properties.

3.3 Chi (χ)

Original Implementation This is the original implementation [10] of the Chi function.

$$P[u, w] = \text{XOR}(Q[u, w, \text{NOT}(Q[u+1, w]) \text{ AND } Q[u+2, w])) \tag{5}$$

$$[u, w = 0, 1, 2, 3, 4]$$

Our Implementation This is our implementation of the Chi function. We added extra XOR and AND operations to increase the inter dependency of the bits.

$$P[u, w] = \text{XOR}(Q[u, w], Q[u-2, w], Q[u+1, w]) \text{ AND } (\text{NOT}(Q[u-1, w])) \tag{11}$$

$$u, w = 0, 1, 2, 3, 4$$

3.4 Iota (ι)

Original Implementation Table 5 [10] shows the original round constants. 24 constants for 24 rounds.

Our Implementation Table 6 shows our modified round constants. 16 constants for 16 rounds.

Table 5 Round constants (original)

I[01] = 1	I[02] = 32898
I[03] = 9223372036854808714	I[04] = 9223372039002292224
I[05] = 32907	I[06] = 2147483649
I[07] = 9223372039002292353	I[08] = 9223372036854808585
I[09] = 138	I[10] = 136
I[11] = 2147516425	I[12] = 2147483658
I[13] = 2147516555	I[14] = 9223372036854775947
I[15] = 9223372036854808713	I[16] = 9223372036854808579
I[17] = 9223372036854808578	I[18] = 9223372036854775936
I[19] = 32778	I[20] = 9223372039002259466
I[21] = 9223372039002292353	I[22] = 9223372036854808704
I[23] = 2147483649	I[24] = 9223372039002292232

Table 6 Round constants (modified)

I[01] = 7948088626323794	I[02] = 6988136757012497
I[03] = 39791030927721416	I[04] = 10809755619314387
I[05] = 8147066428805446	I[06] = 9673464656433063
I[07] = 712894840135913	I[08] = 9856031517555377
I[09] = 7519438105630892	I[10] = 2180132335568229
I[11] = 22694087947679686	I[12] = 4791473861756359
I[13] = 9756078218014334	I[14] = 9910404914141336
I[15] = 7734607116733396	I[16] = 4660076906243047

4 Results

To compensate for the loss of confusion and diffusion factor, we made the internal bits more interdependent. We achieved this by adding more internal methods and changing existing bitwise operations. After trying different combinations of methods, operations and constants, we settled on those which gave us maximum confusion and diffusion.

Table 7 shows the running time of original (24 round implementation) and our modified (16 round implementation) with our changed internals and reduced number of rounds. Table 5 shows an average of 16% reduction in time which was observed for over 1,000,000 sample random inputs.

Table 8 shows some of the hashes computed on both the functions for the same inputs. The diffusion factor can be seen in the last and second last case where just a single change drastically changed the output.

To check the quality of hash being produced, we ran both the implementations on 1000 random inputs of 300+ bytes in size and changed the input by various degrees,

Table 7 Reduction in running time

Size (bytes)	Original (seconds)	Modified (seconds)	% decrease in time
10	0.2071	0.178205	13.9546
100	1.14742	0.920604	19.7676
1000	13.7301	9.46673	31.0512
10,000	120.569	110.511	8.3424
100,000	97.4402	80.9203	16.953
1,000,000	101,573	93,112.5	8.3295

Table 8 Sample output

Message	Original Modified
This is a super secret message	386a35c28e45b7e5783f6cf44ca696d5ae8db06a5eed53e4
	c86a41008bf7dc782156c01c438f29871ad535b009138ac1
Cryptography	cdc9d5c7a7cd902bfc1e6fc142b00aaa9812ea835349bdd2
	28c9409a4db203a5ad445cd3e233e04b983e550f73ce9c3a
Hashing	81d968688b4bbaff13ac3d884bf91f3e0be2708c53247c17
	66a89460f04d70a3f18e28da3ddeffa271d102e179a0aa44
Hashingg	0cd2b4c75f337d88771fd5625228c693013f96aa03df3b80
	3a5049c3bb1c571ef0487cbd153bed70c718d1095283fe59

mapped the corresponding changes in both the implementation. The changes were obtained by calculating the number of bytes being changed when the input is changed by various degrees. Table 9 shows the % difference in both the cases.

5 Conclusions

We were able to produce a 16% reduction in the running time of the schoolbook implementation of SHA-3 without decreasing the confusion and diffusion factor of the algorithm.

Table 9 Changes produced in the output

Percentage change in input (%)	Total changes in original 24 round implementation	Total changes in our modified 16 round implementation	Percentage difference (%)
10	23,913	23,920	−0.07
20	23,916	23,908	+0.08
30	23,887	23,904	−0.17
40	23,898	23,887	+0.11
50	23,914	23,908	+0.06

6 Future Scope and Limitations

This algorithm can be used in Web Browsers. SHA-2 is used in browsers because of its less computation time. This algorithm can replace SHA-2 in Web Browsers or some other light with applications which need secure hashing in less time. KECCAK algorithm is more secure than SHA-2 and is not vulnerable to attacks known for SHA-1 and SHA-2 like length extension attacks. We reduced 8 rounds in the algorithm from original 24. The algorithm is still secure from brute forcing and general crypt-analysis. This is called computational security, because no one has the computation power right now to brute force 16 round SHA-3 algorithm and crack it within the relevant time frame. These changes on internal operations and methods are done on the schoolbook implementation of SHA-3 without using any optimisation techniques. In real world, many space and time optimisation techniques like multi-threading are donc to make the code run faster. Same practices can also be applied on this algorithm as well.

References

1. Sobti, R., Ganesan, G.: Cryptographic hash functions: a review. Int. J. Comput. Sci. Issues **9**, 461–479 (2012). ISSN (Online): 1694-0814
2. Preneel, B.: Cryptographic hash functions: theory and practice. In: Gong, G., Gupta, K.C. (eds.) Progress in Cryptology—INDOCRYPT: INDOCRYPT 2010. Lecture Notes in Computer Science, vol. 6498. Springer, Berlin, Heidelberg (2010)
3. Cryptographic hash function (2014) [online]. Available at: en.wikipedia.org/wiki/Cryptographic_hash_function. Accessed Dec 2019
4. Sahu, A., Ghosh, S.: Review Paper on Secure Hash Algorithm with Its Variants (2017). https://doi.org/10.13140/RG.2.2.13855.05289
5. Moh'd, A., Aslam, N., Marzi, H., Tawalbeh, L.A.: Hardware implementations of secure hashing functions on FPGAs for WSNs. In: Proceedings of the 3rd International Conference on the Applications of Digital Information and Web Technologies (ICADIWT 2010), Istanbul, 12–14 July 2010 (2010)
6. Alfred, M., Oorschot, P., Vanstone, S.: Handbook of Applied Cryptography. CRC Press (1997)

7. Bruce, S.: Applied Cryptography: Protocols, Algorithms and Source Code in C. Wiley, Canada (1996)
8. Stallings, W.: Cryptography and Network Security Principles and Practices. Prentice Hall Press, Upper Saddle River (2010)
9. Bellare, M., Canetti, R., Krawczyk, H.: Keying hash functions for message authentication. In: Koblitz, N. (eds.) Advances in Cryptology—CRYPTO '96. CRYPTO. Lecture Notes in Computer Science, vol. 1109. Springer, Berlin, Heidelberg (1996)
10. Andreeva, E., Bogdanov, A., Mennink, B., et al.: On security arguments of the second round SHA-3 candidates. Int. J. Inf. Secur. **11**, 103–120 (2012). https://doi.org/10.1007/s10207-012-0156-7
11. Stevens, M., Bursztein, E., Karpman, P., Albertini, A., Markov, Y.: The First Collision for Full SHA-1 (2017), pp. 570–596. https://doi.org/10.1007/978-3-319-63688-7_19
12. Wang, X., Yin, Y.L., Yu, H.: Finding collisions in the full SHA-1. In: Shoup, V. (ed.) Advances in Cryptology—CRYPTO: CRYPTO 2005. Lecture Notes in Computer Science, vol. 3621. Springer, Berlin, Heidelberg (2005)
13. Bertoni, G., Daemen, J., Peeters, M., Van Assche, G.: Keccak Sponge Function Family Main Document. NIST, University of California Santa Barbara, Santa Barbara (2009)
14. G. Bertoni, et al.: The Keccak Reference, Version 3.0 (2011). http://keccak.noekeon.org/Keccak-reference-3.0.pdf

Stuck at Fault Testing in Combinational Circuits Using FPGA

Isha Gupta

Abstract VLSI testing has now emerged as a very important and crucial field because it is essential to test circuits before they are proceeded for production. With the increasing complexity of circuits and advancement in technologies, the minimum feature sizes on the chips are diminishing at a very fast pace. Thus, the testing process is becoming difficult and time-consuming. Approximately 40% of the design time is spent on testing. Exhaustive testing is beneficial for maximum accuracy and fault coverage, but it becomes cumbersome as the number of inputs is increased in a circuit. In this paper, a FPGA-based approach for test pattern generation and fault testing of combinational circuits has been presented. Instead of simulation, the concept of emulation is presented for detecting faults in a circuit using the FPGA hardware. Also, the problem of few input and output ports that are available on the FPGA board has been overcome through the approach of multiplexing the switches and push buttons. The FPGA board used in this paper is the Spartan 6 FPGA-based development board developed by CDAC Noida.

Keywords VLSI testing · Stuck at faults · Test pattern generation · FPGA

1 Introduction

Testing was never very popular among the designers in the initial phases of integrated circuit development, and it was considered at the last stage of the design cycle in order to simply verify the correctness of the design. However, with the vast advancement and development in technology we have moved to placing billions of transistors on a single chip. This evolution has lead, testing to emerge as an attractive area of research which needs to be focused on. Thus, the theoretical and practical understanding of the testing subject has now become necessary for designers who are involved in designing complex and compact systems. Also, talking about electronics, there is no great visual inspection sort of activity that can help to check the correctness of

---

I. Gupta (✉)
Education and Training Group, Centre for Development of Advanced Computing (C-DAC), Noida, India
e-mail: ishagupta@cdac.in

A. Noor et al. (eds.), *Proceedings of Emerging Trends and Technologies on Intelligent Systems*, Advances in Intelligent Systems and Computing 1371,
https://doi.org/10.1007/978-981-16-3097-2_23

a circuit. Hence, it is required to have advanced testing techniques and procedures that can help in complete the test process [1].

Faults in a circuit can be of various types, and some of them like the transient errors that usually occur during operation require complex techniques for detection and correction, whereas permanent faults can be detected before the production before the chip is finally introduced to the customers for usage [2]. This fault detection is very important part of the design cycle and needs to be completed in short span of time.

1.1 Stuck at Faults

One such permanent faults are the stuck at faults. Stuck at faults refers to the problem wherein a particular node or nodes in a circuit get stuck to a particular value either logic 0 (stuck at 0 fault) or logic 1 (stuck at 1 fault) [1, 3, 4]. This stuck at fault model is very helpful and common as this model does not have a single reason for its occurrence, but it can be caused due to numerous different reasons.

As illustrated in Fig. 1, there is a stuck at 1 fault on an internal node of the circuit and considering that the inputs are as given values as specified in the figure, this stuck at fault causes a wrong output for the circuit. This is because in the absence of this fault both the internal nodes which are the outputs of the AND gates should have been at logic 0 and this would cause the output of the OR gate also to be 0. But due to the fault, the output is driven to logic 1 which is functionally incorrect. Thus, the fault is detected [1, 3, 4].

Also, if the input combination is depicted in Fig. 2, then the output would be at logic 1 in the presence as well as in the absence of the fault. Thus, the fault is undetected. Hence, it can be concluded that the effect of this fault can be reflected at the output of the circuit based on particular input conditions [1, 3, 4].

In general, circuits can have multiple faults at the same time, but handling multiple faults simultaneously can make our detection algorithm quite complex. This paper focusses on detection of single stuck at faults in a circuit.

The technique for detecting the stuck at faults in a circuit is to apply a set of inputs to the circuit and check the response. The response must be compared with the already known correct response, and any mismatches among these must be reported as faults

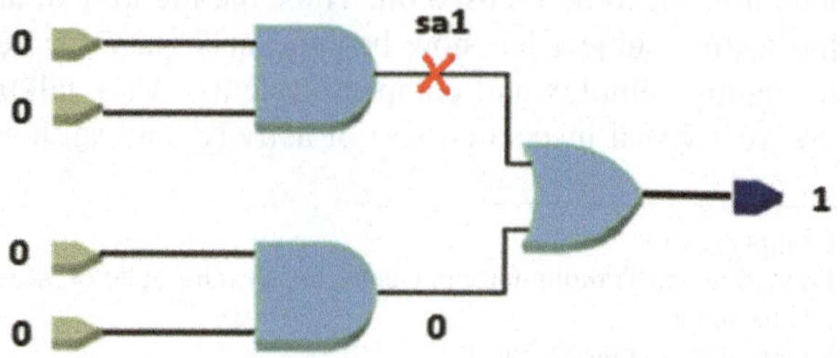

Fig. 1 Stuck at fault model depiction in a circuit and its effect on the output (fault detected) [3]

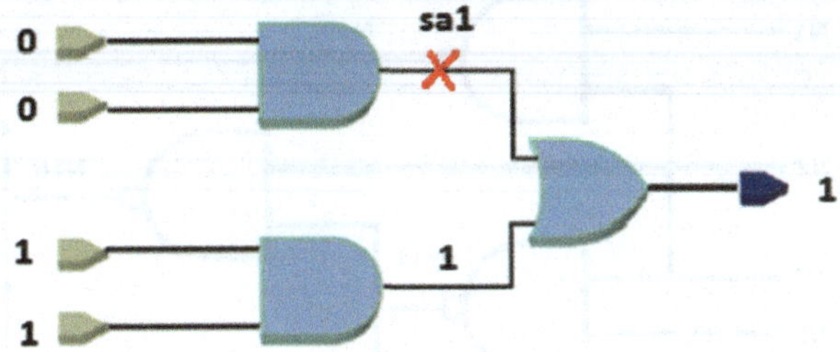

Fig. 2 Stuck at fault model depiction in a circuit and its effect on the output (fault undetected) [3]

in the circuit. In order to detect all possible stuck at faults for each node in the circuit, one obvious way is to apply all possible input combinations and compare the output for each case. This is called exhaustive testing. But as the number of inputs increases for a circuit, this exhaustive testing becomes very difficult as the number of test vectors becomes huge. Handling such volume of data is not preferable. To resolve ATPG (Automatic Test Pattern Generation), techniques are required that can help to find the minimum set of test vectors required to detect all possible faults in a circuit [5].

To generate the required set of minimum test vectors, a various fault simulation software is available, but the need is that the designer must be knowing the complete flow and usage technique for that software in order to get the test vectors. Moreover, it can be time-consuming. Hardware-based fault emulation is thus an attractive solution, and FPGA (field-programmable gate array) can be the possible platform [6]. An FPGA contains a collection of configurable logic blocks and programmable interconnects that can be configured by the designer to fit the design needs [2]. FPGAs do not merely allow implementation of the circuit under test with the stuck at fault model but, additionally, allow the inclusion of test vector generation and output response analysis circuits into a single reconfigurable device [7].

2 Followed Approach

The author in paper 1 in the reference focuses on how the FPGA can be used for fault detection process and proposes the methodology of insertion of multiplexers in the design for fault insertion. This paper attempts to apply this concept and use the Spartan 6-based FPGA board developed by CDAC Noida to implement the c17 benchmark circuit (Fig. 3) and find suitable test vectors to detect the stuck at faults in the c17 circuit. For injecting the faults at the different nodes, 2 × 1 multiplexer is used as depicted in Fig. 4 [2].

Multiple copies of the c17 circuit are implemented, and faults are inserted at the checkpoints using the multiplexer. Then, suitable input combinations are found for each of the inserted faults such that in the presence of the fault, the output changes and hence the fault is detected. Because of multiple copies of the c17 circuit, multiple stuck at faults can be injected in the circuit and suitable test vectors can be found for

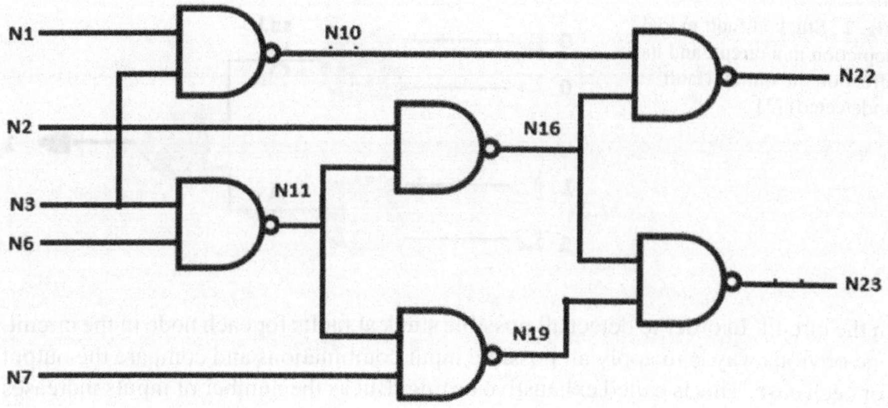

Fig. 3 c17 Benchmark circuit [2]

Fig. 4 Fault insertion using multiplexer [2]

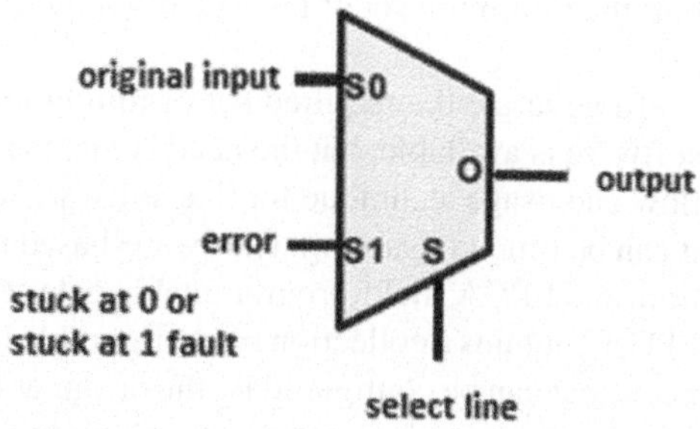

their detection. This paper performs the simulation using Verilog language and also the hardware implementation of the above concept on Spartan 6 FPGA. Different inputs are applied to the circuit, and faults detected are noted as result.

The fault insertion process is based on selecting suitable nodes for insertion of multiplexer based on the check point theorem. The check point theorem focusses on finding the checkpoints in a circuit based on fault equivalence and dominance, and it states that if the stuck at faults is detected for each of the checkpoints, then it automatically detects the stuck at faults for each node in the circuit [3, 8]. The checkpoints in a circuit are the primary inputs and the fan-out branches, and thus, the multiplexer will be inserted at these checkpoints in order to enable fault insertion at these nodes. For the c17 circuit, the checkpoints are shown in Fig. 5. The check point theorem reduces the total nodes from 17 to 11, and hence, the faults to be detected have been reduced from 34 (stuck at 0 or stuck at 1 fault for each node) to 22.

To handle multiple faults at a time on the FPGA, the implementation architecture is shown in Fig. 6. There are three copies of the c17 benchmark circuit implemented on the FPGA. Out of these, one of the circuits is fault free and the other two are implemented with the multiplexers injected at the checkpoints so that faults can be injected into each one of them based on the inputs provided to the select lines. The

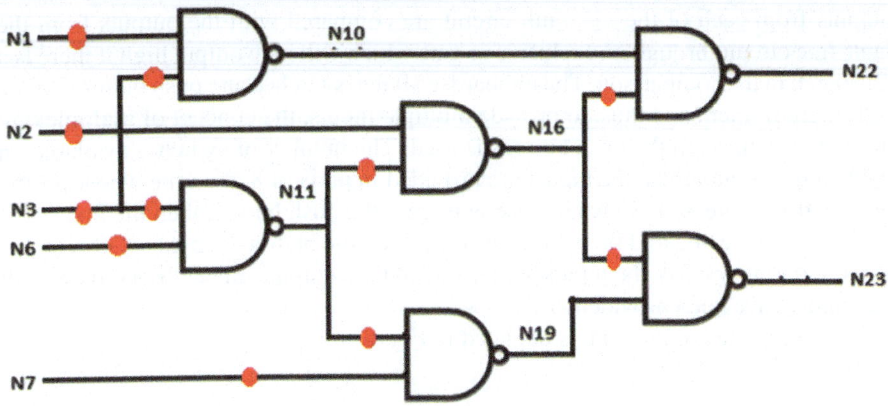

Fig. 5 Checkpoints for the c17 circuit [2]

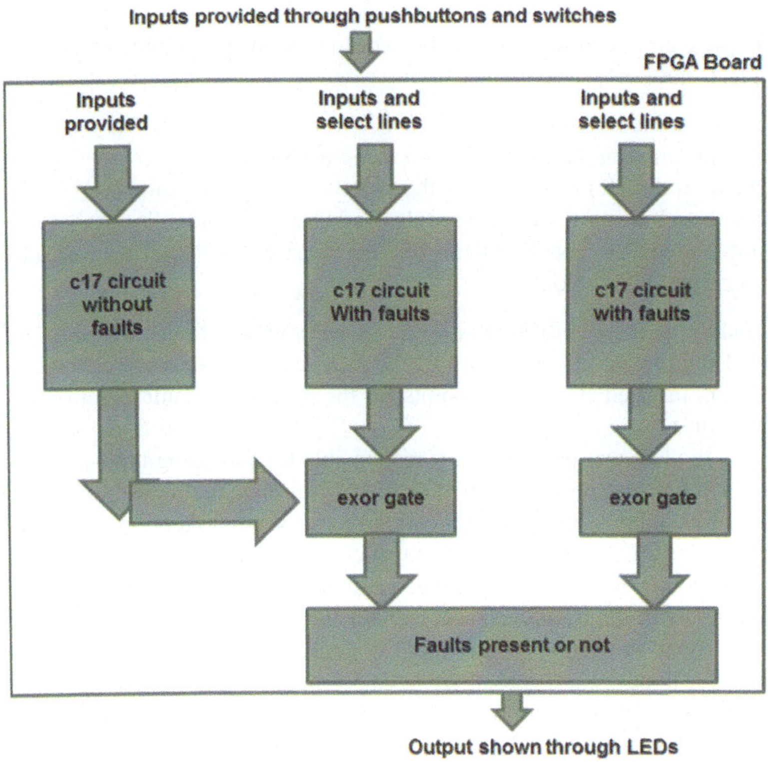

Fig. 6 The complete architecture [2]

outputs from each of the c17 fault circuit are compared with the outputs from the fault-free circuit through an exclusive or gate which will give output high if there is a mismatch in the comparison. The onboard available switches and push buttons on the FPGA are limited, and thus, to provide multiple inputs, the concept of multiplexing the switches through the push buttons is used. The number of switches available on the board is 8, and thus, the inputs are provided in parts of 8 at a time whenever the push button is pressed. So when once you press the push button, then the first set of 8 inputs are received and then on the next push on the button the next set of 8 inputs is send. In this way, all the inputs are passed and the output can be observed once all the inputs have been provided.

The complete architecture is depicted in Fig. 6.

3 Implementation and Simulation Results

The complete architecture was coded in Verilog language with the help of modules created as follows:

(a) c17 benchmark circuit
(b) c17 circuit with the fault injection using multiplexer
(c) Main module for multiplexing the switches and push buttons.
(d) Test bench to stimulate the modules and check functionality
(e) Implementation constraint file for the Spartan 6 FPGA-based development board by CDAC Noida.

A module of the c17 benchmark circuit without any faults was first coded in Verilog. Then two additional copies of the same c17 module were created with the multiplexers inserted at the checkpoints for the purpose of fault injection based on the select lines.

A main module was created that is responsible for providing inputs to the circuit though switches and push buttons. For the c17 circuit, the total number of nodes necessary for fault testing after application of check point theorem is 11 and hence we require checking both stuck at 0 and stuck at 1 fault for each of these nodes. At these checkpoints, we have the multiplexers inserted each of which has a select line. The two inputs to the multiplexers are either the original input or the error signal. This error signal indicates whether there is stuck at 0 or stuck at 1 fault induced at the particular node. Thus, in total we have 5 primary inputs for the c17 circuit, 11 select lines and 1 error signal for both the copies of c17 circuit in our architecture. In total, we need to provide 29 inputs to our design. Since the inputs are provided through the switches on the FPGA board which are limited to 12 in our case, we use the concept of multiplexing the switches through the push buttons.

The state diagram for this switch multiplexing in order to provide the inputs is shown in Fig. 7. At every positive edge of the clock, the state machine is triggered. We begin from state 0. If reset is high through the push button, then we remain in state 0. But if the other push button is pressed, then the state machines start accepting

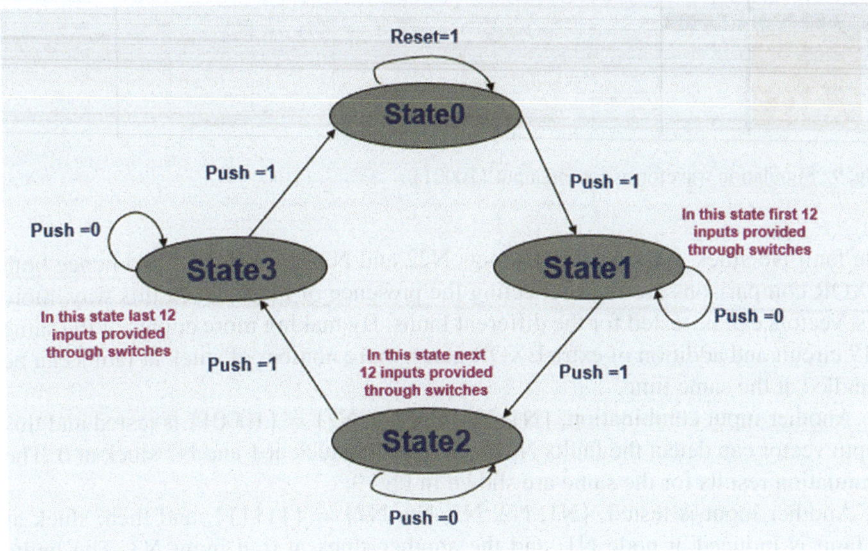

Fig. 7 State diagram

the first 12 inputs through the switches and map them to the modules in our design. We progress to the next state. Again we wait for the push button being pressed again, and then, the next 12 inputs are mapped to our design. In this way, after 3 cycles the complete input sequence is provided and hence we transition back to initial state when push button is pressed fourth time.

The simulation results are shown in Fig. 8.

The waveform shows the clock, inputs given on the switches through the push buttons and the outputs observed. The input provided in the test bench is such that the primary inputs N1, N2, N3, N6 and N7 are given the value 01110 and then in the first copy of the c17 faulty circuit, the select line is made 1 for the input corresponding to N1 stuck at 1 fault and the second copy of the c17 circuit is given inputs at select lines such that the fault N6 stuck at 0 is induced. The output received reflects that in the presence of the fault N1 stuck at 1, the output at primary output N22 changes and hence the EXOR output for this comparison is high, whereas N23 remains same even in the presence of the fault and hence the EXOR output is low for the same. For

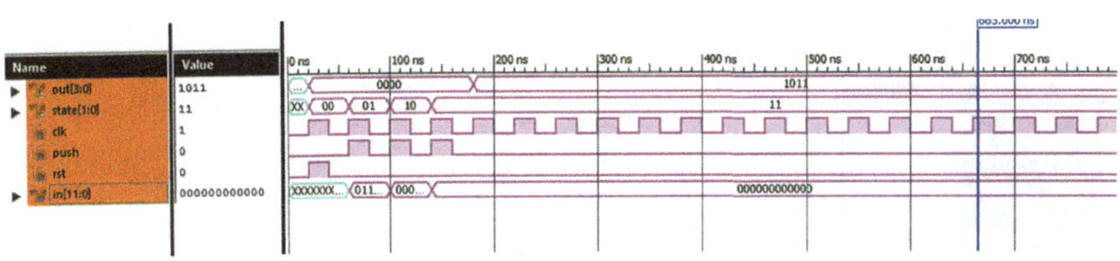

Fig. 8 Simulation waveforms for inputs {01110}

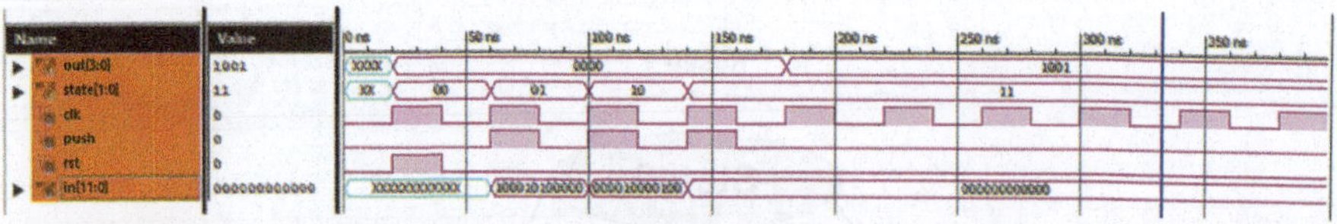

Fig. 9 Simulation waveforms for the input {10001}

the fault N6 stuck at 0 both, the outputs N22 and N23 get effected and hence both EXOR comparisons are high reflecting the presence of the fault. In this way, more test vectors can be tested for the different faults. By making more copies of the same c17 circuit and addition of extra EXOR gates, more number of stuck at faults can be handled at the same time.

Another input combination, {N1, N2, N3, N6, N7} = {10001} is tested and this input vector can detect the faults N2 stuck at 1, N3 stuck at 1 and N7 stuck at 0. The simulation results for the same are shown in Fig. 9.

Another input is tested, {N1, N2, N3, N6, N7} = {11111}, and then, stuck at 0 fault is induced at node N1, and the another stuck at 0 at input N3. The faults are injected by providing suitable value 1 to the select lines for these corresponding inputs. In this case, in the presence of stuck at 0 fault at node N1, the output N22 is changed and in the presence of stuck at 0 fault at N3, the output N23 is changed. The result for the same is shown in Fig. 10.

Another input is tested, {N1, N2, N3, N6, N7} = {01100}, and then, stuck at 0 fault is induced at node N2, and the another stuck at 1 at input N6. The faults are injected by providing suitable value 1 to the select lines for these corresponding inputs. In this case, in the presence of stuck at 0 fault at node N2, both the outputs, N22 and N2, are changed and in the presence of stuck at 1 fault at N6, both the outputs, N22 and N23, are changed. The result for the same is shown in Fig. 11.

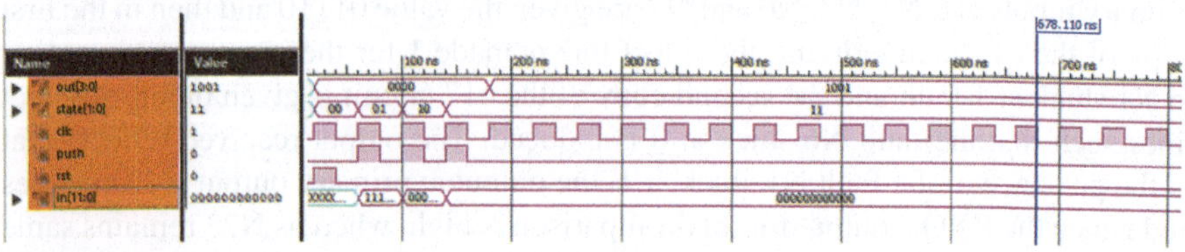

Fig. 10 Simulation waveforms for the input {11111}

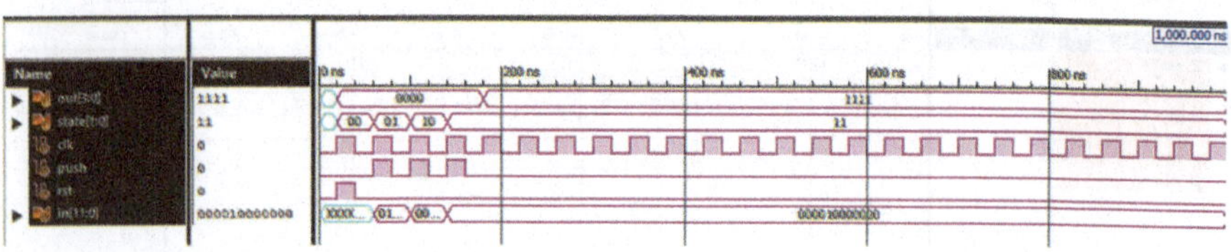

Fig. 11 Simulation waveforms for the input {01100}

Fig. 12 Spartan 6-based FPGA board by CDAC Noida

The same is implemented on the Spartan 6-based FPGA board as shown in Fig. 12. The inputs can be provided through the push buttons and switches. There are 12 switches on this board and 4 push buttons. Every time the push button presses the 12 switches will be mapped to 12 inputs and the process will be repeated n number of times depending on the total number of inputs to be given. Thus, a finite state machine is implemented in order to map this state transition behavior which can tell that how many times the push button has been pressed. The finite state machine is shown in Fig. 7. Once the complete input has been provided, the outputs are executed and faults can be detected. The outputs are reflected through LEDs on the FPGA board. The outputs are the EXOR of the outputs obtained from the fault-free and the faulty circuits. Whenever the LED is on, it depicts that the outputs obtained are different, and hence, it implies that the fault has changed the output. With this testing for every fault induced in the circuit, we can find out the corresponding input test vector that can detect the fault.

The design summary report generated after implementation on FPGA is shown in Table 1.

Table 1 Device utilization summary

Device utilization summary			
Slice logic utilization	Used	Available	Utilization (%)
Number of slice registers	59	4800	1
Number of slice LUTs	80	2400	3
Number used as logic	78	2400	3
Number of occupied slices	26	600	4
Number of MUXCYs used	56	1200	4

4 Conclusion and Future Work

The conclusion is that with the use of FPGA, test vectors for detecting stuck at faults can be derived for combinational circuits. In this work, two copies of the circuit were created and faults were injected in them, but it can be replaced with more number of such copies and with that multiple stuck at faults can be handled simultaneously and test vectors could be detected for them. The same work can be extended to other combinational circuits as well. Also, the method for providing inputs through multiplexing switches and push buttons can be replaced with UART transmission or other modes of providing inputs to the FPGA boards.

References

1. Hurst, S.L.: VLSI testing-digital and mixed analogue/digital techniques. The Institute of Electrical Engineers, London, United Kingdom (1998)
2. Dunbar, C., Nepal, K.: Using platform FPGAs for fault emulation and test-set generation to detect stuck-at faults. JCP **6**, 2335–2344 (2011)
3. Bushnel, M.L., Agrawal, V.D.: Essentials of Electronic Testing for Digital Memory and Mixed-Signal VLSI Circuits. Kluwer Academic Pubs., New York, Boston, Dordrecht, London, Moscow (2002)
4. Wang, L.T., Wu, C.W., Wen, X.: VLSI Test Principles and Architectures: Design for Testability. Elsevier, San Francisco (2006)
5. Parreira, A., Teixeira, J. P., Santos, M.B.: A novel approach to FPGA-based hardware fault modeling and simulation. In: Proceedings of the Design and Diagnostics of Electronic Circuits and Syst. Workshop, pp. 17–24 (2003)
6. Cheng, K.-T., Huang, S.-Y., Dai, W.-J.: Fault emulation: a new methodology for fault grading. IEEE Trans. Comput.-Aided Des. Integr. Circ. Syst. **18**(10), 1487–1495 (1999)
7. Ellervee, P., Raik, J., Tammemae, K., Ubar, R.: FPGA-based fault emulation of synchronous sequential circuits. IET Comput. Digit. Tech. **1**(2), 70–76 (2007)
8. Civera, P., Macchiarulo, L., Rebaudengo, M., Sonza Reorda, M., Violante, A.: Exploiting FPGA-based techniques for fault injection campaigns on VLSI circuits. In: IEEE International Symposium on Defect and Fault Tolerance in VLSI Systems, pp. 250–258 (2001)

How Efficient Is Blockchain While Dealing with Android Malware? A Review Paper

Jagjot Singh Wadali, Sanjay Madan, and Praveen Kumar Khosla

Abstract Android is an open-source mobile operating system that became more popular in recent years than any other mobile operating system. The main reason for its popularity is the wide range of functionality available at an affordable cost. Moreover, millions of android-based free applications are easily available for users on the android market store, that is, Play Store. Because of the expanded user base globally, there is a significant rise in cyber-attacks on android devices in the last decade so as to have large impact. The main goals of the attackers are to gain access to these devices and steal private as well as confidential information stored on these devices. There is a wide variety of android malware specifically designed for these devices with the capability of spreading and propagation. Even some of the android malware code is written in such a way that they have the capability to evade the detection system. Various frameworks and technologies are used for the detection of android malware but still, attackers are finding new ways to evade detection. Meanwhile, blockchain technology is emerging and has proved its worth in various sectors. In this paper, we reviewed the use cases of blockchain technology for the detection of android malware which manifests the capability and effectiveness of this emerging technology to secure ubiquitous mobile cyber-infrastructure.

Keywords Android malware detection · Blockchain · Mobile malware

J. S. Wadali (✉) · S. Madan · P. K. Khosla
Center for Development of Advanced Computing, Mohali, India
e-mail: jagjot@cdac.in

S. Madan
e-mail: msanjay@cdac.in

P. K. Khosla
e-mail: khosla@cdac.in

A. Noor et al. (eds.), *Proceedings of Emerging Trends and Technologies on Intelligent Systems*, Advances in Intelligent Systems and Computing 1371,
https://doi.org/10.1007/978-981-16-3097-2_24

1 Introduction

The enormous development in modern communication technologies in recent years has led to the emergence of mobile devices, which offer intelligent and personalized services. These devices are offering a sophisticated life for humans at an affordable cost. Nowadays, personal and confidential information is stored in these devices. Because of the ever-expanding user base, there has been a tremendous demand for these devices globally. The massive growth in this field has attracted global consumers as well as many cyber-attackers. The android operating system is one of the most preferred platforms globally, and it delivers an exposed market model. Because of being a vulnerable and open source, this platform has been a victim of widespread illegal activities and data thefts.

According to Global Statistics [1], there is a huge growth in smartphone users as shown in Table 1. Ever since the spread of the pandemic, there is enormous growth in the usage of the Internet and a significant rise in cyber-attacks. According to Statista report, around 70–80% of the smartphone are android based, and there are around more than 24,000 variety of embedded devices running on android platform, all over the world [2]. Moreover, there are around 2.87 million android apps available in the Google Play Store making it the largest marketplace for android applications [3]. The smartphone users generally store their personal details including passwords, financial details, etc. on these devices, so because of that and the huge user base, the android-based mobile devices are a promising target for the attackers to steal information or to harm the cyber-infrastructure.

With the rapid growth in android malware and lack of proper security mechanisms for mobiles, hackers are turning toward android devices to steal private and confidential information stored on the mobile phone. The statistics of android malware is shown in Table 2 [4]. Over a period of time, android mobile device popularity is increasing by means to manage many personal as well as confidential information of an individual. Users mostly store specific and classified details on their mobile devices like contact number, date of birth, credit or debit card details, personal information, etc. So, this remains one of the main reasons for attracting attacks on android devices.

There is a significant growth of android malware since 2014 as shown in Fig. 1 [4]. Moreover, an android malicious content writer is nowadays working upon new techniques to evade malware detection systems on the host device. With reference

Table 1 Statistics about android users in 2020s

Parameter	Statistics (in billions)
World population	7.8
Internet users	4.6
Smart phone users	3.5
Android users	2

Table 2 Statistics of android malware in 2020

Parameter (according to 2020)	Statistics (in millions)
Total number of malwares	13.73
Development of android malwares	3.3
Development of potentially unwanted application for android	5.51

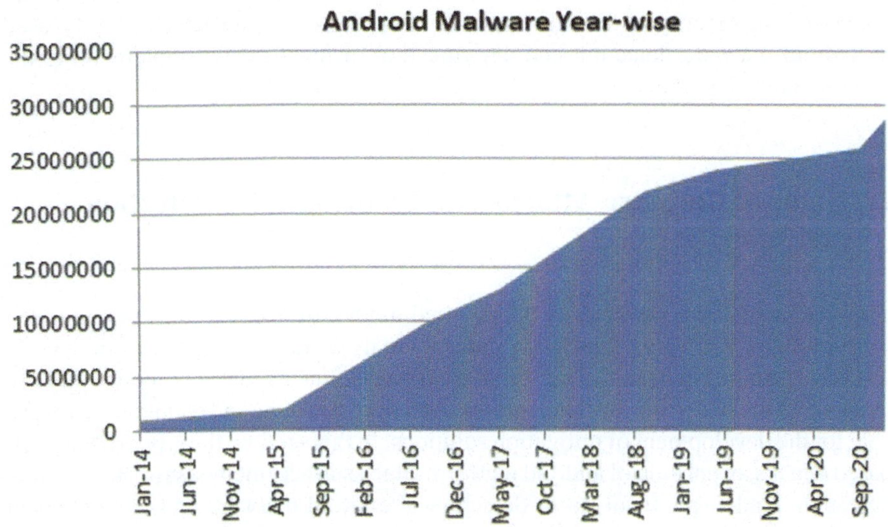

Fig. 1 Statistics of android malware year wise [4]

to the recent threat reports and as mentioned in Table 2, there is significant rise in potentially unwanted application for android platform. Based upon the literature, various approaches for analysis and detection of android malware concludes that the use of hybrid analysis, parallel processing, or multilevel analysis can be quite efficient in the detection of android malware [5, 6].

This huge user base for the active android device users globally gives a great opportunity for cyber-attackers to develop sophisticated techniques for conducting fraudulent activities and financial data theft. The malware program is created and then posted or published online on official or unofficial markets by the attackers. Most of this malware appears to the user as legitimate applications, but they compromise the security of the device upon installation. Even though most of the malicious applications have been removed from the Google Play Store, still android owners are at high risk as some applications remain undetected. Recently, it has been reported that most of the hackers are developing android malware which can bypass detection mechanism and steal private and confidential information [7–9].

Blockchain technology is the composite technology that is used to store and exchange information in a transparent and secured way without any third party or intermediaries [10]. It was created by Satoshi Nakamoto in 2009 for implementing the cryptocurrency, but now, it is emerging as one of the preferred technologies in many other industries as well. The basic fundamental of blockchain lies in maintaining the record which cannot be changed, making it quite useful in different use case scenarios [11]. Blockchain technology finds its use case in cyber-security too. Furthermore, blockchain technology has extensively impacted the mobile industry and is also imperative in security of mobile applications [12–14]. It can be effectively used in android malware detection and classification of malware by implementing the persistent attack records [15].

2 Android Malware Attacks and Challenges for Detection and Analysis

The attacks on the android devices increased in the last decade, so as to rise in the android-based malware variants. In order to analyze the risk pertain to the cyber-infrastructure through these type of android-based attacks, there is requirement to understand the underline mechanisms for detection, analysis, and technologies available for the development of mitigation solutions. In this section, the types of android-based attacks, expansion of android platform attack surface, methods for analysis and detection, challenges involved in detection of android malware and technological growth for detection of android malware.

2.1 Aspects of Android Malware Attacks

Usually, the android attack is carried out by finding the vulnerabilities and exploiting them. And this can be carried out in multiple ways including unauthorized or malicious applications, insecure Internet connection, vulnerabilities in the operating system, etc. As we compare personal computers with mobile devices, we learn that the latter carries more sensitive and confidential information of the user. There are around 2 billion active monthly android users [14], who install and use various applications from different (official and unofficial) sources. As such, most of these devices remain connected to the Internet (WiFi/Cellular Data) and users are all-time logged into their applications, making them a soft target of cyber-threats. This enormous user base, spread around the world, has motivated cyber-criminals to establish different ways of targeting them.

There is a multitude of reasons behind the creation of malware, which include, but are not limited to date theft, device hijacking, financial frauds, etc. Each malware aims for some specific purposes and is therefore classified based on their actions,

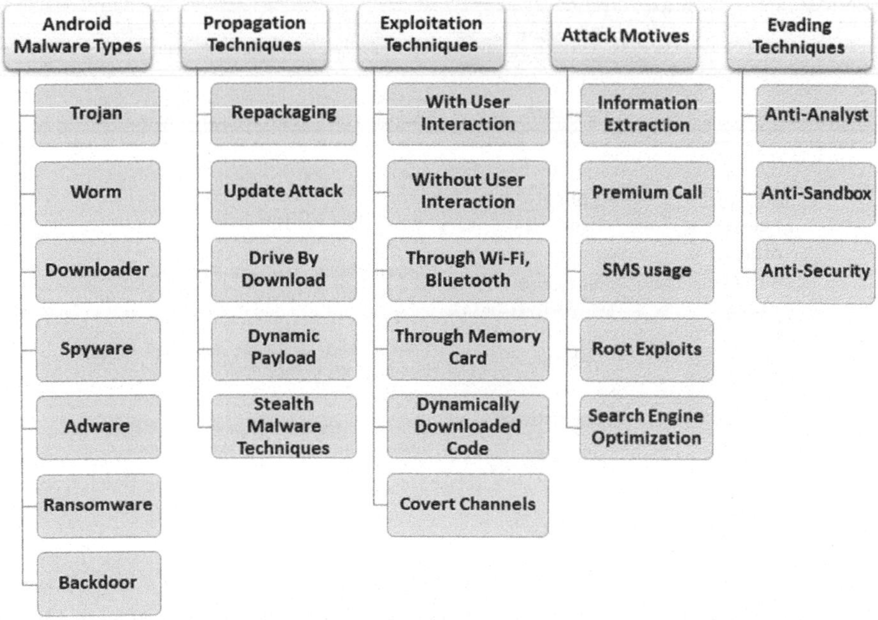

Fig. 2 Aspects of android malware attacks

propagation, behavior, features, etc. [16–18] as shown in Fig. 2. There is a vast variety of android malware, comprising of exclusive characteristics and traits. Each variant uses its unique way of attacking and infecting the device.

2.2 *Android Malware Attack Surface Analysis*

An attack surface is defined as the integration of different exploiting points from where the attacker can get access to the system and can retrieve valuable information. Attack surface for android platform took into consideration of all aspects for the application including (a) application source code; (b) network traffic data for the application; (c) all the data managed by the application; and (d) application interaction with operating platform.

Attack surface analysis is usually done by security professionals for vulnerability accessment or penetration testing, to understand the risk associated with application as well as the impact of the risk involved. It describes about the possibilities of cyber-attack, which helps the developer and solution architect to protect the system or apps from external attacks and to patch the system. With the expanding commercial and

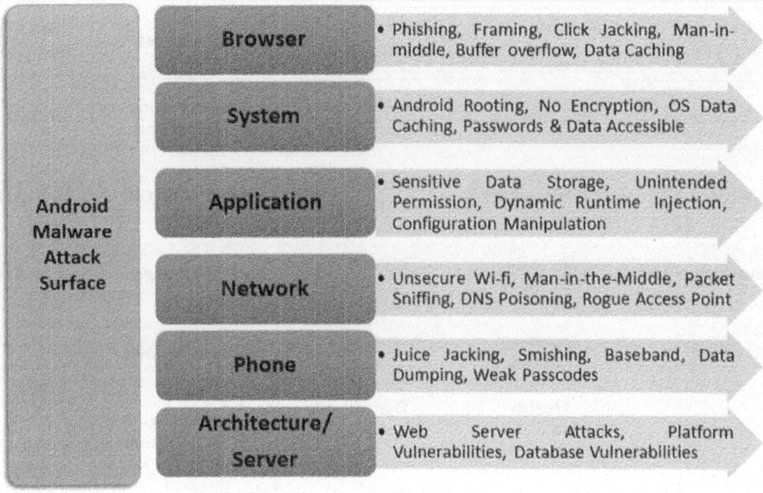

Fig. 3 Android malware attack surface

social use of portable android devices, the attack surface drastically increased and has more possibilities of performing attack on this platform [19]. The possible attack surface for android malware attack is shown in Fig. 3.

2.3 *Android Malware Detection and Analysis*

The malware analysis technique is the process of analyzing the application to find out the malicious content inside or to find out the behavior of the application. The behavior of the application or any malicious content inside it will help in detecting the malware. There are three kinds of malware analysis techniques—static, dynamic, and hybrid analysis as shown in Fig. 4. Each of the types has its kind of strength and weaknesses [17].

Various technologists and researchers concluded that hybrid analysis is better than static and dynamic analysis. Hybrid analysis, parallel processing, or multilevel analysis increases the detection rate and accuracy [6]. It can detect malware that cannot be detected in a static analysis or dynamic analysis.

As security threats evolve, static and dynamic analysis techniques are less capable to identify malware code on their own. Thus, hybrid approaches combine aspects of both static and dynamic analysis [16]. Moreover, different features and algorithms are used to detect different types of malware as shown in Fig. 5.

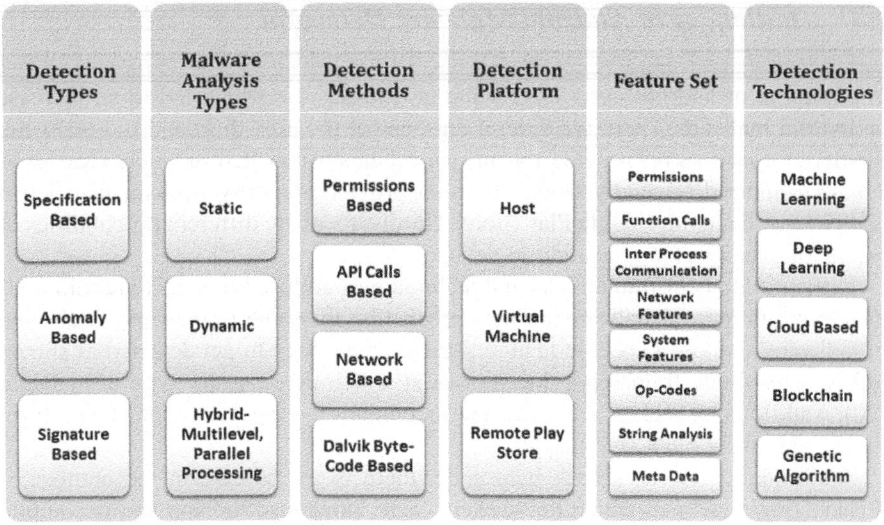

Fig. 4 Android malware detection and analysis

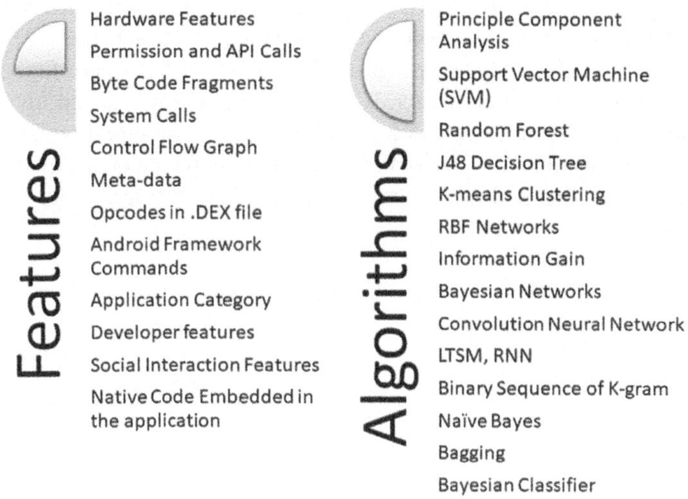

Fig. 5 Features and algorithms used for android malware detection

2.4 Challenges in Android Malware Detection

The number of android users is growing at a faster rate which led to an increase in android malware. There are several reasons for the growth in android malware. Firstly, android does not provide a security mechanism, and there are more chances of attacks in android. Secondly, Google does not provide a security mechanism to detect malicious code in the Google Play Store. Thirdly, there are different probabilities of conducting an attack on the android device.

Different kinds of frameworks and technologies are used to detect android malware but all types of malware are not detected by the same framework. Each type of malware requires a specific feature set and technology to get detected as shown in Fig. 6. A single framework does not exist which detect every type of malware [20]. Additionally, hackers are using new techniques to evade detection and perform illegal activities (Table 3).

Additionally, due to Dalvik byte code, there is an increase in the number of repackaging attacks in which the hackers easily download the app and decompile and again sell it [21]. All the apps are available on Google Play Store and anyone can download and decompile and inject malicious content and again compile it.

There is an increase in mobile Trojans as these Trojans get injected into the apps and work in the background without the user consent or knowledge [22]. A remote access Trojan (RAT) is a malware program that includes a back door for administrative control over the target computer. RATs are usually downloaded invisibly with a user-requested program—such as a game—or sent as an email attachment [23]. One of the examples of Android Trojan is AhMyth which injects malicious code into the genuine application and then operates it from the server and steals confidential information [24].

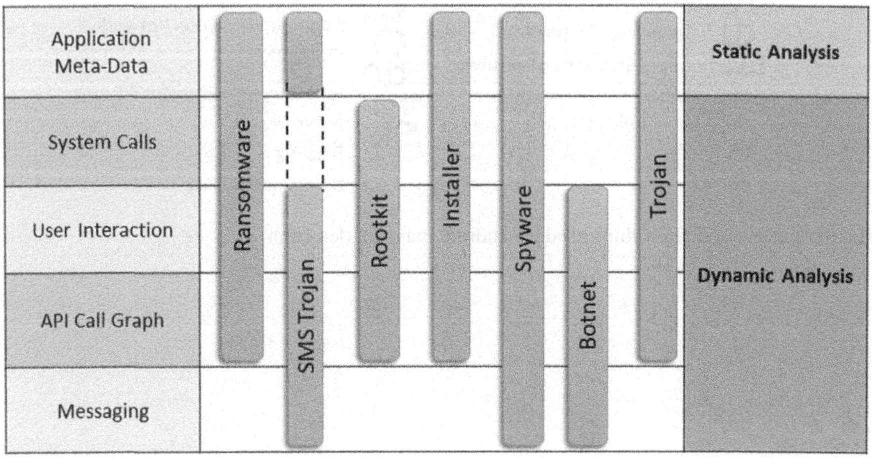

Fig. 6 Android malware versus detection type

Table 3 Challenges for android malware detection and analysis

Challenges	Description
Challenges with respect to android malware detection frameworks	A single framework is unable to detect each and every type of android malwares
	• Different malwares are detected using different set of features
	• Many frameworks are not able to detect dynamically downloaded code and bypass detection mechanism
	• Moreover, scalability, accuracy, and portability of various frameworks is required to detect existing malwares as well as new malwares or zero day malwares
	• Static analysis is effective and scalable however in the case of obfuscation and encryption, dynamic analysis is used but it has less code coverage
	• Hybrid analysis or parallel processing or multilevel analysis gives better accuracy and results
	• Some of the frameworks are not able to detect zero day malwares
Android malware	• Malwares writers are looking for new ways of attacks, such as the bypass detection mechanism including anti-malware, evade sandbox environments, etc.
	• Since the android apps are based on dalvik bytecode, hence the repackaging apps are almost unavoidable
	• Development of more PUA that users are unaware about it
	• Rise of adware and Trojans for SEO purpose and stealing of private information
Others	• Android users are mostly unaware about the attacks
	• Usually, user not bother about the authentication and permissions providing to an apps during installation
	• Weakness in signature-based detection solutions

There is a huge demand or gap area to address security prospects. Users are not aware of the attacks on the smartphone. Mostly, they are aware of the functionality and the application only however they are not aware of the attacks and the way to protect mobile from stealing information. Even much genuine application steals personal data from mobile phone [25]. The main challenge is for correlating the different kinds of android malware open-source datasets available online [21] for further analysis and security product formulation.

2.5 Technological Transformation for Detection of Android Malwares

As the new variants of android malwares are evolving day by day, technologist and researchers are also looking for the new ways to decode the attacking techniques used by the attackers. There is a paradigm shift in designing the new algorithms for detection of new variants of android malware. With the advent of new technologies, researchers are using every possible way to use diversified new different technologies for android malware detection. And now, the android malware detection is not only restricted to machine learning techniques. However, the researchers are using new technologies/methods like deep learning, blockchain, genetic algorithm, etc. to decode the tactis used by attackers for detection of attacks.

So, in our study, we mainly focus upon the use of blockchain technology to identify the various ways through which researchers designed the methodology for detection of android malware attacks along with the advantages and disadvantages of using this technology for the development of more secure mitigation solutions.

3 Blockchain

The blockchain technology market is expected to reach 7.7 billion dollars by 2024 and the financial sector accounting for the major chunk [26]. The reason for the growth in blockchain technology is that it finds its use case in different sectors. Nowadays, almost all companies are dependent upon information technology and information is in the digital format. The most essential thing required is confidentiality, data integrity, privacy, and security. Blockchain not only removes the third party agent but also provides security and confidentiality. This is the prime reason why different companies started adopting this technology. Some of the use cases include:

- **Healthcare**: Blockchain technology can be used to save patient medical records without being tampered with or changed.
- **Cryptocurrency**: Blockchain is the technology behind cryptocurrency and it had already proved the potential.

- **Supply Chain Management**: A supply chain is a network established between business and supplier. The technology's ability to enable the digitization of assets can be beneficial in supply chain management.
- **Intellectual Property**: Blockchain technology can provide a platform upon which clear and accurate ownership of IP assets should be saved.
- **Insurance**: Blockchain-based smart contract can provide a secure and transparent system between insurer and customer.
- **Real Estate**: Transfer land titles can be digitized with blockchain which can make a reliable and transparent system.

3.1 Blockchain in Cyber-Security

Blockchain technology which emerged as the technology behind bitcoin is continuously evolving and it finds its use case in different sectors including banking, healthcare, etc. This technology is spreading due to its various features like reliability, security, and immutability. One of the main objectives of this technology is to make a secure system. It can be used to prevent any kind of data breaches, cyber-attacks, identity thefts, or frauds in transactions [12]. This ensures that the data remains private and secure. Blockchain can be useful in:

- Securing IoT and industrial IoT will help in strengthening the authentication system and enhancing data attribution.
- Maintaining confidentiality and data integrity.
- For securing personal and private information exchanged over messaging applications and social media.
- Improving public key infrastructure.
- To protect against DDoS attacks by integrating blockchain into decentralized solutions.

3.2 Blockchain in Mobile Application Security

Nowadays, mobile devices are more vulnerable than any other device due to the wide variety of attacks. Mobile application security is posing a severe challenge for the developers. The attacker's main purpose is to steal private and confidential information of the mobile for misuse and fraud. Most of the users save their confidential information like bank details and other important documents or files on mobile devices. The problem is that the smartphone user is unaware of these attacks and mostly relies on anti-virus and other protection software to protect vital data. In this scenario, blockchain technology can play a significant role in securing user data [13, 14].

It is quite essential to understand what type of data needs to be secured and how to maintain security before using blockchain technology in the mobile industry. Understanding the mobile application development model and working of the android system will help in developing the correct use case scenario to implement blockchain in providing security. The basic function of blockchain was to secure online transactions through the distributed ledger form. The benefits of using blockchain technology in the mobile application includes:

- No need for password.
- Blockchain does not allow any breach of information.
- Securing the infrastructure.
- Securing the identity.

3.3 Blockchain in Android Malware Detection

Blockchain finds its use case in various domain areas due to several reasons including better transparency and enhanced security. Because of the use of its advantages, it is used in android malware detection also. The number of research articles published in which blockchain technology is used for android malware detection in various ways to improve efficiency or detection rate and creating a reliable source of information.

Researchers had used various techniques for android malware detection like machine learning, deep learning, etc. and now, they had used blockchain technology for the detection of android malware. The blockchain is used in different frameworks with different use cases for creating authentic and reliable data; many research used it for getting authenticated and reliable information.

Furthermore, blockchain alone is not used for the detection of android malware but it is used with other sophisticated technologies including machine learning, deep learning, etc. for host-based or web-based security solutions. It can also be used for storing the information and results so that the user traced backed the authenticated data. The various research articles studied and described here for the use case of the blockchain technology in android malware detection.

Gu et al. [15] used consortium blockchain for detection of android malware. The author created a multi-feature model. Blockchain technology was used to create a fact-based database of android malware. Each block of the data contains the malware name and details about its detection. In this way, it is easy to provide users information about transparent information to the researchers and analysts.

Liu et al. [27] used consortium blockchain to detect malicious code in android malware. Created fact base database of android malware. Each block of data contains the malware name and details about its detection. In this way, it was easy to detect malicious code in malware applications as well as to lower the bogus positive and bogus negative rates. There are several drawbacks to the security mechanism of the Google Play Store. It is unable to detect each and every type of malicious application, and it is also difficult to trace the hackers that hosted the application.

Homayoun et al. [28] presented a blockchain-based framework which detects the illegitimate mobile applications in app stores before it is downloaded by the final user. The framework combines static and dynamic analysis as an integrated system for the detection of malware, which helps in decreasing the false-positive rate of detection. Overall, blockchain technology enables transparency and a reliable source of information leading to optimized solution.

Kumar et al. [29] presented a framework which combines blockchain and machine learning technologies together for malware detection in android-based IoT devices with effective manner. Permissioned blockchain is used in this framework to store authentic information of extracted features in a distributed malware database block that will enable faster detection of malware with accuracy. Furthermore, the information was also available to all the users who joins the blockchain.

Abdellah et al. [30] presented an Android Permissions Scan Registry (ANDROSCANREG) framework based on blockchain technology. In this framework, the requested permissions was extracted and analyzed for the android application through a distributed and decentralized system.

Hwang et al. [21] presented a framework in which they used blockchain technology to identify forgery in mobile applications. In this framework, blockchain technology is used for recording and storing the certified apps in the blocks. And to determine the possibility of forgery, it can be verified from the blockchain whether the application information exists in the blockchain or not. Hence, this design can enable discrimination and verification mechanisms that can identify the fake mobile apps by applying blockchain on Hyperledger Fabric.

Rana et al. [31] developed a framework for the evaluation of various machine learning models for a given android malware dataset using a consortium blockchain network. The competitors who will submit the solutions through training with selected machine learning models in a secure and trustworthy manner, the rewards in the form of incentives was offered to them. This reliable and transparent data in the blockchain is used for analysis and for comparison. This enables various organizations in the network for enhancing or boosting their current malware detection capabilities. The purpose of the decentralized network provides better security and transparency as well as reducing the overhead cost for managing the dedicated data by eliminating the third parties.

Nigam et al. [32] proposed a framework that uses multiple technologies including blockchain and machine learning for the detection of android malware. The purpose of the research is to shift from the traditional method of detection of android malware to new ways by using advance technologies. The purpose of this research is to include blockchain technology that helps in improving the false-positive and false-negative rates to increase the accuracy of malware detection in devices. Moreover, the significant importance of this study is to pin down a new type of malware that are not identified yet using the signature-based method or pattern analysis method. The experimental result shows better accuracy and precision for the detection of malware.

Lee et al. [33] constructed a discrimination Decentralized Application (DApp) framework, which detects forged Android APK by storing and maintaining the genuine android application in the consortium blockchain framework. The author had used Hyperledger Fabric for recording the database of legitimate applications. The main purpose of this design is to prevent the forgery and modification of the application from being installed on the user's device. Additionally, users can get access to genuine and verified apps only.

Jan et al. [34] proposed a state-of-the-art model for the detection of android malware and in the paper explained the traditional ways for detection of android malware, which can be used for the small dataset. However, the new approaches or technologies should be explored for detection in large datasets. Generally, for the detection of android malware, various machine learning techniques are used which are based on static, dynamic, and hybrid analysis; however, these approaches are not feasible for large-scale Android malware analysis. Deep neural architectures enable analyzing large-scale static details of the applications. Although static analysis techniques ignore the malicious behaviors of applications during execution. This study reviews various approaches for the detection of malware and proposes the conventional and state-of-the-art models, developed for analysis, which facilitates the provision of basic insights for researchers working in this malware analysis domain. It also provides a dynamic approach that employs deep neural network models for the detection of malware. Furthermore, this research also uses the android permissions as a parameter to measure the dynamic behavior of around 16,900 benign and malicious applications. Blockchain technology is used to preserve the integrity of the dataset and the results of the analysis.

3.4 Discussion

As the android applications are easy to decompile, there are more chances of repackaging attacks. The attacker download the applications and inject the malicious code after decompilation. Blockchain can be used to create a database of apps to avoid repackaging attacks. Users can rely on a single source of data to avoid installing malicious applications.

Android applications access important information through permission. Android permissions are declared in the android manifest file as well as in the source code. Usually, android malware injects malicious code in the applications and uses dangerous permissions in the source code but however does not use it in the android manifest file. Blockchain can be used to create a transparent and reliable database of android applications along with the permissions declared by it so that the user can trace the original and authenticated source of information. Before installing applications from the Play Store, the end-user can match the hash value of APK file with the stored value in the blockchain. If the signature is not present in the blockchain of the hash value, the alert for the user is generated that the specific application might be malicious. From there, the user can get the information whether the app is genuine or tampered. Blockchain can be used to create a transparent and reliable database of android applications along with the permissions declared by it so that the user can trace and track the original and authenticated source of information.

There are a different various types of possible android attacks and the system cannot be able to detect every type of attack. Blockchain can be used in creating the fact base database of android malware, which can help researchers and malware analysts to trace and track the malware related information. The developed pool of information is available to all, which is transparent, reliable, and easily traceable.

4 Conclusion

The conclusion of this study is that because of the expanded attack surface of the android-based devices due to increase usage of commercial and social applications, there is huge possibility of exploiting these devices and performing attach. Now, the researchers are shifting their detection methodology techniques from traditional methods of detection to the new technological transformation ways for cost-effective detection techniques. Moreover, the evading techniques used by attackers to bypass detection mechanism forced to design the new mitigation frameworks/solutions, which uses the diversified modern technologies in more efficient and effective way to detect these new types of malwares. Blockchain is an emerging technology, which proven its strength in various domain areas for solving complex problems and professionals are working on it to use in every possible way because of transparency and reliability. Android malware analysis is the promising domain area, in which researchers are also using it in different ways to deal with android malware detection. Various blockchain-based frameworks developed in last decade to deal with android malware. It helps in detection of android malware in various ways including (a) reducing the false positive and false-negative rate, (b) better transparency and reliability of malware information, (c) better efficiency in handling of large data or traces of android malware, and (d) easiest way to trace the information. Although, blockchain is quiet a complex technology and it is being used in the technological frameworks for authentic and authorized information persistence, reliability of information, and security of data is on high priority.

References

1. Global digital overview—datareportal—global digital insights. https://datareportal.com/global-digital-overview. Accessed: 27 Oct 2020
2. Liu, S.: Topic: android. https://www.statista.com/topics/876/android/
3. AppBrain: Number of android applications on the google play store. https://www.appbrain.com/stats/number-of-android-apps. Accessed: 28 Dec 2020
4. AVTest: Malware statistics trends report: Av-test. https://www.av-test.org/en/statistics/malware/. Accessed: 28 Dec 2020
5. Zachariah, R., Akash, K., Yousef, M.S., Chacko, A.M.: Android malware detection a survey. In: 2017 IEEE International Conference on Circuits and Systems (ICCS), pp. 238–244. IEEE (2017)

6. Tam, K., Feizollah, A., Anuar, N.B., Salleh, R., Cavallaro, L.: The evolution of android malware and android analysis techniques. ACM Comput. Surv. (CSUR) **49**(4), 1–41 (2017)
7. Google play store spews malware onto 9 million droids. https://forums.theregister.com/forum/all/2019/01/09/google_play_store_malware_onto_9m_droids/
8. September 2018's most wanted malware: cryptomining attacks against apple devices on the rise (2018). https://blog.checkpoint.com/2018/10/15/september-2018s-most-wanted-malware-cryptomining-attacks-against-apple-devices-on-the-rise/
9. Elise, A.: 5 types of android malware that made headlines in 2017 (2017). https://www.kcci.com/article/5-types-of-android-malware-that-made-headlines-in-2017/14508001
10. Rosic, A.: Blockgeeks: what is blockchain technology? a step-by-step guide for beginners (2020). https://blockgeeks.com/guides/what-is-blockchain-technology/
11. Blockchain 101. https://www.coindesk.com/learn/blockchain-101/what-is-blockchain-technology
12. Sharma, T.K.: The future of cyber security: blockchain technology (2018). https://www.blockchain-council.org/blockchain/the-future-of-cyber-security-blockchain-technology/
13. Srivastav, S.: How is blockchain revolutionizing mobile app economy (2020). https://appinventiv.com/blog/blockchain-technology-redefine-mobile-economy
14. Rajput, M.: Mehul Rajput (2018). https://customerthink.com/how-does-blockchain-boost-the-security-of-mobile-app/
15. Gu, J., Sun, B., Du, X., Wang, J., Zhuang, Y., Wang, Z.: Consortium blockchain-based malware detection in mobile devices. IEEE Access **6**, 12118–12128 (2018)
16. Amro, B.: Malware detection techniques for mobile devices. Int. J. Mob. Netw. Commun. Telematics (IJMNCT) **7** (2017)
17. Raveendranath, R., Rajamani, V., Babu, A.J., Datta, S.K.: Android malware attacks and countermeasures: current and future directions. In: 2014 International Conference on Control, Instrumentation, Communication and Computational Technologies (ICCICCT), pp. 137–143. IEEE (2014)
18. Sawle, M.P.D., Gadicha, A.: Analysis of malware detection techniques in android. Int. J. Comput. Sci. Mob. Comput. **3**(3) (2014)
19. Rodrìguez-Mota, A., Salinas-Rosales, P.J.E.A.A.: Malware analysis and detection on android: the big challenge (2017). https://www.intechopen.com/books/smartphones-from-an-applied-research-perspective/malware-analysis-and-detection-on-android-the-big-challenge
20. Saracino, A., Sgandurra, D., Dini, G., Martinelli, F.: Madam: effective and efficient behavior-based android malware detection and prevention. IEEE Trans. Dependable Secure Comput **15**(1), 83–97 (2016)
21. Hwang, S., Lee, H.W.: Identification of counterfeit android malware apps using hyperledger fabric blockchain. J Int Comput Serv **20**(2), 61–68 (2019)
22. Ltd., Q.H.T.: The growing threat of mobile malware: top android malware families of 2012. https://blogs.quickheal.com/the-growing-threat-of-mobile-malware-top-android-malware-families-of-2012/. Accessed: 27 Oct 2020
23. Contributors, T.: What is rat (remote access trojan)?—definition from whatis.com. https://searchsecurity.techtarget.com/definition/RAT-remote-access-Trojan. Accessed: 27 Nov 2020
24. AhMyth: Ahmyth/ahmyth-android-rat. https://github.com/AhMyth/AhMyth-Android-RAT. Accessed: 27 Nov 2020
25. Conner, K.: Over 1,000 android apps were found to steal your data. Here's what you can do. https://www.cnet.com/how-to/over-1000-android-apps-were-found-to-steal-your-data-heres-what-you-can-do/. Accessed: 27 Nov 2020
26. Willms, J.: Report: blockchain technology market to reach $7.7 billion by 2024. https://bitcoinmagazine.com/articles/report-blockchain-technology-market-reach-77-billion-2024/. Accessed: 27 Oct 2020
27. Liu, K., Xu, S., Xu, G., Zhang, M., Sun, D., Liu, H.: A review of android malware detection approaches based on machine learning. IEEE Access **8**, 124, 579–124, 607 (2020)
28. Homayoun, S., Dehghantanha, A., Parizi, R.M., Choo, K.K.R.: A blockchain-based framework for detecting malicious mobile applications in app stores. In: 2019 IEEE Canadian Conference of Electrical and Computer Engineering (CCECE), pp. 1–4. IEEE (2019)

29. Kumar, R., Zhang, X., Wang, W., Khan, R.U., Kumar, J., Sharif, A.: A multimodal malware detection technique for android IoT devices using various features. IEEE Access **7**, 64411–64430 (2019)
30. Ouaguid, A., Abghour, N., Ouzzif, M.: A novel security framework for managing android permissions using blockchain technology. Int. J. Cloud Appl. Comput. (IJCAC) **8**(1), 55–79 (2018)
31. Rana, M.S., Gudla, C., Sung, A.H.: Evaluating machine learning models on the ethereum blockchain for android malware detection. In: Intelligent Computing-Proceedings of the Computing Conference, pp. 446–461. Springer (2019)
32. Nigam, R., Pathak, R.K., Kumar, A., Prakash, S.: PCP framework to expose malware in devices. In: 2020 International Conference on Electronics and Sustainable Communication Systems (ICESC), pp. 1–6. IEEE (2020)
33. Lee, H.W., Lee, H.: Consortium blockchain based forgery android APK discrimination DApp using hyperledger composer. J Int Comput Serv **20**(5), 9–18 (2019)
34. Jan, S., Musa, S., Ali, T., Nauman, M., Anwar, S., Ali Tanveer, T., Shah, B.: Integrity verification and behavioral classification of a large dataset applications pertaining smart OS via blockchain and generative models. Exp. Syst. e12611 (2020)

Performance Evaluation of Parameters for Wi-Fi Network for Oil Pipeline Management Using OPNET Simulator

Chavala Lakshmi Narayana, Rajesh Singh, and Anita Gehlot

Abstract In this world, oil industry is perhaps the greatest business. The working of this for the most part relies upon the transportation of unrefined petroleum from source to destination. Pipelines are generally appropriate for the transportation of oil. In the event that pipelines are not well maintained, it will give a negative effect on the climate. Oil transmission pipelines are vitally significant for safe activity. On the off chance that any irregular issues like holes are recognized, it is important to send data to the end-client by means of a correspondence convention. Wi-Fi correspondence convention assumes a conspicuous function in communicating data to end user. Wireless Fidelity (Wi-Fi) network is made on the IEEE 802.11 standards with the developing interest just as infiltration of wireless services, the clients of the wireless network presently expect quality of service (QoS). The aim of this paper is to assess their quality of service (QoS) exhibitions regarding Wi-Fi. The proposed structure will be dissected utilizing the OPNET modeler reenactment apparatus with the basic boundaries. Finally, we accomplished a streamline throughput, the base measure of delay just as zero tolerance on the packet loss.

Keywords Wi-Fi · QoS · Oil pipeline · OPNET

1 Introduction

Oil pipeline networks are the most financial and most secure method of shipping these fuel sources. As a method for transportation, pipelines need to satisfy high requests of wellbeing, dependability and proficiency [1]. It has been demonstrated that oil and gas pipeline systems are the most conservative and most secure mean of shipping rough oils and they satisfy a popularity for proficiency and unwavering quality [2, 3]. The explanations behind the failures are either purposeful (like disfigurement) or material damages and so on harms [4, 5]. It is essential to communicate pipeline abnormal parameters data to end user and alert local workers. In this regard, Wi-Fi network

C. Lakshmi Narayana (✉) · R. Singh · A. Gehlot
School of Electrical and Electronics Engineering, Lovely Professional University, Phagwara, Punjab, India

A. Noor et al. (eds.), *Proceedings of Emerging Trends and Technologies on Intelligent Systems*, Advances in Intelligent Systems and Computing 1371,
https://doi.org/10.1007/978-981-16-3097-2_25

plays a prominent role to accomplish this task. All these pipeline consequences will be communicated by Wi-Fi communication protocol to end users. Toward the last part of the 1990s, the WLAN turned into the most favored decision of correspondence among individual users [6]. The first form of the IEEE 802.11 WLAN standard is intended to work just with the best exertion administrations, which implies there is no QoS allot on the assigned applications [7, 8]. It has been reported that from 2004 to 2012, the usage popularity of Wi-Fi around the world has increased by over 4 times [9].

2 Proposed Architecture

The proposed architecture is divided various sections as shown in Fig. 1. In the pipeline monitoring section, all the parameters of the pipeline are monitored and by using IoT sensors and send data to microcontroller for analysis whether the information collected from sensors is normal or abnormal and displayed in LCD. If abnormal data found. Wi-Fi communication protocol is activated, and hooter is activated and alerts the pipeline monitoring people. Moreover, the data are sent to application server, i.e., end user.

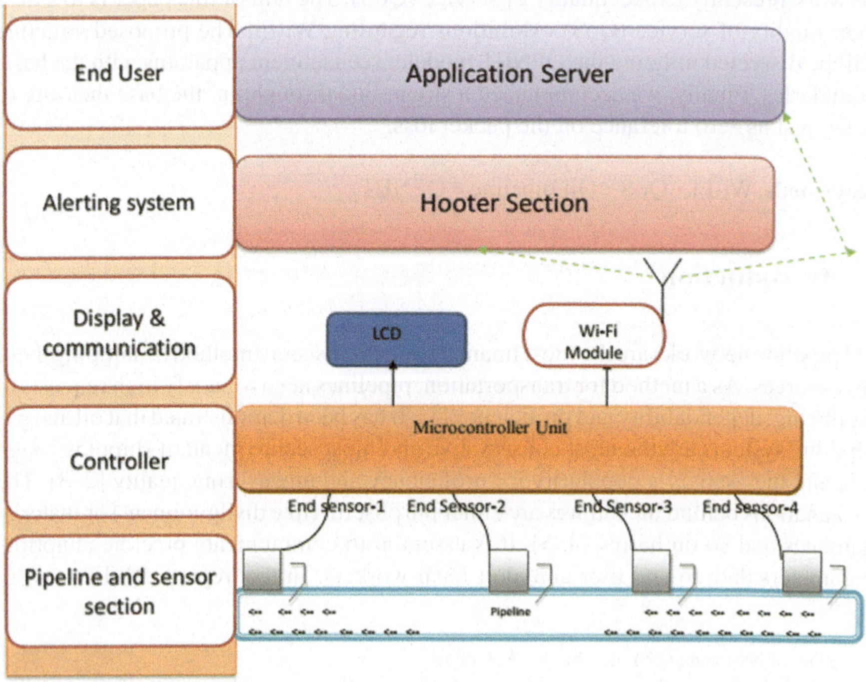

Fig. 1 Architecture of oil pipeline monitoring using Wi-Fi module

Fig. 2 Hooter system

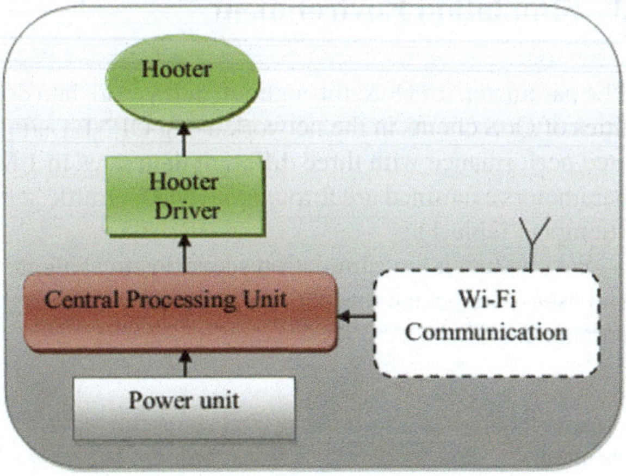

2.1 *Hooter Section*

Figure 2 shows the hooter system, which is used to alert the local pipeline monitoring people. The Zigbee receives signal and sent to controller gives instructions to the hooter driver is activated and gives alarm.

2.2 *End User*

All the real-time monitoring data are transferred to the application server using Wi-Fi communication protocol as shown in Fig. 3 which also called end-user section.

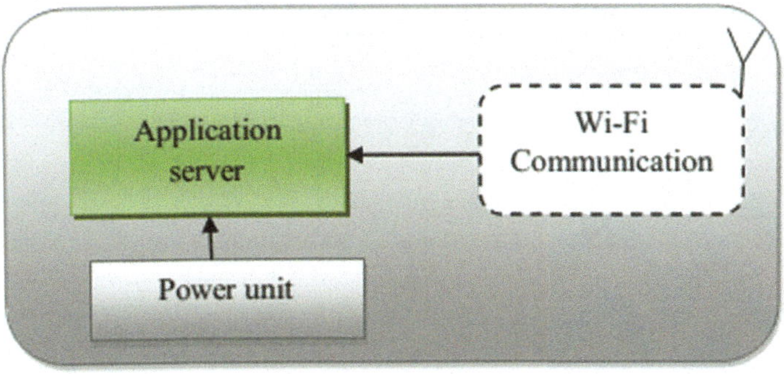

Fig. 3 End-user section

3 Simulation Environment

The parameters of QoS: throughput, delay, and data dropped including various quantities of QoS clients in the network using OPNET simulator [10–12]. We also evaluated performance with three different data rates in 1 Mbps, 12 Mbps, 54 Mbps. The parameters examined are throughput, MAC, traffic sent, received and retransmission attempts (Table 1).

We conduct a few simulation scenarios to evaluate the proposed method. In this, end user is connected to an access point with a 100 m range and evaluated the various parameters (Fig. 4).

Table 1 Simulation parameters

Parameters	Description
No. of wireless nodes	10
No. of access points	1
No. of servers	1
Range	100 m * 100 m
Antenna height	6 ft
BSS identifier	Auto assigned
Transmit power (W)	2 W
Configure PHY	802.11g
Data rates	1 Mbps, 12 Mbps, 54 Mbps

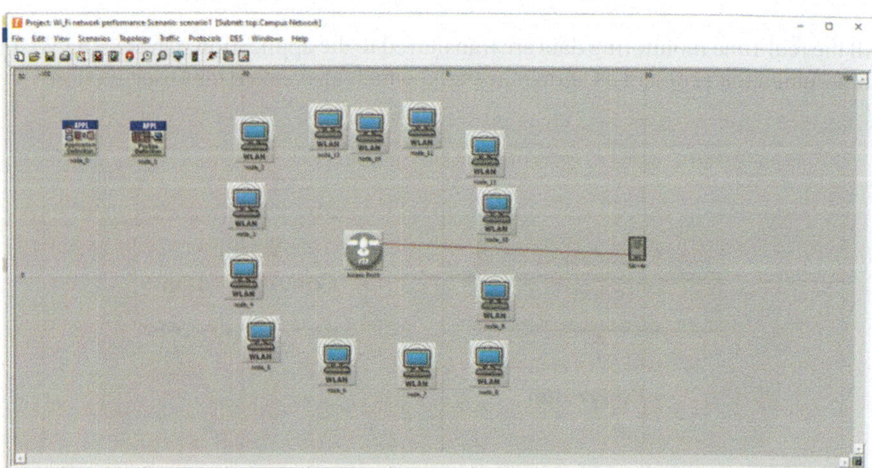

Fig. 4 Wi-Fi performance evaluation scenario using OPNET

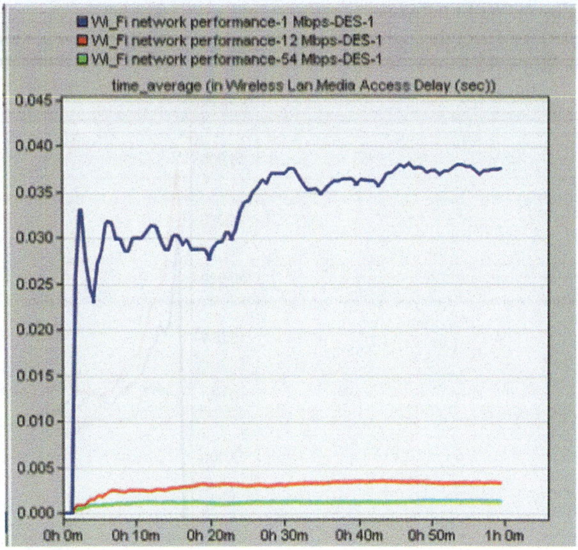

Fig. 5 Wi-Fi network performance: media access delay for 1 Mbps, 12 Mbps, 54 Mbps

4 Results and Discussion

4.1 Media Access Delay

In Fig. 5, we evaluated MAC with 1 Mbps, 12 Mbps, 54 Mbps data rates. With 1 Mbps data rate less delay is noticed, and remaining data rates almost delay is negligible. The average delay of all data rates is in milliseconds.

4.2 Throughput

The throughput for all data rates is average 10,000 bits/s which are noticed in Fig. 6.

4.3 Data Traffic

In Fig. 7, the traffic sent is highest in beginning, i.e., 3200 bits/s in 54 Mbps data rate and less traffic noticed in other data rates. So traffic sent in less data rates is good in performance.

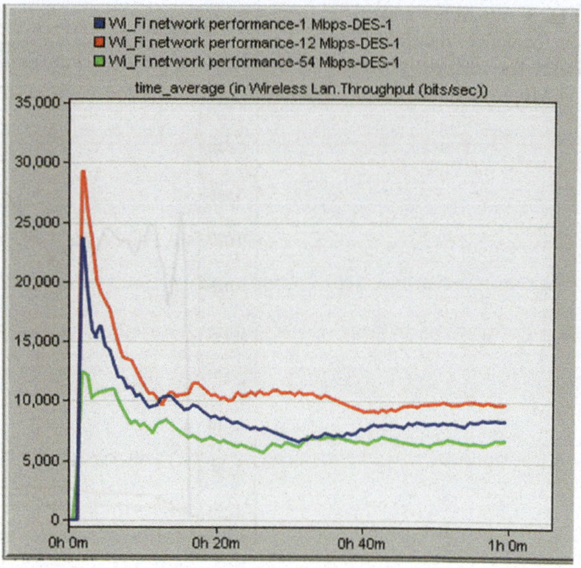

Fig. 6 Wi-Fi network performance: throughput for 1 Mbps, 12 Mbps, 54 Mbps

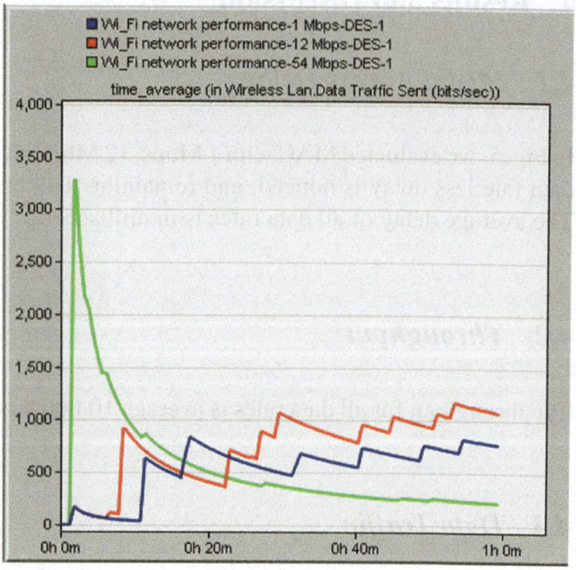

Fig. 7 Wi-Fi network performance: data traffic sent for 1 Mbps, 12 Mbps, 54 Mbps

4.4 Retransmission Attempts

Retransmission attempts are considerably less in 1 Mbps rate when compare to other data rates as shown in Fig. 8. The average retransmission rate for all data rates is 1 packet (Table 2).

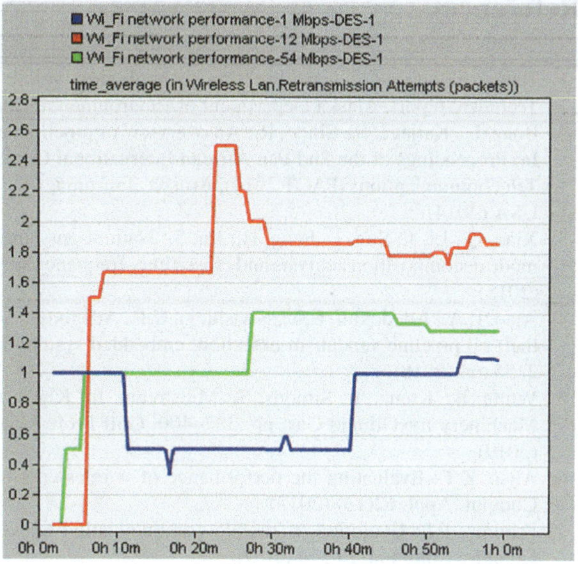

Fig. 8 Wi-Fi network performance: retransmission attempts for 1 Mbps, 12 Mbps, 54 Mbps

Table 2 Average performance table of Wi-Fi network with various data rates

Parameters	1 Mbps data rate	12 Mbps data rate	54 Mbps data rate
Media access delay (MAC)	0.03 (s)	0.003 (s)	0.001 (s)
Throughput (bits/s)	15,000 (bits/s)	15,000 (bits/s)	10,000 (bits/s)
Data traffic sent	1000 (bits/s)	700 (bits/s)	500 (bits/s)
Retransmission attempts	0.5 (packets)	1.6 (packets)	1.2 (packets)

5 Conclusion

As we know that oil and gas pipeline systems are the most conservative and most secure means of shipping rough oils and they satisfy popularity for proficiency and unwavering quality. If pipelines are not well maintained, they will negatively impact the human and natural environment. So is highly essential to monitor oil pipelines without failure. Thus, the paper gives the proposed methodology of pipeline management using a Wi-Fi communication protocol. Finally, all the simulation results show the optimized scenario of the Wi-Fi network gives the performance of various parameters with different data rates. In this scenario, the overall performance of Wi-Fi network gives retransmission attempts rate shows good performance. Improvement needed to get consistent throughput, and MAC delay gives a good performance. Consistent performance in MAC management traffic sent data, however, improvement needed in received data in 54 Mbps data rate.

References

1. Technical review of leak detection technologies, vol. I: crude oil transmission pipelines. Technical report. Alaska Department of Environmental Conservation (1999)
2. Boaz, L., Kaijage, S., Sinde, R.: An overview of pipeline leak detection and location systems. In: Proceedings of the 2nd Pan African International Conference on Science, Computing and Telecommunications (PACT 2014), Arusha, Tanzania, 14–18 July 2014, IEEE, Piscataway, NJ, USA (2014)
3. Xiao, Q., Li, J., Sun, J., Feng, H., Jin, S.: Natural-gas pipeline leak location using variational mode decomposition analysis and cross-time–frequency spectrum. Measurement **124**, 163–172 (2018)
4. Ajao, L.A., Adedokun, E.A., Nwishieyi, C.P., Adegboye, M.A., Agajo, J., Kolo, J.G.: An anti-theft oil pipeline vandalism detection: embedded system development. Int. J. Eng. Sci. Appl. **2**, 55–64 (2018)
5. White, B., Kreuz, T., Simons, S.: Midstream. In: Klaus, B., Rainer, K. (eds.) Compression Machinery for Oil and Gas, pp. 387–400. Gulf Professional Publishing, Houston, TX, USA (2019)
6. Alisa, Z.T.: Evaluating the performance of wireless network using OPNET modeler. Int. J. Comput. Appl. **62**(13) (2013)
7. Prinima, P.J.: Evolution of mobile communication network, 1G to 5G. Int. J. Innov. Res. Comput. Commun. Eng. **4** (2016)
8. Ramia, B., Amin, B., Osman, A.: A comparison between IEEE 802.11 a, b, g, n and ac standards. Int. Comput. Eng. **7**(5) (2015)
9. Trimintzios, P., Georgiou, G.: WiFi and WiMAX secure deployments. J. Comput. Syst. Netw. Commun. Article ID 423281 (2010)
10. Measuring delay, jitter, and packet loss with Cisco IOS SAA and RTTMON—Cisco. [Online]. Available: http://www.cisco.com (2016)
11. Jitter. [Online]. Available: http://www.wikipedia.com (2014)
12. Alawi, M.A., Alsaqour, R.A., Sundarajan, E., Ismail, M.: Prediction model for offloading in vehicular WiFi network. Int. J. Adv. Sci. Eng. Inf. Technol. **6**(6). ISSN 2088-5534

Design and Development of Industrial IoT Gateway for Robotic Arm Control

K. Hariharan and M. Rajesh

Abstract Modern-day robots are playing a greater role in wider application areas, thereby driving increased performance and efficiency. For these robots, Industrial Internet-of-Things (IIoT) provide a way for driving operational efficiency in Industrial Applications, with the help of Analytics, Automation, and Connectivity. On the other hand, Cyber-Physical Systems (or CPS) consist of a 'Physical Entity' and a 'Cyber-Twin,' which could virtually replicate the behavior of the Physical Entity under various conditions. In this project, an existing Mitsubishi's Multifunctional Reprogrammable Manipulator (or simply an 'Industrial Robot') is considered. With proper study, efforts to add features like Graphical User Interface (GUI), remote accessibility, wireless teach pendant, etc. are attempted. A Gateway is developed on a Single Board Computer to add IIoT features that make the Industrial Robot targeting Cyber-Physical System (CPS) applications. Currently, the robot learns its movements with a Wired Teach Pendant and could be programmed with a DOS Control Software, which would be upgraded into an Interactive HMI and also wireless teach pendant will be developed as a by-product. The Robot supports 5 Degrees of Freedom (DOF) with the basic application as 'Pick and Place Robot.'

Keywords Internet of things · Mitsubishi robots · Cyber-physical system · Human-to-machine interface · Linux

1 Introduction

When a connected computer reported the status from a remote Inventory in 1982, it was marked as the dawn of IoT Technology. Over the years, the growth of IoT has taken an exponential growth and now it is playing a vital role and wider coverage

K. Hariharan (✉)
Centre for Development of Advanced Computing (C-DAC), Chennai, India
e-mail: hariharank@cdac.in

M. Rajesh
National Institute of Electronics and Information Technology (NIELIT), Calicut, India
e-mail: rajesh@calicut.nielit.in

A. Noor et al. (eds.), *Proceedings of Emerging Trends and Technologies on Intelligent Systems*, Advances in Intelligent Systems and Computing 1371,
https://doi.org/10.1007/978-981-16-3097-2_26

in Applications ranging from DIY Projects, Smart Home, Industrial Automation, Connected Factories, Surveillance Systems, Manufacturing Plants and Aerial Robotics, etc. Also, this discipline is taking different forms according to application use cases viz., Narrow Band-IoT (NBIoT), Industrial IoT (IIoT), Real-Time IoT (RT-IoT), Consumer IoT (CIoT), Internet of Medical Things (IoMT), etc. In Industry 4.0's trend toward manufacturing technologies and processes, Cyber-Physical Systems (CPS) play a greater role that facilitates automation, connectivity, and data exchange and analytics [1]. CPS emerged around 2006 and was coined by Helen Gill from the National Science Foundation of the USA. When a 'Physical Entity' combines with a 'Cyber-Twin,' it makes a CPS. Here, 'Cyber-Twin' refers to a 'computational relational model' or 'coupled-model' approach, derived from the physical entity, which is also capable of virtually replicating the entire system [2]. A simple example to understand CPS would be the 3D Wheel Alignment for Cars.

In this paper, attempts to modify and upgrade an existing Industrial Robot System (which is meant for pick and place, material handling, etc.) with IIoT-CPS, to enhance the existing functionalities are described. This paper provides details starting from studying an old Mitsubishi Robot to attempting some features through Five Sections. The Introduction section, where a brief note on IoT is discussed. The second section is the Existing System, where the robot's study phase results are shown. It also includes existing technical problems and some solutions to them. Following that, the Proposed System comes in the Third Section, where the formulated solution to mitigate the problems are dealt with. The Design Stage of the proposed methodology comes in section four, under the System Design part which is followed by actual implementation in the fifth section and the discussion concludes in section six.

2 Existing System

Mitsubishi's RV-M1 (Movemaster EX) is a reprogrammable multifunctional manipulator and a 'vertical articulated robot' that is categorized under the Industrial Micro-Robot System. Block Diagram of the existing system is shown in Fig. 1. Programming and Controlling the Robot takes place with a teach pendant and a PC. Centronics Cable (Parallel Port) is used as a standard interface between DOS-PC and the robot. RV-M1 was designed and released in the early 1990s complying with Centronics Interface standards. The existing workflow for this robot is defined in Fig. 2. Initially, when a user defines a specific pick-and-place application, co-ordinate axes/points must be taught with the support of Teach Pendant so that the robot learns its position and region of operation. After teaching successfully, new programs can be loaded from the DOS-PC and the robot will move in a pre-planned path as per the instructions in the program.

A complete methodology of Teaching, Instruction set, and Controlling the robot is found in Programming Manual [3]. A total of 63 Instructions classified into five categories as, Position/Motion Control Instructions—24, Program Control Instructions—19, Hand Control Instructions—4, I/O Control Instructions—6, RS-232C

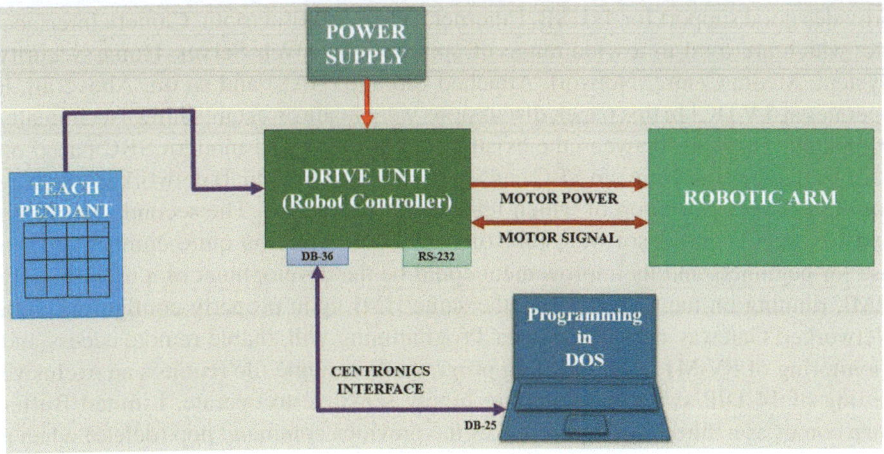

Fig. 1 Existing System

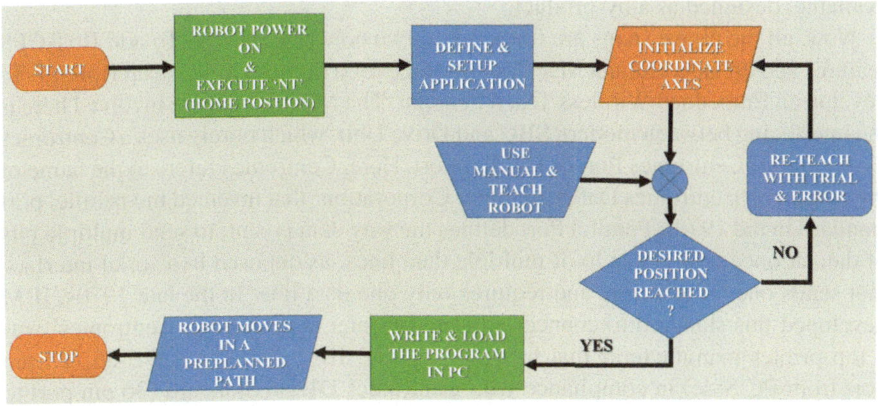

Fig. 2 Existing Workflow

Read Instructions—6, and Miscellaneous Instructions—4. On the other side, there are some snags observed in the existing system which could be improved with modern technologies. Firstly, Programs to RV-M1 (written in Q-BASIC Language) are sent from a DOS-PC, which uses a bulky CPU to communicate with the Drive Unit. (Drive Unit, then actuates the robotic arm in the desired path.) A good alternative for this bulky CPU would be a modern-day Single Board Computer (SBC) that is capable of greater performance, smaller form factor, good interface support, network connectivity, power efficiency, etc. All these features when properly analyzed and technically made compatible with RV-M1's Drive Unit will elevate its capabilities to another level. One such example of an SBC could be the series called 'The Raspberry Pi' (RPi) which is well-known by the name '$35 credit-card-sized computer' provides good support for 4xUSB, Ethernet, Wi-Fi and Bluetooth, Camera Interface, etc. which are used in a wide range of applications—Web Server, Home Security System, Media Centre, Network Attached Storage (NAS) and so on. Above all, it operates at 5 V DC! In this paper, discussions were made on establishing a technically compatible interface between the existing Drive Unit and a modern SBC based on various parameters. Here, an SBC can act as a (multifunctional) networked gateway for the Drive Unit, on top of which features will be added. The second snag is the text-based PC control software (existing method), which is quite cumbersome to use for beginners and its improvement could be the development of a user-friendly HMI, running on the Gateway [4], the same HMI upon properly configuring with Networked Gateway and Web Server Programming will enable remote access and monitoring of RV-M1. Also, sending programs as a single file requires an exclusive setting of 14 DIP switches which are highly sensitive to operate. Limited Buffer Size comes as a third snag which makes the previous command pop (delete) when a new command arrives, which is again a problem for Looping and Conditional statements. The fourth and final improvement is the addition of Wireless Teach Pendant using Embedded Microcontrollers in addition to the default wireless teach pendant available, designed as a by-product.

Now, all the above snags are framed into two core objectives—To add IIoT-CPS features and create a Human Machine Interface for a non-GUI Industrial Robot—To develop an Embedded Wireless Teach Pendant. The Main Challenge involved here is the interfacing between modern SBC and Drive Unit, which purely uses a Centronics Connector or Centronics Port or Parallel Port. Here, Centronics refers to the name of the company 'Centronics Data Computer Corporation' that invented the parallel port standard in the 1970s. Parallel Port defines the way data is sent; to send multiple bits of data at once with the help of multiple data lines, as opposed to a serial interface that sends one bit at a time and requires only one data line. In the late 1970s, IBM developed this standard to connect a PC and Printer. Apparently, 'Centronics' was a top printer manufacturer that had DB-36 Port. IBM configured DB-25 (25 pin port from PC Side) in compliance with Centronics DB-36 Standard (36 pin port to Printer Side). For this reason, all other printer manufacturers started adapting to the Centronics Standard and were called the IEEE 1284 standard port that supports a data rate of 50–100 kbit/s. Enhanced Parallel Port (EPP) and Extended Capability Port (ECP) support data rates up to 2 and 2.5 MB/s, respectively [5]. The mapping of DB-25 to DB-36 is shown in Fig. 3.

Operational Signal Levels are between 0 and +5 V DC and maximum voltage can go up to 5 V DC. On a PC side, Detailed Pinout for a DB-25 connector is classified into Data Pins, Control Pins, and Status Pins and showing pin direction—In, Out and In-Out is shown in Fig. 4.

Line	DB-25 Male (Computer)	DB-36 Centronics	Port Direction
Strobe	1	1	⇒
Data Bit 0	2	2	⇒
Data Bit 1	3	3	⇒
Data Bit 2	4	4	⇒
Data Bit 3	5	5	⇒
Data Bit 4	6	6	⇒
Data Bit 5	7	7	⇒
Data Bit 6	8	8	⇒
Data Bit 7	9	9	⇒
Acknowledge	10	10	⇐
Busy	11	11	⇐
Paper Out	12	12	⇐
Select	13	13	⇐
Line Feed	14	14	⇒
Error	15	32	⇐
Reset	16	31	⇒
Select Printer	17	36	⇒
Signal Ground	18	33	⇔
Signal Ground	19	19 + 20	⇔
Signal Ground	20	21 + 22	⇔
Signal Ground	21	23 + 24	⇔
Signal Ground	22	25 + 26	⇔
Signal Ground	23	27	⇔
Signal Ground	24	28 + 29	⇔
Signal Ground	25	16 + 30	⇔
Shield	Cover	Cover + 17	⇔

Fig. 3 DB-25 to DB-36 pin mapping

3 Proposed Solution

With all the above details from the existing solution, a novel solution is proposed as shown in Fig. 5. In the proposed system, an SBC will be interfaced to the Drive Unit of the Robot with the help of an appropriate communication protocol. Since the Drive Unit uses Centronics Interface and modern SBC lacks Centronics port, the main challenge occurs at this stage in the form of interfacing between these two systems. But, this is a very important section and has to be achieved. On top of this SBC Gateway, an HMI will be developed which will be followed by the addition of IIoT-CPS Features. Finally, a wireless teach pendant will be developed as a by-product. HMI Design Approach can be done with 'Qt' (pronounced as "Cute") which is a free and open-source Toolkit for designing Graphical User Interface (GUI) and Cross-Platform Applications. Applications developed in 'Qt' can run on various

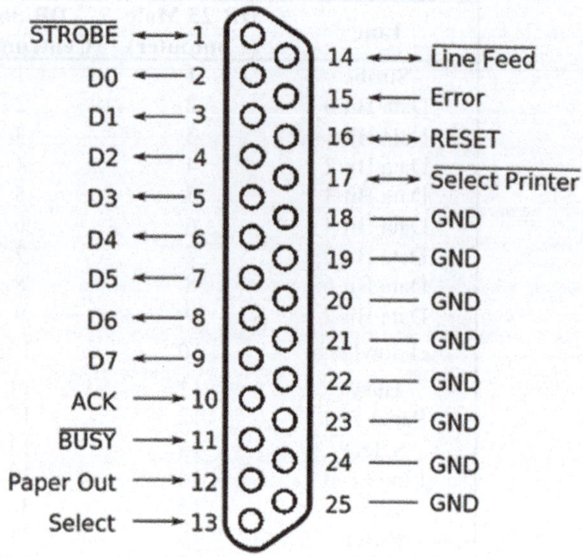

Fig. 4 A Standard DB-25 connector

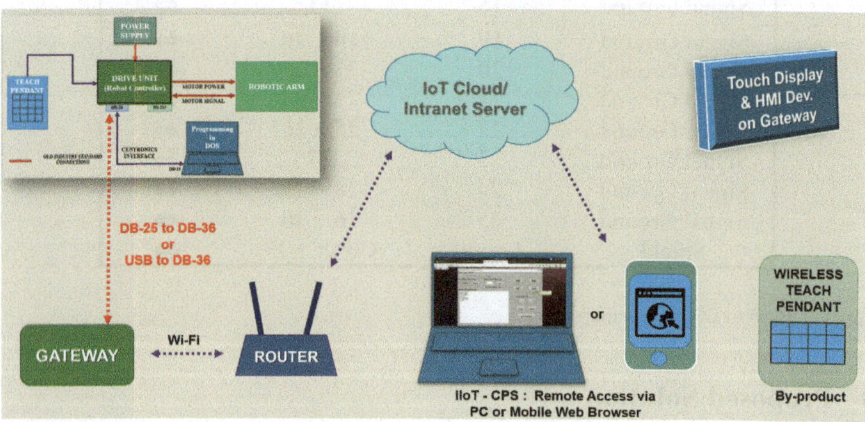

Fig. 5 Proposed System

Hardware and Software Platforms such as Linux, Windows, Mac OS, Android, and even in some Embedded Devices. Availability of a wide variety of Widgets, Online Documentation, and Community Support, etc. are some of the advantages of 'Qt'. Many real-world projects viz. VLC media player, TeamViewer, EAGLE and Google Earth, etc. are developed using the 'Qt' Framework. A simple HMI frontend for RV-M1 is shown in Fig. 6.

In our context, the Initial stages of HMI development, i.e., coding, testing, debugging will be done in 'Qt' running on a Host Computer (Kubuntu OS in Linux Environment) and later ported to Target (Gateway) [6].

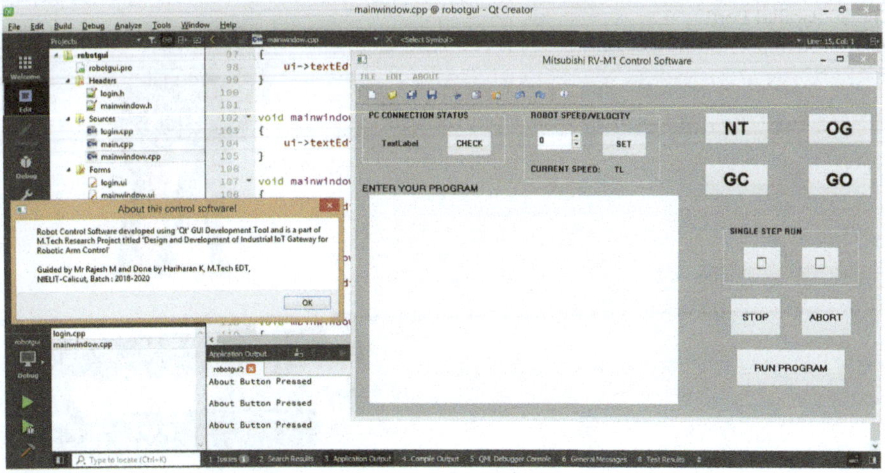

Fig. 6 'Qt' GUI frontend

4 System Design

The system design is initially approached with three different Use Case Diagrams—Teach Pendant, Human-to-Machine Interface, and Remote Accessibility with different use case names/parameters such as Use Case ID, Description, Pre-Condition, Trigger, Sample Program, Exception Paths, etc. For Example, if the 'Nest + Enter' use case under Teach Pendant is considered, then all the use case names must be defined properly to understand the design as shown in Fig. 7. Also, a simple circuit was designed with 10 segments LED Array Bar Graph (as shown in Fig. 8)

Use Case Name	'NT' - Nest and 'ENTER'
Use Case ID	#01
Description	NT (Nest) command followed by ENTER makes the robot to the Origin Position.
Pre-condition	Command causes the robot to return to origin, which must be performed immediately after the power is turned ON
Trigger	User presses the ' NT ' button on the Teach Pendant
Sample Program	*10 LPRINT "NT"* ;*QBASIC COMMANDS* ***(will be in HLL)*** *20 LPRINT "MO 10"* ;*Move to Position 10*
Exception Paths	User may stall the operation at any time by pressing the Emergency Stop Button on the Teach Pendant

Fig. 7 Use-case description for 'NT + Enter'

Fig. 8 Parallel Port Analysis Setup

which is one of the simple setups to read the states of Parallel Port Signals. Subsequent Parallel Port analysis schematics were designed in EasyEDA—a web-based interactive circuit design tool. Following that, Test Setup(s) to study the operation of Parallel Port Devices in Linux SBC's are designed. In Linux File System, all the Read/Write operations are done on the `/dev` directory, where it contains special device files, and `/dev/lp0` is the first parallel printer device. Our test setup is a Host-Target Model, where the host is a CPU running Kubuntu OS (Gateway); Target is a generic parallel port device. To interface the host with the target, two different approaches were proposed: A Standard DB-25 to DB-36 Cable and A USB to DB-36 Cable with CH340G Driver Chip [7]. Upon connecting, when `echo "NT" > > /dev/lp0` command (or `/dev/usb/lp0`, in case of CH340G) was executed from the terminal, the target prints 'NT' for one time. For the command to work properly, user-level permissions must be changed every time as the write operation is done in the `/dev` directory. Instead, a device manager for the Linux Kernel known as `'udev'` is available. A custom `udev` script containing a range of inode values, action, and user-level must be enabled, so that when a device is plugged in, it triggers the script in the backend and there's no necessity of changing user-level permissions every time in the frontend [8]. The same test setup was connected to the robot and at the basic level, when two different robot commands were sent from the terminal, the robot started moving accordingly. It marks the completion of the first step in interfacing the proposed Gateway with the robot. With that information, a more feasible interface approach was designed, as shown in Fig. 9, which is a slight modification of Fig. 5. Here, the Gateway itself is a CPU running Kubuntu OS which is connected with a standard DB-25 to DB-36 cable. On this Gateway, HMI development is done and when it is configured through LAN/Intranet along with some Software tools, it adds remote connectivity to the robot.

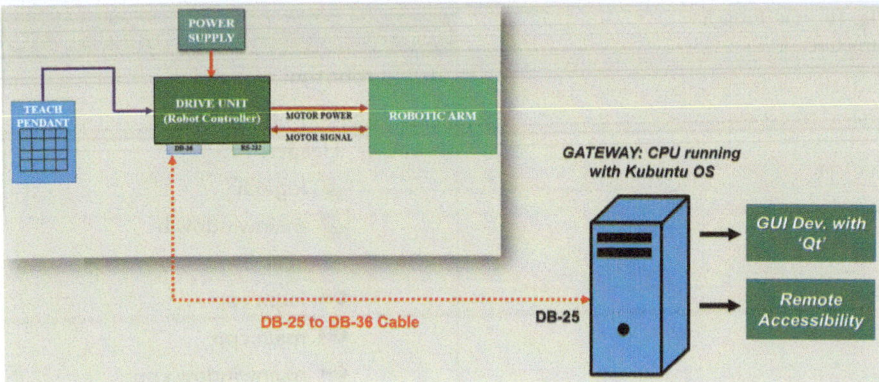

Fig. 9 Final Interface Approach

5 Implementation

Initially, the HMI Implementation was made with 'Qt' and it is followed by configuring the remote accessibility with the 'Xpra' tool. In this context, Human-to-Machine Interface (HMI)/Graphical User Interface (GUI)/Control Software refers to the novel software tool that was developed and used to control the robot. As specified earlier, there are around 63 total instructions available for the robot and only these instructions need to be sent to the robot's drive unit from the Gateway. Also, all these instructions have their syntax, boundary condition, limits, etc. and all these conditions are kept in mind before designing the new HMI in 'Qt' [9]. A typical 'Qt' Project Structure is shown in Fig. 10.

While designing the GUI, the two important design logic must be addressed. One is the logic for the SW Push Button and the other is the logic for GT (Go To) Commands. Entire HMI development is done in C++ and additional support of Linux System Call Programming is used for four robot commands—NT (Nest), OG (Origin), GC (Gripper Close), GO (Gripper Open) as 'pushbutton' objects using 'QPushButton' class. A flowchart is shown in Fig. 11.

On the other hand, Robot Programming Manual contains some looping and conditional statements, which also should be designed at the software level. For Example, the 'GT' command stands for 'Go-To' and it permits the program sequence to jump to a specified line number unconditionally (inside the program). The approach goes as: When the program is typed in the 'QTextEdit' widget, the 'Run Program' button is clicked and while parsing through the strings line-by-line, the system has to check whether the 'GT' command exists or not. If not, then continue further execution. If 'GT' and its associated syntax is found, then go to the line number and execute from that line. Furthermore, the block diagram for Remote Accessibility is shown in Fig. 12.

Here, two new terms are introduced: Remote Host and Remote Target. This scenario can be assumed as two different PCs connected through Intranet in an

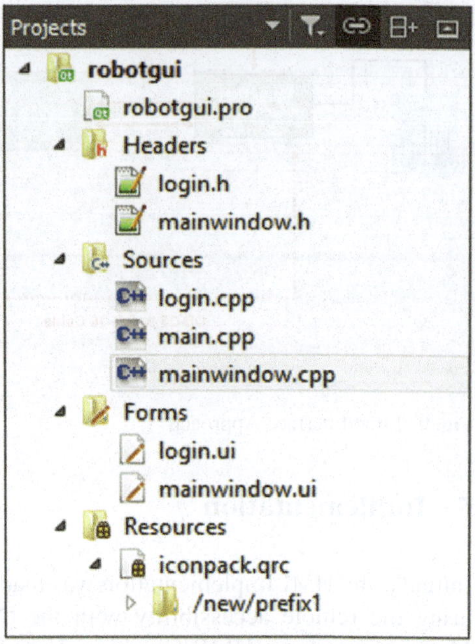

Fig. 10 'Qt' Project Structure

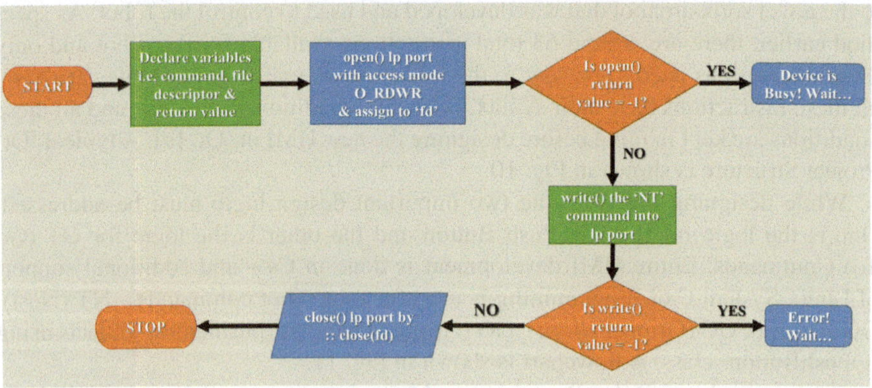

Fig. 11 Push Button Logic

Fig. 12 Remote Accessibility

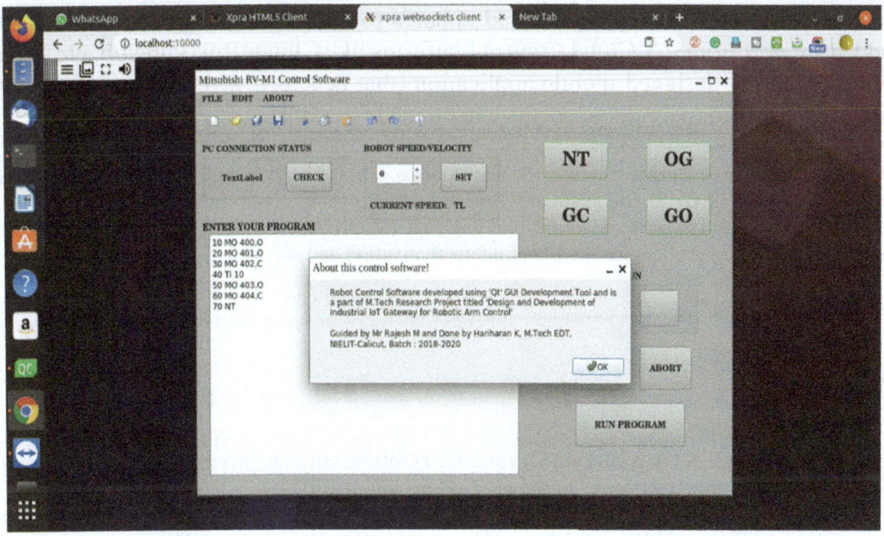

Fig. 13 GUI on running on Gateway Web Browser

Organization. As mentioned earlier, ‘Xpra’ (X Persistent Remote Applications) was used, which is an open-source multi-platform persistent remote display server and client for forwarding applications and desktop screens [10]. Its source code is written in Python and it supports UNIX-like, Windows, and Mac OS. In this case, the two goals are—Accessing the robot GUI on Gateway PC’s Web Browser (Lab B) and accessing the robot GUI on Remote PC/Mobile Web Browser (Lab A). For the first case, ‘Xpra’ must be configured on the Gateway and to start a new session, enter the command followed by the name of the control software executable as `xpra start --start=./robotgui --bindtcp=0.0.0.0:10000`. This will start a new session and to view the result, go to the web browser of the gateway and enter `http://localhost:10000`. The same GUI can be viewed on the web browser and to stop the session, enter `fuser -k 10000/tcp`. Results are shown in Fig. 13. For the Second case, start a new session on Xpra, copy the IP Address of the Gateway to the remote PC and the same GUI is available on the Remote PC as well.

6 Future Scope

Currently, the proposed system is successfully tested with Intranet. Additionally, a Camera module could be added to the system along with a video server, to monitor the positions of the robot in its region of operation in a closed loop. Apart from the Pick and Place application, Robot use-cases can be extended by changing the End Effector Module for Welding, Industry Inspection Purposes, etc. Under these cases, more algorithms can be designed and updated to the proposed system. Also, with added support from IoT Cloud Control Platforms like Cloud4Rpi, PubNub, etc., and with a securely designed mobile application (that acts as a wireless teach pendant too!), fully functional remote access can be made possible.

Acknowledgements This research work was done as a Master's Research Project at Embedded Systems Group (ESG) at the National Institute of Electronics and Information Technology (NIELIT), Calicut. The authors would like to thank CAD-CAM Group, ES Group, and all the members of NIELIT Calicut for their continuous motivation and support.

References

1. Patel, A., Azadi, S., Babee, et al.: Significance of robotics in manufacturing, energy, goods and transport sector in the internet of things (IoT) paradigm (2018)
2. Reis, J., Pinto, R., Gonçalves, G.: Human-centered application using cyber-physical production system. In: IECON 2017—43rd Annual Conference of the IEEE Industrial Electronics Society, pp. 8634–8639 (2017)
3. Mitsubishi RV-M1 instruction manual, https://manualzz.com/doc/8635155/mitsuibishi-rv-m1-instructionmanual. Accessed 15 Nov 2019
4. Świder, J., Foit, K., et al.: The off-line programming and simulation software for the Mitsubishi Movemaster RV-M1 robot. J. Achievements Mater. Manuf. Eng. **1**(20), 499–502 (2007)
5. Diamond, S.L.: A new PC parallel interface standard [IEEE Std 1284–1994]. IEEE Micro **14**, 3 (1994). https://doi.org/10.1109/40.296145
6. Cross compilation for RPi3, https://gist.github.com/hariharank97/b37af2d776d1bc3441974 1dda9b30c20. Accessed 11 Jan 2020
7. SparkFun electronics serial basic breakout CH340G, https://www.sparkfun.com/products/ 14050. Accessed 30 Dec 2020
8. Hackaday (2009) How to write Udev rules, https://hackaday.com/2009/09/18/how-to-write-udev-rules/. Accessed on 02 Nov 2019
9. Getting started programming with 'Qt' widgets, https://doc.qt.io/qt-5/qtwidgets-tutorials-not epad-example.html. Accessed 16 Nov 2019
10. Xpra home page, https://www.xpra.org/. Accessed 30 Dec 2020

On the Development of a Mobile TurtleBot3 Burger Multi-robot System for Manufacturing Environment Monitorization

Carmen Nica, Mihaela Oprea, and Alexandru-Călin Stan

Abstract The paper presents a research study on developing a mobile multi-robot system for autonomous exploration of a manufacturing environment. The experimental mobile multi-robot system is composed of three TurtleBot3 Burger mobile robots. A simulation of the experimental system performed in Gazebo is described in detail. The results of the simulation are used for a real-world monitorization for a flexible manufacturing line control room.

Keywords Manufacturing environment exploration · Flexible manufacturing line · Multi-robot system · SLAM · Turtlebot3 · ROS

1 Introduction

In the last decades, the manufacturing industry has gone through continuous changes from technologization point of view. The first step, transition from manual work to the automatization of manufacturing process, was led by the advantages that came along with them: increase in efficiency, flexibility, quality, etc. More and more company that have reached the highest point in terms of automation level are permanent searching for methods of improving the productivity. Artificial intelligence application for manufacturing process represents the key for achieving these goals.

Multi-robot systems can perform efficiently several tasks in manufacturing environments, such as manufacturing tasks, work environment exploration, and indoor air pollution monitoring. One of the challenging problems that a multi-robot system must solve is exploring unknown or partially known dynamic environment and building a 2D or 3D map of it.

Our research goal is to develop a multi-robot system for environment monitoring of a flexible manufacturing line using simultaneous localization and mapping

C. Nica (✉) · M. Oprea · A.-C. Stan
Petroleum-Gas University of Ploiești, Ploiești, Romania
e-mail: carmen.nica1991@yahoo.ro

M. Oprea
e-mail: mihaela@upg-ploiesti.ro

A. Noor et al. (eds.), *Proceedings of Emerging Trends and Technologies on Intelligent Systems*, Advances in Intelligent Systems and Computing 1371,
https://doi.org/10.1007/978-981-16-3097-2_27

(SLAM). In order to do that, first, we will test the system using the Gazebo simulator to study the multi-robot system behaviour and to discover the weakness before the real-world test. After we complete the simulator test, we will make an experiment using a mobile multi-robot system to explore and construct a 2D map of a manufacturing area, in this case, the control room for a flexible manufacturing line.

For this research work, we chose to use for multi-robot implementation, the TurtleBot3 Burger robot, a small, cost efficient and perfect equipped for SLAM applications, programmable using ROS software.

The paper is structured as follows. Section 2 presents a brief overview on multi-robot system modelling and simulation. The simultaneous localization and mapping approach are discussed in Sect. 3. The experimental mobile multi-robot system is presented in Sect. 4, the first part presents the simulator test and the second part describes the experiment using a real multi-robot system into an industrial environment. The last section concludes the paper.

2 Modelling and Simulation of Multi-robot Systems

The development of a multi-robot system involves three main stages: design or choose/adapt a model for the multi-robot system, implement the model as a simulation and test it for various scenarios while doing model adaptation to the simulation results, and implement and test the final version of the model in a real-world multi-robots system. Each stage tackles in detail the requirements related to robots coordination, robots communication and control and so on, according to the multi-robot system goal achievement (e.g. industrial work environment exploration, indoor air pollution monitoring or specific industrial manufacturing tasks). In this section, we make a short overview on multi-robot system modelling and simulation.

2.1 Workspace Presentation: Flexible Manufacturing Line

The modern manufacturing industry confronted with new challenges in terms of volume demand, production cost, quality assurance, labour cost, production efficiency, etc.

The appearance of flexible line manufacturing gave the manufacturing industries the needed tools to achieve the flexibility and productivity [1] of the production process.

A flexible manufacturing line represents a traditional production line but having in its structure multiple adjustable machines that are linked by automatic transport systems.

The architecture of a flexible production line has in its structure: workstations, material handling/storage systems (automatic transport system) and a central computer [1]. The automatic transport system is used for transferring the products from one workstation to another and the central computer is responsible for controlling the entire process, the performance of the machines and the automatic transport system/material handling.

A flexible manufacturing system can be designed in different configurations and layouts, depending on the field of production, strategies, or preferences [2].

Figure 1 shows the schematic representation of a flexible manufacturing line. The workstation 1–6 represents the stages in the production process, linked by transportation devices (automatic conveyors). The control room is the central computer that controls and coordinates all the manufacturing process of a flexible line.

The aim of this paper is to monitor the HVAC (heating, ventilation, and air conditioning), smoke detection, hot spots, etc. of the environment in which the control room is located. In this purpose, a multi-robot system is used. In the past, a person was dedicated to manual check the parameters of the control room.

From the variety of multi-robot systems models that were proposed in the literature, we mention three types: the multi-agent model, the Petri Nets based model, and swarm intelligence-based model. A very short overview of some selected papers that use such models is made in this section.

Intelligent agents can easily model mobile robots due to their basic characteristics: autonomy, pro-activity, reactivity, and social ability. Moreover, multi-agent systems which are systems composed of minimum two intelligent agents that have a common global goal and are embedded in a dynamic, usually, unknown environment, represent a straightforward model for multi-robot systems being distributed

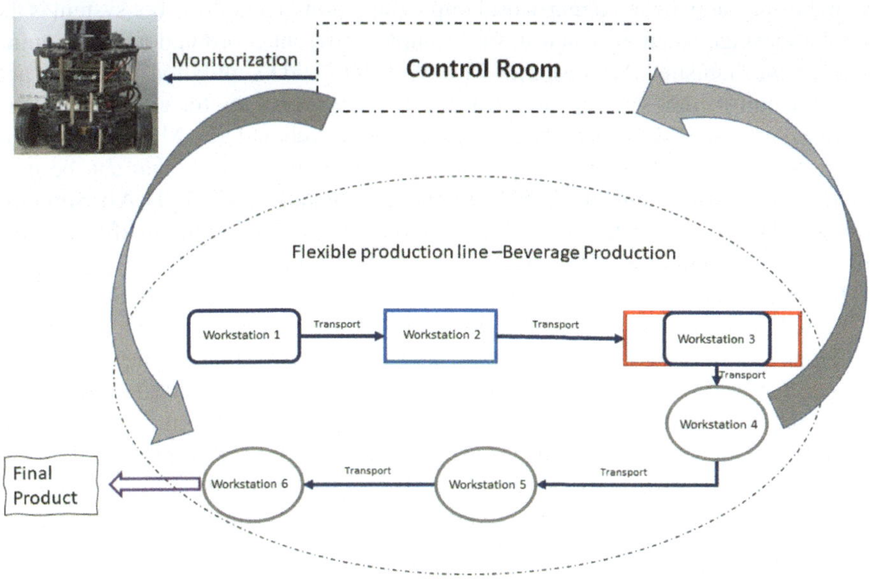

Fig. 1 The schematic representation of a flexible manufacturing line

systems. Several research papers on modelling a multi-robot system as a multi-agent system were reported in the literature. For example, the main benefits of using multi-agent robot systems are analysed in [3], starting from the trends in robotics and distributed autonomous systems development, while a cognitive multi-agent approach is proposed in [4] for cooperative mobile robots. An agent-based model for multi-robot systems, including the agent adaptation ability was reported in [5]. The author proposed the use of a common ontology for knowledge sharing and of a machine learning technique, reinforcement learning. As one of the challenging problems in multi-robot systems is their control for navigation tasks that include the exploration of unknown and unstructured environments, we have selected two papers that describe two types of approaches: a multi-agent-based coordination approach (detailed in [6]) and hierarchical fuzzy and sliding mode controllers within a multi-agent architecture (as shown in [7], proper for partially unknown environments and dynamic obstacle avoidance).

Comparing with traditional SCADA that uses fixed sensors with wired communication, this type of system presents the advantage of scalability due to the fact that it is based on decentralized system composed from mobile cooperative robots in which the number of entities can be increased without need to modify the main control.

2.2 *Simulation of Multi-robot Systems*

An important step toward a real-world implementation of a multi-robot system is the simulation stage when we can test, for example, different coordination and communication mechanisms. At present, a variety of 2D or 3D simulation platforms and tools for multi-robot systems are provided for developers, as for example, V-REP [9] (currently named CoppeliaSim), Gazebo [10], Webots [11], MORSE [12] and SwarmSimX [13]. Some common examples of simulation tools that can be used for multi-robots systems are LabView Robotics module and MATLAB Simulink package. Depending on the type of application a proper simulation platform or tool for 2D or 3D simulation of the multi-robots system is chosen.

In the next sections, we focus on the environment exploration problem solving by a mobile multi-robot system and we present the approach we have used for simultaneous localization and mapping and the experimental mobile multi-robot system that was tested as a simulation in Gazebo and as a real-world system in an industrial environment. Our previous research work related to multi-robot systems modelling is detailed in [5] (for the multi-agent model), and in [8] (for the Petri Nets model). Also, we have described in [14] a case study of multi-robot system coordination using Particle Swarm Optimization that was simulated in Webots.

3 Simultaneous Localization and Mapping in Manufacturing Environment

In the last few years, the needs for control of mobile robots in uncertain environment have grown exponentially. The central problem is to develop a method in order to obtain precise information about robot position and environment. In the literature, this method is called Simultaneous Localization and Mapping (SLAM), a method that allows to create a map while the robot system explores the unknown space and detects its surroundings, using a laser scanner or a deep camera, and each robot estimates its current location.

In the last decade, a number of papers have been published, trying to solve these problems such as Whyte and Beiley [15] using SLAM for a single robot system or Sasaoka, Kimoto, Kishimoto, Takaba and Nakashima [16] that takes the concept even further and propose a method for multi-robot system SLAM using Kalman Filters. We have adapted this method for performing SLAM for exploring an industrial environment using a multi-robot system composed of three TurtleBot3 Burger mobile robots.

First of all, the robot needs to be able to measure its position and orientation, such as the GSP measures the position of the personal car to determine the position on the map. However, for our mobile robot system, the GSP is not possible to use because GSP has large errors for estimation (more than 5 m) and for our robotic system needs precise measurement. Currently, the most widely used indoor position estimation method for robotic systems is dead reckoning, described in ROS Robot Programming book [17] that use wheels encoders and accelerometer in order to estimate with good precision the robot position and will be described next in this paper.

Assuming that the robot travels a distance during time T_e, the rotational speed of the left and right wheel can be determined as follows:

$$v_l = \frac{E_{lc} - E_{lp}}{T_e} * \frac{\pi}{180} \text{ (rad/s)} \tag{1}$$

$$v_r = \frac{E_{rc} - E_{rp}}{T_e} * \frac{\pi}{180} \text{ (rad/s)} \tag{2}$$

where

v_l	is the rotational speed of the left wheel.
v_r	is the rotational speed of the right wheel.
E_{lc}, E_{rc}	is the current value of the encoder of left/right wheel.
E_{lp}, E_{rp}	is the previous value of the encoder of left/right wheel.

From relations (1) and (2), we can calculate the linear speed of each wheel knowing that wheel radius of Turtlebot3 Burger is 65 mm.

$$V_l = v_l * 65 \text{ (m/s)} \tag{3}$$

$$V_r = v_r * 65 \text{ (m/s)} \quad (4)$$

$$\omega_k = \frac{V_r - V_l}{D} \text{ (rad/s)} \quad (5)$$

$$v_t = \frac{V_r + V_l}{2} \text{ (m/s)} \quad (6)$$

where

V_l, V_r	is linear speed of the left/right wheel.
ω_k	is the angular velocity of the robot.
v_t	is the linear speed of the robot.

Due to construction limitation of Turtlebot3 Burger, presented in next section, the maximum linear speed allowed is 0.22 m/s and the maximum angular velocity is 2.84 rad/s.

Using relations (1)–(5), we can obtain the position and the orientation of the robot, as given by relations (7)–(11):

$$\Delta d = v_t T_e \quad (7)$$

$$\Delta\theta = \omega_k T_e \quad (8)$$

$$x_{(k+1)} = x_k + \Delta d \cos\left(\theta_k + \frac{\Delta\theta}{2}\right) \quad (9)$$

$$y_{(k+1)} = y_k + \Delta d \cos\left(\theta_k + \frac{\Delta\theta}{2}\right) \quad (10)$$

$$\theta_{k+1} = \theta_k + \Delta\theta \quad (11)$$

where

Δd	is the distance travelled by the robot.
$\Delta\theta$	is the orientation done by the robot.
x_k, y_k	is the actual position of the robot.
θ_k	is the actual orientation of the robot.
$x_{(k+1)}, y_{(k+1)}$	is the next position of the robot after Δd is done.
θ_{k+1}	is the next orientation of the robot after $\Delta\theta$ is done.

4 The Experimental Mobile Multi-robot System

4.1 Description of the Experimental Multi-robot System

A multi-robot system can be viewed as a group of robots that have a joint goal and cooperate in order to accomplish it, increasing the utility of the whole system.

In this scenario, we use Turtlebot3 Burger mobile robots. TurtleBot3 is a small, affordable, programmable, ROS-based mobile robot for use in education, research, hobby, and product prototyping.

The TurtleBot3 can be customized into various ways depending on the application where is used. TurtleBot3's core technology is SLAM, Navigation and Manipulation, making it suitable for indoor and outdoor environment exploration. The TurtleBot3 robot can run SLAM (simultaneous localization and mapping) algorithms to build a map and can drive around autonomous that makes this robot a perfect choice for our application. Figure 2 shows our experimental Turtlebot3 Burger robot, developed by Robotis.

Taking in consideration this spatial restrains, our approach is to use a multi-robot system composed of three Turtlebot3 robots for better and faster exploration. In order to do that, we will need a method of exploration that allows the robots to cooperate in the construction of the environment map.

Fig. 2 The experimental Turtlebot3 Burger robot

Next in this paper, we will present the SLAM method described in section III, implemented in a virtual environment using Gazebo simulator for a multi-robot system using TurtleBot3 Burger robots.

4.2 SLAM Simulation Using Mobile Multi-robot System

First, we will implement the multi-robot in Gazebo simulator, and each robot will be described by a unique ID and a parameter file. The virtual system will be composed of three TurtleBot3 Burger mobile robots that will need to cooperate in order to achieve the global goal, i.e. the construction of the environment map.

The robots will be placed in a virtual environment that describes an office floor. The environment shape is a maze shape with different static obstacles.

Figure 3 shows the work environment for the multi-robot system and the initial position of each robot, where the environment was generated in Gazebo simulator.

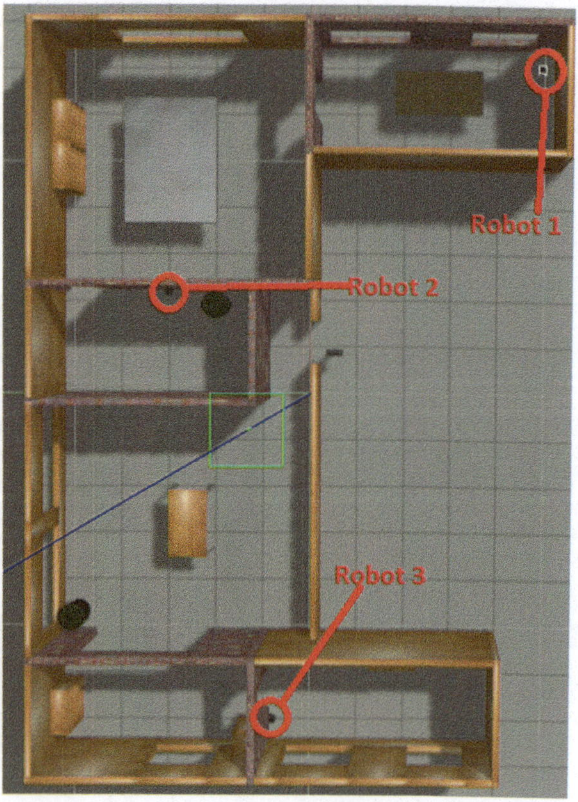

Fig. 3 The experimental Turtlebot3 Burger robot

Table 1 Simulation parameters

Maximum linear speed	0.22 m/s
Max Backwards linear speed	−0.22 m/s
Maximum angular velocity	2.84 rad/s
Linear acceleration limit	2.5 m/s^2
Angular acceleration limit	2 rad/s^2
Map resolution	0.05 m/pixel
Map type	2-dimensional map

Our main objective is to obtain the map of the environment using the data received from multiple robots. The map will be drawn based on robots odometry and scan information when the robot moves.

In the first step, we established the starting position and the orientation of each robot. In our case, we chose the positions shown in Fig. 3.

The second step is to navigate the environment and collect data from each robot. Robots and map parameters used for this simulation are described in Table 1.

For simulation, we have used the maximum speed and acceleration possible for a Turtlebot3 Burger mobile robot because we consider an ideal surface for moving (no drift or drag and flat).

Using the "turtlebot3_gazebo multi_map_merge" package, we construct the global map by merging the map provided by each robot.

Figure 4 shows the initial map that was constructed, represented in Rviz.

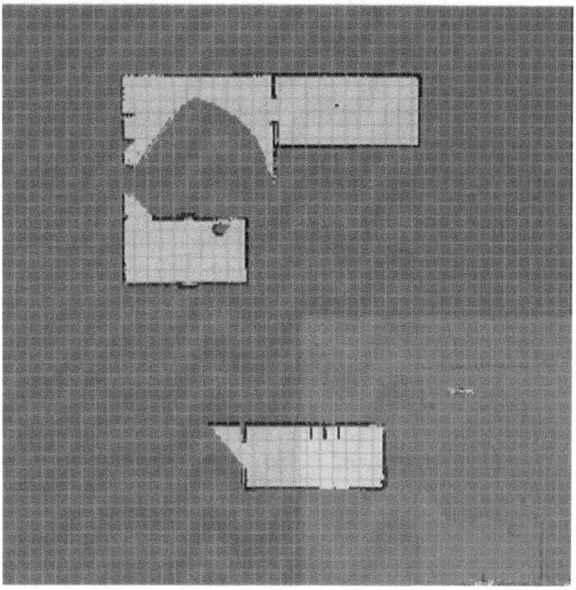

Fig. 4 Initial map constructed—represented in Rviz

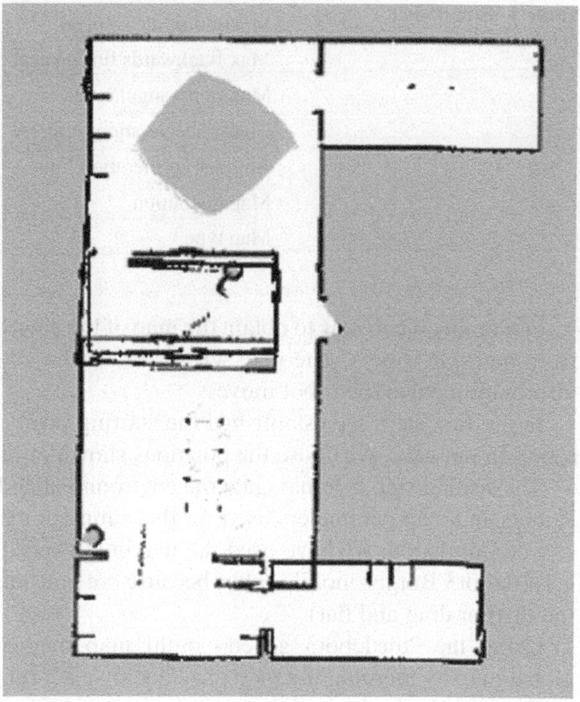

Fig. 5 Final global map of experimental environment

The map generated is a 2-dimensional map with the origin in bottom left corner defined by $x = 0$, $y = 0$ coordinates. White colour indicates free area in which the robot can move, black colour specifies the occupied area in which the robot cannot move, and grey colour represents the unknown area.

Figure 5 shows the final global map obtained through merging the maps created by each robot.

Comparing Fig. 3, in which is presented the experimental environment, and Fig. 5 in which it is represented the final global map obtained by using the SLAM method adopted in this paper applied to the multi-robot system, we can see that the results are good, and we can proceed with testing in an real environment.

4.3 Simultaneous Localization and Mapping Approach for Mapping and Navigate into an Industrial Environment Using Turlebot3 Burger Robot

In this experiment, we have tested the SLAM method using Turtlebot3 Burger robot into a manufacturing environment, the control room of a flexible manufacturing line (Fig. 6). The scope of this experiment is to test if we can construct the map of the

Fig. 6 Real experiment manufacturing environment

entire room and to navigate freely in this environment using the map obtained by the robot system.

We have started the experiment using the same parameters that we use for the virtual simulation, but because of the construction of the control room floor, slippery panels with joints between them and rubber isolation carpets 15 mm high, we saw the drift and the drag of the robot wheels on the surface is too high.

Due to this fact, we test different parameters, and we establish the limitation for the robot system as given in Table 2.

The parameters used for the robotic system during the experiment are given in Table 2. Comparing with the parameters used for virtual environment, we can see

Table 2 Real robotic system parameters

Maximum linear speed	0.16 m/s
Max Backwards linear speed	−0.16 m/s
Maximum angular velocity	2.31 rad/s
Linear acceleration limit	1.6 m/s^2
Angular acceleration limit	1.8 rad/s^2
Map resolution	0.05 m/pixel
Map type	2-dimensional map

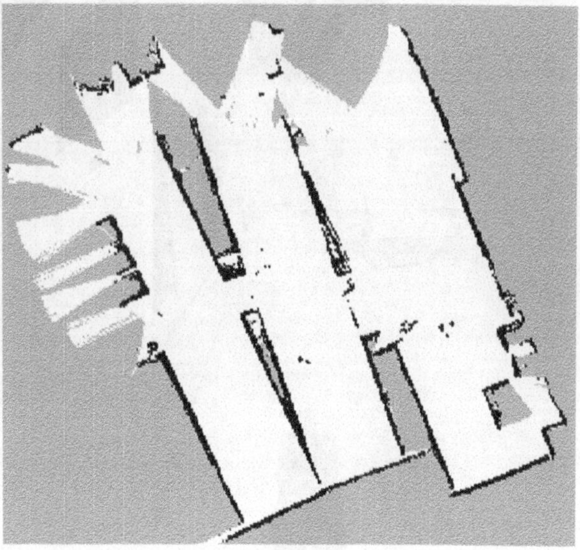

Fig. 7 Global map of the control room

a 27% of reduction in term of linear speed and 36% in term of linear acceleration, because of the surface constraints.

Figure 7 presents the map of the environment created by the mobile robot system.

After the map was created, the next step was to use the map for autonomous driving of the mobile robot system form an initial starting position to a final position on the map.

For this, we used ROS turtlebot_navigation package and RViz for visualisation. For navigation, we use the same parameters for the robot system, presented in Table 2, and for creation of the 2D cost map, we tried different values for inflation radius and cost scaling factor, but from safety reasons, to avoid getting to close to electrical panels we choose:

Inflation radius = 0.5;
Cost scaling factor = 8;

In Fig. 8, it is presented the cost map used for mobile robot navigation generated using the map contracted using SLAM method presented in this paper. As we can see, the Turtlebot3 Burger can navigates freely through the environment from an initial position to a requested final position using the cost map and the local sensors.

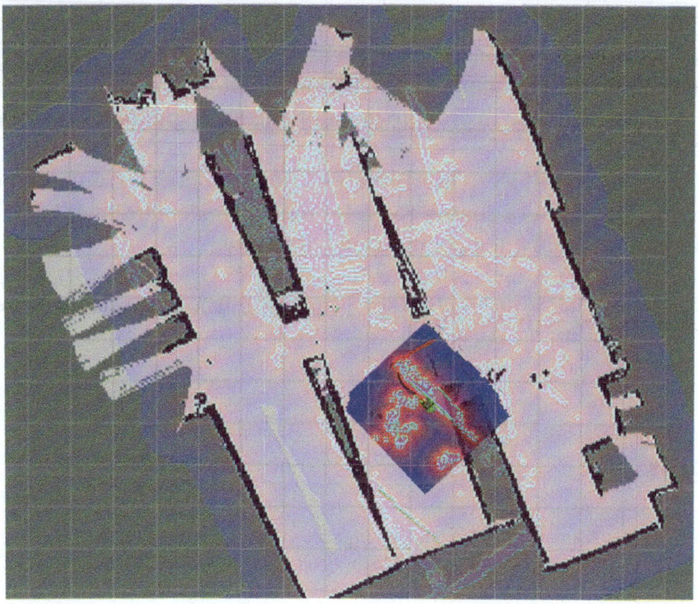

Fig. 8 Cost map used for Turtlebot3 Burger navigation presented in Rviz

4.4 Manufacturing Environment Monitorization—Data Collection

The data gathered by the robotic system are collected into a database. The database serves as an input for an intelligent diagnostic system and forecast software that provides heatmaps for the control room. Based on those heatmaps, we can make a predictive maintenance for electrical panels and electronic devices due to the specific signature temperature for each of the devices.

5 Conclusion and Future Work

In conclusion, we can say that the Simultaneous Localization and Mapping method presented in this paper can be applied successfully to a Turtlebot3 multi-robot system and can be used for mapping and navigation through an unknown environment taking in consideration the specific constrains of the environment for parameterisation of the robot system.

We proved that the construction of the map is possible using only the Turtlebot3 Burger odometry and the laser distance sensor. Furthermore, the global map can be constructed using the data obtained from multiple robots that allow mapping to be faster and more accurate.

The main advantage using this type of system in industrial environment is that does not longer needs a person to monitor a high risks environment such as a control room is and is used a combination of intelligent monitoring system a mobile robotic system to create a heat map in order to monitor and apply predictive maintenance to the control room.

In future work, we will try to use Turtlebot3 Burger robots for autonomous survey of industrial areas such as a control room for monitoring environmental parameters, as for example, air quality, temperature, and humidity.

Acknowledgements The research work of the first author is funded by a doctoral research grant received at Petroleum-Gas University of Ploiești.

References

1. Manu, G., Vijay Kumar, M., Nagesh, H., Jagadeesh, D., Gowtham, M.B.: Flexible Manufacturing Systems (FMS): a review. Int. J. Mech. Prod. Eng. Res. Dev. (IJMPERD) **8**(2), 323–336 (2018)
2. Lafou, M., Mathieu, L., Pois, S., Alochet, M.: Manufacturing system configuration: flexibility analysis for automotive mixed-model assembly lines. IFAC hosting by Elsevier **48–3**, 094–099 (2015)
3. Ota, J.: Multi-agent robot systems as distributed autonomous systems. Adv. Eng. Inform. **20**(1), 59–70 (2006)
4. Da Costa, A.L., Bittencourt, G.: Cooperative mobile robots: a cognitive multi-agent approach. In: Proceedings of the 1st IFAC Workshop on Multivehicle Systems, pp. 77–82 (2006)
5. Oprea, M.: Agent-based modelling of multi-robot systems. In: The 8th International Conference on ACME, vol. 444, p. 052026. IOP Publishing (2018)
6. Mouad, M., Adouane, L., Schmitt, P., Khadraoui, D., Martinet, P.: Architecture controlling multi-robot system using multi-agent based coordination approach. In: Proceedings of the 8th International Conference on Informatics in Control, Automation and Robotics, July 2011
7. Ayedi, D., Boujelben, M., Rekik, C.: A multi-agent architecture for mobile robot navigation using heirarchical fuzzy and sliding mode controllers. Mathematical Problems in Engineering (2018)
8. Stan, A.C., Oprea, M.: Petri Nets based coordination mechanism for cooperative multi-robot system. J. Electr. Eng. Electron. Control Comput. Sci. **6**(20), 7–14 (2020)
9. Freese, M., Singh, S., Ozaki, F., Matsuhira, N.: Virtual robot experimentation platform V-REP: a versatile 3D robot simulator. In: Ando, N., et al. (eds) SIMPAR 2010, LNAI 6472, r, pp. 51–62. Springe (2010)
10. Koenig, N., Howard, A.: Design and use paradigms for Gazebo. An open-source multi-robot simulator. In: Proceedings of IEEE/RSJ International Conference on Intelligent Robots and Systems, vol. 3, pp. 2149–2154 (2004)
11. Michel, O.: Webots: professional mobile robot simulation. Int. Journal Adv. Robot. Syst. **1**, 39–42 (2004)
12. Echeverria, G., Lassabe, N., Degroote, A., Lemaignan, S.: Modular open robots simulation engine: MORSE. In: Proceedings of IEEE International Conference on Robotics and Automation, pp. 46–51 (2011)
13. Lachele, J., Franchi, A., Bulthoff, H.H., Giordano, P.R.: SwarmSimX: real-time simulation environment for multi-robot systems. In: Noda, I., et al. (eds.) SIMPAR 2012, LNAI 7628, pp. 375–387. Springer (2012)

14. Stan, A.C., Oprea, M.: A case study of multi-robot systems coordination using PSO simulated in webots. In: The 11th International Conference Electronics, Computers and Artificial Intelligence, Pitești, Romania (2019)
15. Whyte, H., Bailey, T.: Simultaneous localisation and mapping (SLAM): Part I The essential algorithms. IEEE Robot. Autom. Mag. **13**(2) (2006)
16. Sasaoka, T., Kimoto, I., Kishimoto, Y., Takaba, K., Nakashima, H.: Multi-robot SLAM via information fusion extended Kalman filters. IFAC-PapersOnLine **49**(22), 303–308 (2016)
17. Pyo, Y., Cho, H., Jung, R., Lim, T.: ROS Robot Programming From the Basic Concept to Practical Programming and Robot Application. Robotis Co Ltd. (2017)

Automated Public Transport System Using IoT

Prabha Selvaraj, Hari Kishan Kondaveeti, Arghya Biswas, S. Gaurav, and Shubham S. Mane

Abstract The adaption of new technologies and application of IoT has solved many critical problems. Create an efficient and data-driven ecosystem. This paper aims to provide commuters with an organized, efficient, and cashless public transportation system. This work aims to give commuters a clear idea about the arrival and departure of public buses. It also makes the payment system efficient, effortless, and cashless. The buses equipped with this technology will provide more options to plan organized trips. Fleets will experience less disorganization due to the data-driven technology they will use.

Keywords Public transport · IoT · Barcodes · QR codes · Cashless transaction · Mobile application · Android

1 Introduction

Transportation is one of the major systems that need attention. So it has taken its way to transportation in order to change from manual ticketing to automated ticketing using smart e-tokens. More accidents due to footfall and increased volume of passenger initiated the thought of taking a change from closed-loop fare media to new tech-based ticketing methodologies, such as open loop, account-based ticketing, near-field communication (NFC), quick response (QR) code, and mobile ticketing. New technologies such as the National Common Mobility Card (NCMC) are enabling us to standardize and unify multi-modal transit within a city and can be gradually extended to inter-state and inter-nation transit as well. In the future, biometric-based ticketing like face and fingerprint recognition may evolve which will reduce the operational and management costs. It also increases the convenience of passengers.

P. Selvaraj · H. K. Kondaveeti · A. Biswas · S. Gaurav (✉) · S. S. Mane
VIT-AP University, Amaravati, Andhra Pradesh, India
e-mail: shivaprasad.gaurav@gmail.com

A. Noor et al. (eds.), *Proceedings of Emerging Trends and Technologies on Intelligent Systems*, Advances in Intelligent Systems and Computing 1371,
https://doi.org/10.1007/978-981-16-3097-2_28

2 Literature Study

Inconvenience makes the consumers dissatisfied and unhappy. Sometimes, it leads to loss of business. India is a diverse country with many states and languages. If anyone migrates from one state to another, the language which they speak differs from the language spoken by the local people [1]. Most of the banners and boards have the information written in the regional languages. It will be difficult for them to figure out the details of the bus and its route.

Forming queue and waiting to get the ticket is a hectic process. Taking your wallet out, counting and calculating the amount of money you are supposed to pay and the due that you are supposed to get, and more often than not getting the due back because the person at the other end of the counter does not have the 'change' has been a frustrating affair for most of the commuters. The cash transaction for small amount can be digitized into cashless which needs focus [2].

For long, the lack of technological feasibility and a conservative regulatory regime stalled the adoption of mass cashless transit payment ecosystems. Since the profit is very less in semi-closed-loop smart card, it had less scope to take action. Imo Transport Company transport system design and integration with other services is explained in [3].

In the absence of an efficient and feasible business system for issuing tickets, getting profit in the closed environment has several restrictions on implementing an interoperable solution for public transport system. IoT in ITS using RFID and how it can be applied in transport system are discussed [4]. The real time in which bus arrives in a particular location based on traffic is explained [5]. The use of GPS and crowd-sourced data positioning using smartphone is explained [6].

Cashless travel on public transport systems is slowly gaining momentum in a developing country like India yet. Just 2% of transactions were non-cash a few years ago, but now 13–14% of people prefer to pay the bus fare using cashless mode [7, 8]. Others still use cash as a preferred method for payment. During peak hours, overcrowding of passengers occurs; in this case, the conductor may not ask for a ticket or the passenger may forget as well. The Metropolitan Transportation Authority says nearly 22% of bus riders do not pay, compared with 3–4% of subway riders. Theft identification using IoT and GPS system is said [9].

An overview of different activities using IoT improved intelligence in transport system is analysed [10]. The method to provide bus details to the user with bus number along with the number of passengers and arrival time is explained [11]. The use of smartphone in ITS is studied [12]. The use of smartphone to provide information to drivers to have efficient experience in driving is described [13]. The use of RFID and GPS for tracking messages [14] and updating details to the user using IoT for public transportation is discussed [15].

3 Proposed Work

This work aims to make inter-city public transport more efficient and organized. One of the objectives of this work is to develop an automated bus stop announcement system to provide commuters with a precise idea about the route of a bus arriving at a bus station. The second objective is to develop a user-friendly mobile application for fast and secure payments. This application consists of two different user interfaces: one for the driver/conductor and second for the commuters. This application simplifies the process of commuters in buying a ticket. It also reduces the burden of the conductors in checking the payment status in real time.

3.1 *Automated Bus Stops Announcement System*

In general, manual announcements are made at the bus stops about the arrival and departure of buses and their routes. An automated system to reduce the burden of human operators is developed. This system makes announcements automatically to the commuters waiting in bus stops about the arrival and departure of a bus along with the route in which bus travels. Each bus is provided with a barcode, and barcode scanners are placed on the premises of the bus stops. These scanners scan the barcode when a bus arrives and sends the information to the system. The system matches the information scanned with the information stored in its database. Then, the system makes announcements to the commuters using text to speech (Fig. 1).

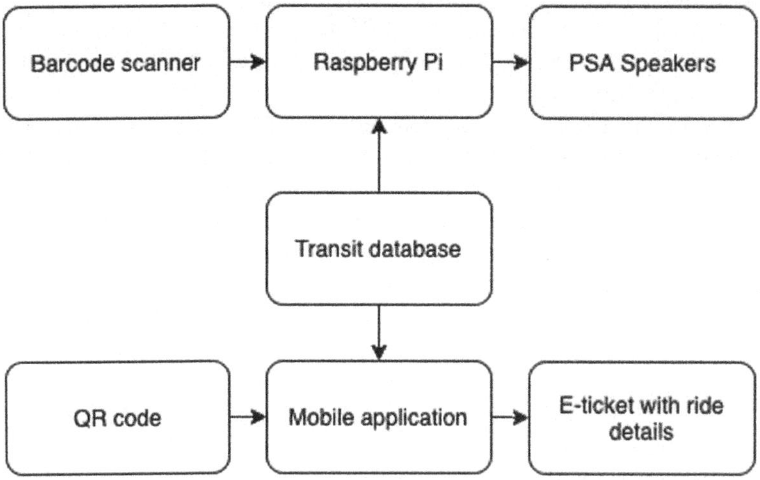

Fig. 1 System architecture

3.2 Ticket Booking Application

A mobile application that allows the commuters to scan the QR code available inside the bus and enter their drop location to get a virtual ticket is implemented. They can make the payment of the bus fare directly through the app, which makes the transaction cashless and effortless and get an e-receipt in the app, which they can show the conductors before getting off at their destinations.

The proper, coordinated working of all the individual components together leads to the efficient functioning of our ecosystem as a whole. This section intends to define the overall working of the project as a whole. It has a barcode reader, which is placed at a bit before the entrance to a bus stop where the bus arrives. A barcode stuck to the side of the bus facing towards the bus stand. When the bus arrives at the bus stop, the barcode reader scans the barcode, and the scanned value is received by the Raspberry Pi. Each bus has a unique value on its barcode. Based on the unique value the Raspberry Pi scans, it plays the corresponding audio in multiple languages (English, the national language, regional language) about the bus details and its route through a public announcement system at the bus stop. This alerts the passengers regarding the arrival of the specific bus.

On the other hand, in the mobile application, an authentication module for all users and bus conductors is provided. The commuters are expected to scan the barcode attached to the back of the seat of the bus they are travelling in. After scanning the barcode, the app would direct them to a page where the departure and destination points are available as a drop-down list. The commuter can choose his/her 'from' and 'to' bus stops accordingly. Based on his/her departure and arrival destinations, the fare is automatically calculated and displayed on the screen. Then, he/she can go ahead with the payment of the same through the app itself. Once the payment is completed, he/she gets an e-receipt. He/she can even track the bus route, get details about the bus stops on-route, and get an estimate of the travel time by using the maps integrated into the app (Fig. 2).

Once the payment is done by the commuter and the e-receipt is generated in the app, the conductor also gets a list of passengers who paid their fares, their names, and their boarding and departing details. When the passenger wants to get down at the destination, they just need to show the e-ticket generated in the app and the conductor will verify by checking them off the list and they are allowed to leave the bus (Figs. 3, 4, 5, and 6).

4 Claims and Result

4.1 Principal Claims

1. Helping migrant commuters to announce the arrival of a bus in multiple languages.

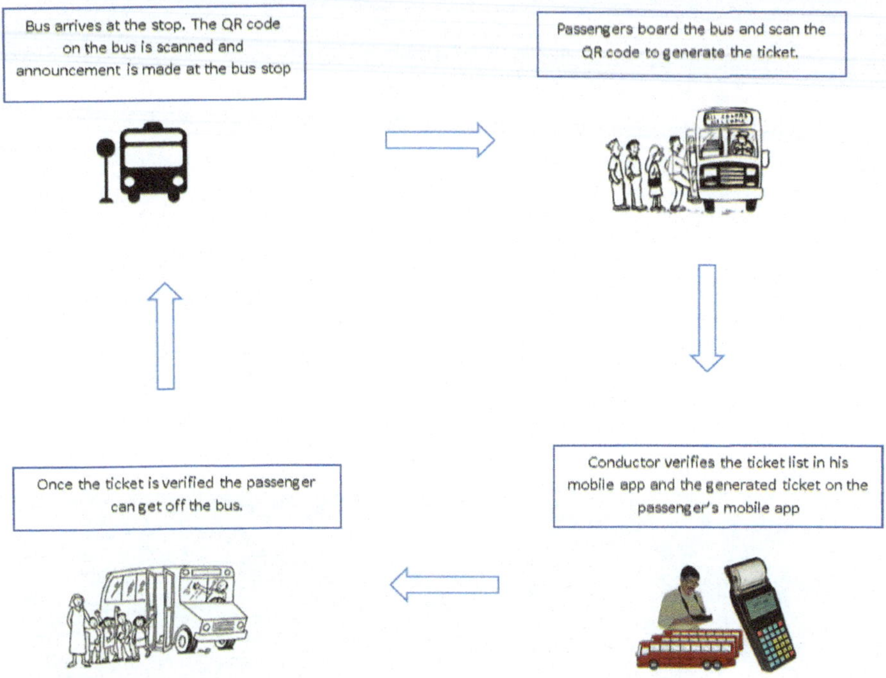

Fig. 2 System flow

2. Having an automated and cashless payment system that reduces human effort and contact.

4.2 Dependent Claim

1. Minimizing the effort of the conductor and making sure everyone pays for the public transport system.
2. At the time of the pandemic, the online payment method in public transport is a much safer way than the transaction with cash, which reduces human interaction.

5 Conclusion

In conclusion, our work suggests that a more sustainable future lies in the creation of an automated transport system. The computerized system will help to reduce human efforts and will also reduce the cash transaction and encourage people to go for an online payment method. By giving the option to pay through online mode by scanning QR code would help to make growth payment cashless.

Fig. 3 Authentication

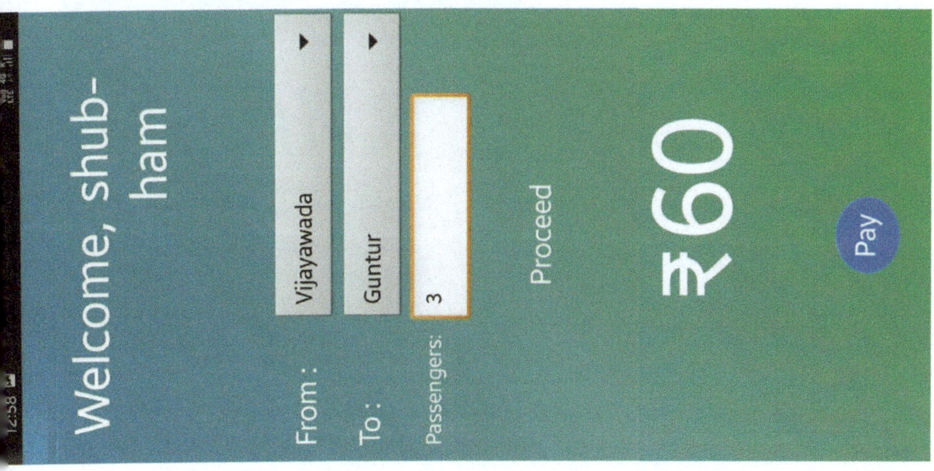

Fig. 4 Travel details

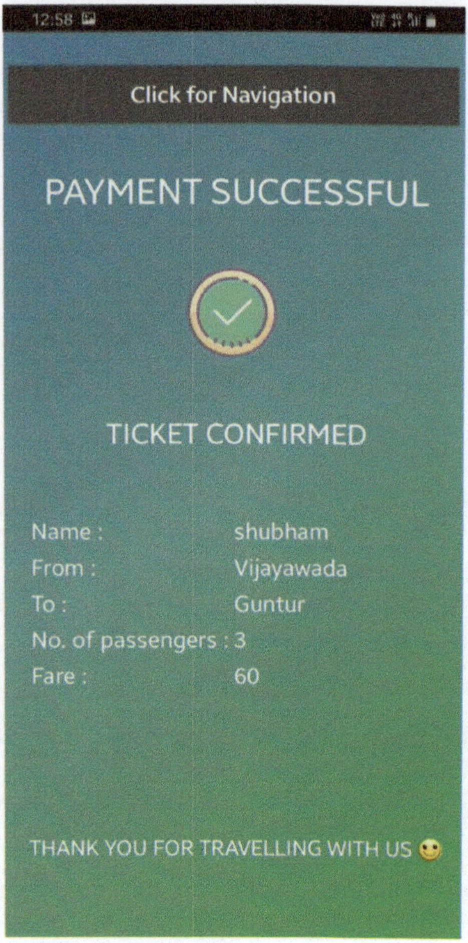

Fig. 5 Ticket

The following features can be embedded to improve our model further by using more sophisticated technology like GPS to detect when the bus is arriving at the bus stop, issue virtual 'smart cards'/accounts with prepaid passes, and use NFC tags for payments, instead of scanning QR codes and better gate-keeping and ticket avoidance features for the conductor's model.

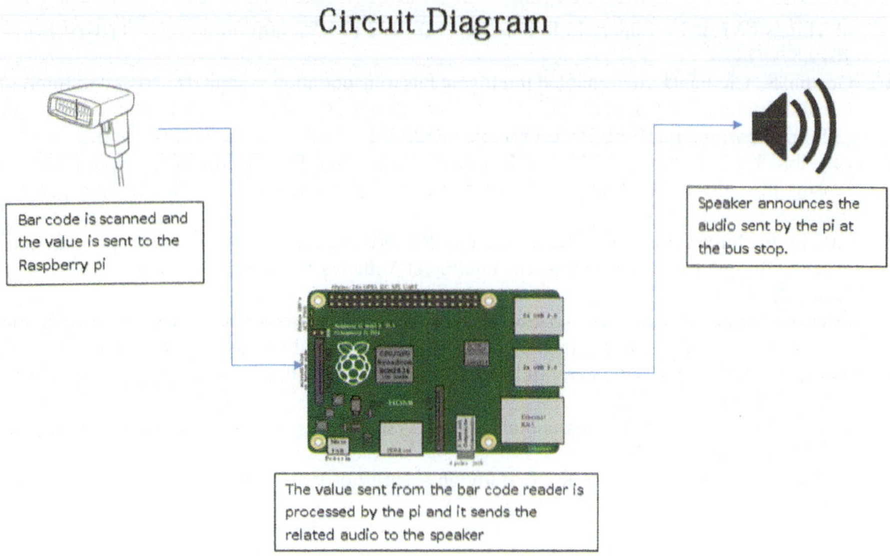

Fig. 6 Hardware components

References

1. Sharma, K.: India has 139 million internal migrants. They must not be forgotten. World Economic Forum, 2020. [Online]. Available: https://www.weforum.org/agenda/2017/10/india-has-139-million-internal-migrants-we-must-not-forget-them/ (2020)
2. Manikandan, A.: Digitising transit payments key to a cashless economy. The Economic Times, 2020. [Online]. Available: https://economictimes.indiatimes.com/industry/banking/finance/banking/fastag-digitising-transit-payments-key-to-a-cashless-economy/articleshow/74467929.cms (2020)
3. Nwakanma, I., Etus, C., Ajere, I., Agomuo, U.: Online bus ticket reservation system. Stat. Comput. **1**(2) (2015)
4. Lohokare, J., Dani, R., Sontakke, S., Adhao, R.: Scalable tracking system for public buses using IoT technologies. In: 2017 International Conference on Emerging Trends & Innovation in ICT (ICEI), pp. 104–109. IEEE (2017)
5. Mukheja, P., Velaga, N.R., Sharmila, R.B.: Smartphone-based crowdsourcing for position estimation of public transport vehicles. IET Intell. Transp. Syst. **11**(9), 588–595 (2017)
6. Vijayalakshmi, R., Premalatha, G., Dhivya, K.: IOT based wireless controlled smart transportation system. In: 2019 1st International Conference on Innovations in Information and Communication Technology (ICIICT), Chennai, pp. 1–6, India (2019). http://doi.org/10.1109/ICIICT1.2019.8741507
7. Performance analysis of India public bus sector 2015–16 | UITP India (2020). [Online]. Available: https://india.uitp.org/articles/performance-analysis-of-india-public-bus-sector. Accessed: 15-June-2020
8. Dave, P.: Cashless travel on public transport in India: waiting for the future | Smart Cities Dive (2020). [Online]. Available: https://www.smartcitiesdive.com/ex/sustainablecitiescollective/cashless-travel-public-transport-systems-india-future-be-awaited/190691/. Accessed: 15-June-2020
9. Brincat, A., Pacifici, F., Martinaglia, S., Mazzola, F.: The internet of things for intelligent transportation systems in real smart cities scenarios. In: 2019 IEEE 5th World Forum on Internet

of Things (WF-IoT), Limerick, Ireland, pp. 128–132 (2019). http://doi.org/10.1109/WF-IoT.2019.8767247
10. Geetha, S., Cicilia, D.: IoT enabled intelligent bus transportation system. In: 2017 2nd International Conference on Communication and Electronics Systems (ICCES), Coimbatore, pp. 7–11 (2017). http://doi.org/10.1109/CESYS.2017.8321235
11. Engelbrecht, J., Booysen, M.J., van Rooyen, G.-J., Bruwer, F.J.: Survey of smartphone-based sensing in vehicles for intelligent transportation system applications. IET Intell. Transp. Syst. **9**(10), 924–935 (2015)
12. Araujo, R., Igreja, A., de Castro, R., Araujo, R.: Driving coach: a smartphone application to evaluate driving efficient patterns. In: Intelligent Vehicles Symposium (IV), pp. 1005–1010. IEEE (2012)
13. Deebika Shree, A., Anusuya, J., Malathy, S.: Real time bus tracking and location updation system. In: 2019 5th International Conference on Advanced Computing & Communication Systems (ICACCS), Coimbatore, India, pp. 242–245 (2019). http://doi.org/10.1109/ICACCS.2019.8728353
14. Padmapriya, S.: Real time bus announcement system using RFID. In: Conference: Recent Trends of Computing (2012)
15. Chepuru, A., Venugopal Rao, K.: A study on security of IoT in intelligent transport systems applications (2015)

USWSBS: User-Centric Sensor and Web Service Search for IoT Application Using Bagging and Sunflower Optimization

Deepak Surya, Gerard Deepak, and Santhanavijayan

Abstract An enormous volume of data is generated by the sensors as the sensors incorporated in an IoT infrastructure rise rapidly. So, it becomes a daunting task to determine the best sensor for an intended application. A search for the most suitable sensor has become a crucial requirement. This paper proposes an approach based on the bagging classification method and sunflower optimization to suggest sensors that can generate the most appropriate data for the application. The query terms are collected from the user, and terms from IoT and sensor thesaurus are merged. Relevant query terms are aggregated, and the semantic similarity among them is measured using Lin Similarity. Then the terms are classified using random forest and support vector classification techniques. Upon classification, the top 10% of the semantically similar terms are merged with the annotated dataset. The sunflower optimization algorithm is applied to recommend the most relevant sensors. Web repositories such as WSDL and UDDI are exploited to obtain web service metadata. The annotated dataset and the metadata obtained are then classified and optimized to suggest web services for the IoT application. The proposed framework achieves an accuracy of 94.04%.

Keywords Bagging classification · Lin Similarity · Semantic sensor network · Sunflower optimization

1 Introduction

Sensors are a vital component of an IoT system. They provide the IoT systems data by detecting the variations in physical phenomena such as pressure, temperature, and humidity. Sensors capture the environmental variations and convert them into digital or analog signals that can be interpreted by machines or humans. The sensors cannot just detect an environment's physical properties, but they can also capture the human senses. Thermistors, Odo sensors, Imaging sensors, and Sound pressure

D. Surya · G. Deepak (✉) · Santhanavijayan
Department of Computer Science and Engineering, National Institute of Technology Tiruchirappalli, Tiruchirappalli, India

A. Noor et al. (eds.), *Proceedings of Emerging Trends and Technologies on Intelligent Systems*, Advances in Intelligent Systems and Computing 1371,
https://doi.org/10.1007/978-981-16-3097-2_29

sensors are included among the sensors to detect the human senses. Although IoT is about building a network of entities, objects, users, applications and devices, WoT is encompasses them into the Web. Semantic Search facilitates search engines to retrieve the most appropriate web documents by considering the contextual meaning of the key word. Also, an effective management and representation strategy is vital for information systems as they have an enormous volume of data to process. Semantically driven recommendation systems utilize a knowledge base to confront these issues. It analyzes the user profiles to reflect the long-term information requirements and preferences of users.

Motivation: In the present-day time, there is a need for semantically compliant strategies for the recommendation. The World Wide Web is gradually adapting the standards of Web 3.0. As a result, the IoT-driven Web is transforming into the Web of Things (WoT). Also, recommending sensors along with web services is a challenge in the era of WoT integrated Web 3.0. Owing to this reason, an approach that facilitates both sensor search and web service recommendation is proposed. The queries collected from the user are preprocessed, and the relevant query terms are aggregated from the IoT and sensor thesaurus. Lin Similarity is applied to the aggregated terms to compute semantic similarity. Web service metadata are obtained from the web repositories such as UDDI and WSDL. Then the terms are classified by bagging method using random forest and support vector clustering. Upon classification, semantic similarity is calculated among the classified terms with concepts in IoT and sensor ontologies. The top 10% of the similar terms are merged with the annotated dataset and classified using the sunflower optimization algorithm to suggest sensors and web services.

Contribution: In this paper, a novel technique is put forth to suggest sensors and web services for an IoT application. The proposed approach integrates agent-based classification, agent-based ontology integration, and agent-based semantic similarity computation using sunflower optimization. The annotated dataset, along with sensor and IoT ontologies, is used to suggest sensors. The model uses query terms with web repositories for web service suggestions. The bagging method based on random forest and support vector clustering is used for classification. Finally, the query facets enrichment is performed based on the combination of Lin Similarity, and an meta-heuristic optimization algorithm is exploited to further optimize the computation of semantic similarity.

Organization: The organization of the remainder of the paper is as follows. Section 2 comprises Related Work, and Sect. 3 encompasses the suggested framework. The implementation and performance evaluated is depicted in Sect. 4, while Sect. 5 contains the results. Finally, Sect. 6 concludes the paper.

2 Related Work

Gomes et al. [1] introduced a customized version of the optimization technique for sunflower detection of damage in complex structures. In the altered optimizer algorithm, the terms are implemented as root velocity, and the pollination provides robustness. A problem based on an inverse technique to detect defects in composite laminated plates has been used in the suggested methodology. By using the lip flight algorithm with the sunflower optimization algorithm, Raslan et al. [2] suggested a technique to enhance the life of wireless sensor networks. Jiang et al. [3] proposed a reliable novel multi-deployable slope one algorithm developed by merging reliable data and user similarity approach. The algorithm applied in their work has better accuracy than the conventional slope one algorithm. Elsaleh et al. [4] introduced a novel model to illustrate the semantic model's use and lists examples of instantiation of the system for various use cases. Based on microservices, middleware, web services and other common IoT architectures, the system framework has been built. Their study also indicates frameworks which are required to ingest and annotate IoT-Stream-labeled data. Iwendi et al. [5] proposed a TF-IDF based on the Louvain temporal strategy to evaluate text collected from numerous sensing systems. The proposed work enables analysts to make reliable decisions by classing documents into hierarchical structures and presenting the relationships among the variables in the document. Sejdiu et al. [6] enriched the raw sensor data by adding semantic annotations based on the concepts from the ontologies to facilitate knowledge discovery by describing the data more expressively. Their work also examines the solutions based on attaching semantic annotations to the sensor data, current stream data annotation models, and trending domains in IoT that exploit semantics. A novel approach to integrating the output streams of the services included in the Data Acquisition Plan and the Domain Ontology to present a semantic specification for its ultimate result was proposed by Valtolina et al. [7]. Speiser et al. [8] highlight different techniques to determine the variables for the random forest algorithm to identify suitable techniques based on intended applications and intelligent systems. Lin et al. [9] implement a method of similarity to calculate the similarity between two sets of documents commonly employed in text clustering and classification. Mansour et al. [10] developed an ontology to extend the reusable Semantic Sensor Ontology by considering several SSN platforms. Their work also extended the Ontological representation of sensed data, sensors, and deployment environments. Deepak and Santhanavijayan [11] proposed the Ontological Best Fit RDF Driven Scheme for Semantic Search. Semantic Latent analysis was incorporated by Pushpa et al. [12] to measure the ontological relevance. Content-based filtering was used to improve the quality of the proposed framework. These tools have been used to increase the overall efficiency of Semantic Web Content recovery. In [13–22] several ontological models have been discussed in correlation to the proposed model.

3 Proposed Work

Queries that represent the users' interest are collected and preprocessed. Upon query preprocessing, the relevant query terms are aggregated based on the IoT and sensor Thesaurus. IoT Ontologies and Sensor Ontologies are also appended to the aggregated terms, and the semantic similarity among the concepts is calculated using Lin Similarity. Lin Similarity is employed to approximate the degree of the semantic relationship between units of concepts, language, and instances. The Lin Similarity is measured as the ratio of the similarity between the terms upon the difference between them, i.e., the commonality and difference ratio. It can be calculated, as shown in Eq. (1).

$$\mathrm{sim}_{\mathrm{Lin}}(A, B) = \frac{\log P(\mathrm{comman}(A, B))}{\log P(\mathrm{description}(A, B))} \tag{1}$$

The workflow for the proposed architecture is presented in Fig. 1. Metadata of web services is obtained from web repositories such as UDDI and WSDL. The bagging method is employed to classify the concepts from Ontologies and the metadata of web services from the repositories. Random forest algorithm with support vector clustering is employed in the bagging process to classify the concepts based on the sensor and IoT ontologies. Random forest can handle multi-output problems, i.e., both numerical and categorical data. Random forest has been chosen for various

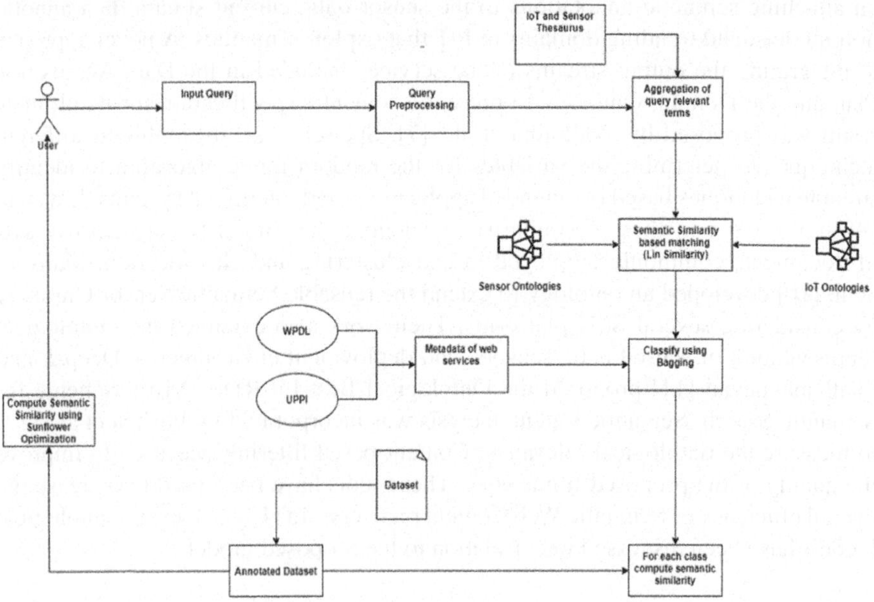

Fig. 1 Proposed USBSWS algorithm for sensor and web service suggestion for IoT application

reasons. As the final prediction is based on average of all the predictions made by each decision tree, it reduces the risk of overfitting and reduces the training data time, thus increasing the model's performance. It predicts with a very high accuracy even when the large proportion of data set is missing. Thereby, it can run even without hyperparameter tuning. It can also be run efficiently on large datasets with high dimensionality. It is not affected by the non-linear relationships between the parameters. However, if the growth of an individual decision tree is not limited, it overfits the training data and its performance decreases.

4 Implementation and Performance Evaluation

4.1 Dataset Preparation

The proposed framework's implementation is performed in Google's Colaboratory environment on a computer with an i7 processor and 16 GB RAM. Environment-based datasets such as MesoWest and Air Quality Sensor Dataset are exploited for the experimentation objectives and evaluation. The former dataset comprises weather observations data, while the latter contains air quality estimation reading. The semantic sensor network (SSN) ontology is utilized for modeling the sensor data.

The SSN ontology is used to portray sensors and their noticed properties. The ontology is built based on vertical and horizontal modularization design that encompasses a lightweight yet self-contained core ontology termed Sensor–Observation–Sampling–Actuator ontology (SOSA) for its rudimentary concepts and properties. With various axiomatization levels, SSN and SOSA can uphold a broad scope of utilization, including social detecting, observing enormous industrial and scientific infrastructures, Web of Things, and Ontology Engineering.

4.2 Implementation

The dataset is annotated, and the semantic similarity is calculated between the concepts in the dataset and the top 10% of the terms obtained from the previous step. The concept similarity is computed using the sunflower optimization algorithm. The sunflower optimization algorithm is developed based on inspiration from the sunflowers' reproduction, sunflower movement towards the Sun, and the inverse square law. It is a nature-inspired population-based meta-heuristic method used to determine the global optimum without getting stuck in the local optima. It can also be applied to find a solution for tasks that are multi-modal and multi-dimensional. The inverse square law states that the intensity of the radiation emitted is inversely proportional to the square of the distance of separation. As the Sun rises, the sunflower faces it, and pollination tends to occur with a nearby flower. The algorithm for the proposed architecture is presented in Algorithm 1. The intensity of radiation or the solar energy received by a sunflower is calculated based on the inverse square law, as shown in Eq. (2).

Algorithm 1: The proposed Sensor and Web Service Recommendation

Input: User query, IoT and Sensor ontologies, and Web repositories such as UDDI and WSDL.
Output: Sensors and Web services that best serve the purpose of the IoT application.
Start
Step 1: Collect the queries that represents the user intention.
Step 2: Query preprocessing
Step 3: Aggregation of relevant query terms based on sensor and IoT ontologies.
Step 4: Compute semantic similarity using Lin Similarity.
Step 5: Employ random forest and SVC to classify the data set.
Set the SVC kernel that defines the linear hyperplane when training to the algorithm.
Input the dataset to the fit function.
Benchmark Import Accuracy
Step 6: Apply sunflower optimization the classified results.
Set a random n-flower population.

Determine the Sun in the population at start (best answer S*).
Orient plants in the direction of the Sun.
While (K < Maximum Days):
For each individual plant, determine the position vector.
Further out from the Sun, eliminate m% of individuals.
Compute a step for every plant in the population.
Top P individuals shall pollinate around the Sun.
Assess the new plants.
Update the Sun if the new plant is the global best.

End While
The optimum values are obtained.
Step 7: Repeat the previous step to recommend the web services that are most relevant to the IoT application by incorporating web service repositories such as UDDI and WSDL.
End

The sunflower changes its direction to receive the maximum solar radiation, as depicted in Eq. (3). The movement taken by each individual sunflower towards the

Sun is calculated as presented in Eq. (4), and the movement is restricted to a threshold value. The threshold value is calculated based on the formula shown in Eq. (5). Based on the sunflower's movement and the direction towards the Sun, a new generation of sunflowers is produced due to pollination which is illustrated in Eq. (6).

$$I = \frac{P}{4\pi d^2} \tag{2}$$

$$s_i = \frac{X^* - X_i}{||X^* - X_i||} \quad i = 1, 2, 3, \ldots, n \tag{3}$$

$$d_i = \gamma * P_i(||X_i + X_{i-1}|| * ||X_i + X_{i-1}||) \tag{4}$$

$$d_{\max} = \frac{||X_{\max} - X_{\min}||}{2 * n} \tag{5}$$

$$X_{i+1} = X_i + d_i * s_i \tag{6}$$

In Eq. (3), P is the power emitted by the Sun and d is the distance of separation between the Sun and sunflower. In Eq. (4), X^* and X represent the global best and current solution. The $X_{\max}$, $X_{\min}$, and n represent the upper bound, lower bound and the population size. The newly generated individual's position is represented by X_{i+1}. Finally, based on the similarities obtained, sensors and web services are suggested for the intended IoT application.

5 Results and Performance Evaluation

The queries which capture the need of the user are obtained as input and preprocessed. The relevant terms among the preprocessed queries, sensor, and IoT Ontologies are aggregated. Concept similarity among the classes and terms are measured by employing Lin Similarity. Then Bagging technique is applied to classify the terms. The dataset is annotated, and sunflower optimization is operated on the annotated dataset to facilitate sensor recommendation. Web service repositories like WSDL and UDDI, along with the annotated dataset, are exploited to suggest the most relevant web services. Table 1 compares the performance of the proposed approach with the baseline model and other frameworks. The performance is evaluated by considering Precision, Recall, Accuracy, F- Measure, False Discovery Rate (FDR) as potential metrics. Precision, Recall, and Accuracy are calculated as depicted in the Eqs. (7), (8), and (9), respectively. The F-measure and FDR are computed as illustrated in Eqs. (10) and (11), respectively, and Fig. 2 compares the accuracy distribution of the baseline models with USWSBS.

Table 1 Performance comparison of the proposed approach with the baseline model and other variants

Metric	AntCluster [1]	COUCSFC [2]	Lin Similarity + Classification using bagging + IoT ontologies + Sensor ontologies (LSBSWSO)	Proposed approach for sensor recommendation	Proposed approach for recommendation of web services
Precision%	80.01	83.53	87.27	92.83	93.32
Recall%	83.17	89.74	93.47	95.17	95.21
Accuracy%	81.34	87.42	90.43	93.81	94.27
F-measure%	81.55	86.52	90.26	93.98	94.55
FDR	0.19	0.16	0.12	0.07	0.06

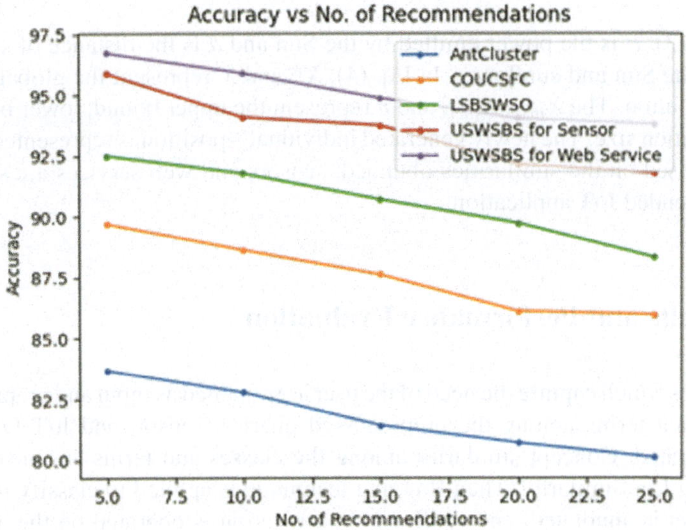

Fig. 2 Accuracy distribution graph

$$\text{Precision} = \frac{\text{Retreived} \cap \text{Relevant}}{\text{Retreived}} \tag{7}$$

$$\text{Recall} = \frac{\text{Retreived} \cap \text{Relevant}}{\text{Relevant}} \tag{8}$$

$$\text{Accuracy} = \frac{\text{Proportion correct of each query passing ground truth test}}{\text{Total No. of Queries}} \tag{9}$$

$$F\text{-measure} = \frac{2 * \text{Precision} * \text{Recall}}{\text{Precision} + \text{Recall}} \tag{10}$$

$$\text{FDR} = 1 - \text{Precision} \tag{11}$$

It is evident from Table 1 that the proposed framework USWSBS operates better than AntCluster [1], COUCSFC [2], and the other approach that utilizes Lin Similarity and Bagging Classification with IoT Ontologies Sensor Ontologies (LSBSWSO). The USWSBS has a better score for precision, F-measure, FDR, and accuracy. The F-measure value of USWSBS for web service recommendation is greater than AntCluster, COUCSFC, and LSBSWSO by 13.00%, 8.03%, and 4.49%, respectively. The accuracy percentage of USWSBS for web service recommendation is more significant than AntCluster by 12.93%, COUCSFC by 6.85%, and LSBSWSO by 3.84%. The proposed model's precision to recommend web service is 93.32%, while that of AntCluster, COUCSFC, and LSBSWSO are 80.01%, 83.53%, and 87.27%, respectively. The accuracy of the web services recommended by USWSBS is greater than that of AntCluster, COUCSFC, and LSBSWSO by 12.04%, 1.74%, and 5.47%. The recall of the proposed approach for sensor recommendation and web service recommendation are better than AntCluster, COUCSFC, and LSBSWSO in percentage by 13.83, 7.75, 4.47, and 12.1, 5.53, 1.8. The precision of the proposed approach for sensor recommendation and web service recommendation are better than AntCluster, COUCSFC, and LSBSWSO in percentage by 12.82, 9.3, 5.56, and 13.31, 9.79, 6.05. The FDR values of AntCluster, COUCSFC, and LSBSWSO are greater than USWSBS by 0.13, 0.10, 0.6 for the web service recommendation and 0.12, 0.09, 0.5 for the sensor recommendation.

6 Conclusions

The efficacy of the results obtained for the proposed framework validates its potential to be exploited for IoT applications. The queries obtained are preprocessed to obtain the query terms. Domain Knowledge, Sensor and Web Service Ontologies, Web service repositories like UDDI and WSDL are incorporated into the dataset to improvise the classification accuracy. Upon classification, sunflower optimization is employed on the classified results to recommend web services and sensors. The proposed USBSWS framework yields an average accuracy of 94.04% with a very low FDR of 0.65, which owes to the generation of metadata from web repositories and incorporation of sensor and IoT ontologies. Also, the sunflower optimization algorithm has been employed on the classified results to yield more accurate results. The results validate that USBSWS is the best-in-class approach to recommend sensors and web services for an IoT architecture.

References

1. Gomes, G.F., da Cunha, S.S., Ancelotti, A.C.: A sunflower optimization (SFO) algorithm applied to damage identification on laminated composite plates. Eng. Comput. **35**(2), 619–626 (2019)
2. Raslan, A.F., Ali, A.F., Darwish, A.: A modified sunflower optimization algorithm for wireless sensor networks. In: Joint European-US Workshop on Applications of Invariance in Computer Vision, pp. 213–222. Springer, Cham (2020)
3. Jiang, L., Cheng, Y., Yang, L., Li, J., Yan, H., Wang, X.: A comprehensive study of web usage mining. In: Symposium on Colossal Data Analysis and Networking Trust-Based Collaborative Filtering Algorithm for E-commerce Recommendation System. J. Ambient Intell. Humanized Comput. **10**(8), 3023–3034 (2016)
4. Elsaleh, T., Enshaeifar, S., Rezvani, R., Acton, S.T., Janeiko, V., Bermudez-Edo, M.: IoT-stream: a lightweight ontology for internet of things data streams and its use with data analytics and event detection services. **20**(4), 953 (2020)
5. Iwendi, C., Ponnan, S., Munirathinam, R., Srinivasan, K., Chang, C.-Y.: An efficient and unique TF/IDF algorithmic model-based data analysis for handling applications with big data streaming. Electronics **8**(11), 1331 (2019)
6. Sejdiu, B., Ismaili, F., Ahmedi, L.: Integration of semantics into sensor data for the IoT: a systematic literature review. Int. J. Semant. Web Inf. Syst. (IJSWIS) **16**(4), 1–25 (2020)
7. Valtolina, S., Ferrari, L., Mesiti, M.: Ontology-based consistent specification of sensor data acquisition plans in cross-domain IoT platforms. IEEE Access **7**, 176141–176169 (2019)
8. Speiser, J.L., Miller, M.E., Tooze, J., Ip, E.: A comparison of random forest variable selection methods for classification prediction modeling. Expert Syst. Appl. **15**(134), 93–101 (2019)
9. Lin, Y.S., Jiang, J.Y., Lee, S.J.: A similarity measure for text classification and clustering. IEEE Trans. Knowl. Data Eng. **26**(7), 1575–1590 (2013)
10. Mansour, E., Chbeir, R., Arnould, P.: An ontology for hybrid semantic sensor networks. In: Proceedings of the 23rd International Database Applications & Engineering Symposium, pp. 1–10 (2019)
11. Deepak, G., Santhanavijayan, A.: OntoBestFit: a best-fit occurrence estimation strategy for RDF driven faceted semantic search. Comput. Commun. **160**, 284–298 (2020)
12. Pushpa, C.N., Deepak, G., Kumar, A., Thriveni, J., Venugopal, K.R.: Improving web service discovery by hybridization of ontology focused concept clustering and interface semantics. In: 2020 IEEE International Conference on Electronics, Computing and Communication Technologies (CONECCT), pp. 1–5. IEEE (2020)
13. Deepak, G., Kasaraneni, D.: ONTOCOMMERCE: an ontology focused semantic framework for personalised product recommendation for user targeted e-commerce. Int. J. Comput. Aided Eng. Technol. **11**(4–5), 449–466 (2019)
14. Gulzar, Z., Leema, A.A., Deepak, G.: Pcrs: Personalized course recommender system based on hybrid approach. Procedia Comput. Sci. **125**, 518–524 (2018)
15. Deepak, G., Teja, V., Santhanavijayan, A.: A novel firefly driven scheme for resume parsing and matching based on entity linking paradigm. J. Discrete Math. Sci. Crypt. **23**(1), 157–165 (2020)
16. Haribabu, S., Kumar, P.S., Padhy, S., Deepak, G., Santhanavijayan, A., Kumar, N.: A novel approach for ontology focused inter-domain personalized search based on semantic set expansion. In: 2019 Fifteenth International Conference on Information Processing (ICINPRO), pp. 1–5. IEEE (2019)
17. Deepak, G., Kumar, N., Bharadwaj, G.V.S.Y., Santhanavijayan, A.: OntoQuest: an ontological strategy for automatic question generation for e-assessment using static and dynamic knowledge. In: 2019 Fifteenth International Conference on Information Processing (ICINPRO), pp. 1–6. IEEE (2019)
18. Santhanavijayan, A., Kumar, D.N., Deepak, G.: A semantic-aware strategy for automatic speech recognition incorporating deep learning models. In: Intelligent System Design, pp. 247–254. Springer, Singapore (2021)

19. Deepak, G., Kumar, A.A., Santhanavijayan, A., Prakash, N.: Design and evaluation of conceptual ontologies for electrochemistry as a domain. In: 2019 IEEE International WIE Conference on Electrical and Computer Engineering (WIECON-ECE), pp. 1–4. IEEE (2019)
20. Deepak, G., Priyadarshini, J.S.: Personalized and enhanced hybridized semantic algorithm for web image retrieval incorporating ontology classification, strategic query expansion, and content-based analysis. Comput. Electr. Eng. **72**, 14–25 (2018)
21. Ebrahimi, M., ShafieiBavani, E., Wong, R.K., Fong, S., Fiaidhi, J.: An adaptive meta-heuristic search for the internet of things. Future Gener. Comput. Syst. **76**, 486–494 (2017)
22. Pattar, S., Sandhya, C.R., Vala, D., Chouhan, D., Buyya, R., Venugopal, K.R., Iyengar, S.S., Patnaik, L.M.: Context-oriented user-centric search system for the IoT based on fuzzy clustering. In: International Conference on Computational Intelligence, Security and Internet of Things, pp. 343–356. Springer, Singapore (2019)

Author Index

A. Noor et al. (eds.), *Proceedings of Emerging Trends and Technologies on Intelligent Systems*, Advances in Intelligent Systems and Computing 1371,
https://doi.org/10.1007/978-981-16-3097-2

The manufacturer's authorised representative in the EU is Springer Nature Customer Service Centre GmbH, Europaplatz 3, 69115 Heidelberg, Germany. If you have any concerns regarding our products, please contact ProductSafety@springernature.com

Printed and bound by CPI Group (UK) Ltd, Croydon, CR0 4YY
30/09/2026
02225251-0011